European Consortium for
Mathematics in Industry 6

Wacker and Zulehner (Eds.)

Proceedings of the Fourth European Conference
on Mathematics in Industry

European Consortium for Mathematics in Industry

Edited by
Michiel Hazewinkel, Amsterdam
Helmut Neunzert, Kaiserslautern
Alan Tayler, Oxford
Hansjörg Wacker, Linz

ECMI Vol. 6

Within Europe a number of academic groups have accepted their responsibility towards European industry and have proposed to found a European Consortium for Mathematics in Industry (ECMI) as an expression of this responsibility.

One of the activities of ECMI is the publication of books, which reflect its general philosophy; the texts of the series will help in promoting the use of mathematics in industry and in educating mathematicians for industry. They will consider different fields of applications, present casestudies, introduce new mathematical concepts in their relation to practical applications. They shall also represent the variety of the European mathematical traditions, for example practical asymptotics and differential equations in Britain, sophisticated numerical analysis from France, powerful computation in Germany, novel discrete mathematics in Holland, elegant real analysis from Italy. They will demonstrate that all these branches of mathematics are applicable to real problems, and industry and universities in any country can clearly benefit from the skills of the complete range of European applied mathematics.

Proceedings of the

Fourth European Conference on Mathematics in Industry

May 29-June 3, 1989 Strobl

Edited by

Hansjörg Wacker

and

Walter Zulehner

Institute for Mathematics,
Johannes Kepler University, Linz, Austria

 B. G. Teubner Stuttgart

KLUWER ACADEMIC PUBLISHERS
DORDRECHT / BOSTON / LONDON

Library of Congress Cataloging-in-Publication Data

European Conference on Mathematics in Industry (4th : 1989 : Strobl,
  Austria)
   Proceedings of the Fourth European Conference on Mathematics in
  Industry, May 29-June 3, 1989, Strobl / edited by Hansjörg Wacker,
  Walter Zulehner.
      p. cm. -- (European Consortiun for Mathematics in Industry :
  ECMI vol. 6)
    ISBN-13:978-94-010-6802-4   e-ISBN-13:978-94-009-0703-4
    DOI:10.1007/978-94-009-0703-4

    1. Engineering mathematics--Congresses.   I. Wacker, Hansjörg,
  1939-   . II. Zulehner, Walter, 1955-   . III. Title. IV. Series:
  European Consortium for Mathematics in Industry (Series) ; vol. 6.
  TA329.E96 1989
  620'.00151--dc20   90-19153

ISBN-13:978-94-010-6802-4

CIP-Titelaufnahme der Deutschen Bibliothek

European Conference on Mathematics in Industry 04, 1989,
Strobl :
Proceedings of the Fourth European Conference on
Mathematics in Industry: May 29 - June 3, 1989 Strobl / ed. by
Hansjörg Wacker; Walter Zulehner. - Stuttgart: Teubner;
Dordrecht; Boston; London: Kluwer Acad. Publ., 1990
  (European Consortium for Mathematics in Industry, Vol. 6)
  ISBN-13:978-94-010-6802-4
NE: Wacker, Hansjörg (Hrsg.); European Consortium for Mathematics
  in Industry: European Consortium for ...

| WG: 27,35 | DBN 90.127499.2 | 90.09.06 |
| 7172 | wai | |

ISBN-13:978-94-010-6802-4

Sold and distributed in Continental Europe (excluding U.K.)
by B. G. Teubner GmbH, P.O. Box 801069, D-7000 Stuttgart-80

Sold and distributed in the U.S.A. and Canada
by Kluwer Academic Publishers,
101 Philip Drive, Norwell, MA 02061, U.S.A.

Kluwer Academic Publishers incorporates
the publishing programmes of
D. Reidel, Martinus Nijhoff, Dr W. Junk and MTP Press.

In all other countries (including U.K.), sold and distributed
by Kluwer Academic Publishers Group,
P.O. Box 322, 3300 AH Dordrecht, The Netherlands.

Printed on acid-free paper

Preface

The Fourth ECMI Conference on Industrial Mathematics took place at Strobl in Austria, May 29–June 2, 1989. The conference was devoted to the exchange of ideas, models and methods from various fields of industrial applications of mathematics. About 140 people from 21 countries attended the meeting. The aim was to bring together people from industry and from university. In this respect the organizers were only partly successfull. The participance of about 20 people from industry shows that there is still much work to be done to increase the acceptance from this side.

72 speakers presented their results as invited or contributed lectures, or in the frame of 2 minisymposia. One minisymposium was organized by Heinz W. Engl and focused on steel processing, the other one, organized by Hansjörg Wacker, dealt with chemical engineering.

These proceedings consist of 56 papers. The articles within each of the sections: Invited Lectures, Minisymposium Steel Processing, Minisymposium Chemical Engineering, and Contributed Lectures are in alphabetical order of the first author. Exept for the contributions to the minisymposia, which clearly concentrate on the corresponding topics, it is hard to find a reasonable classification of the papers . This, we believe, is typical for industrial mathematics and underlines the vast variety of fields where mathematics could be used to support problem solving.

We would like to acknowlegde the valuable work of the referees of the articles who certainly helped to improve the quality of this volume.

Linz, Austria   Hansjörg Wacker, Walter Zulehner

Table of Contents

Minisymposium: Chemical Engineering

Contributed Lectures

Invited Lectures

ON A MATHEMATICAL MODEL FOR THE CRYSTALLIZATION OF POLYMERS

D.Andreucci,A.Fasano,M.Primicerio, Università di Firenze,Italy

1. Introduction

Injection moulding is a largely used process in polymer manifacture which consists of two major stages: first the mould is filled with melted polymer and solidification begins at the mould walls; in the second stage — "packing and cooling" — the temperature is decreased while pressure is increased up to 100-150 M Pa in order to complete the filling, compensating for the change in volume due to solidification (see [9]).

This problem was brought to our attention by Dr. S. Mazzullo (Himont, Italia) in the framework of an ongoing cooperation. The aim of the research is to describe the process in a mathematical form so that:

a) the mechanical properties of the solid polymer can be predicted;

b) the quantities governing the process (such as the duration of the pressure pulses, the heat extraction rate, etc.) can be controlled in order to achieve a desired final state.

Here, we present some preliminary results concerning a model of crystallization of a polymer at rest. Effects of motion, of integral stresses etc. will be taken into account in future studies.

This phenomenon exhibits two main differences with respect to the solidification of metals:

(i) solidification takes place over a wide range of temperatures (θ_g, θ_m) (e. g. for Nylon 6, $\theta_g = 333°K$, $\theta_m = 501°K$);

(ii) solidification can occur in two different ways: crystallization (with release of latent heat) and transition to

3

Hj. Wacker and W. Zulehner (eds.),
Proceedings of the Fourth European Conference on Mathematics in Industry, 3–16.
© 1991 B.G. Teubner Stuttgart and Kluwer Academic Publishers.

an amorphous glassy phase. More precisely this transition is by no means a phase-change process but merely a very large increase of viscosity.

A remarkable feature of the phenomenon is that the crystallization fraction of the final product, as well as the average dimensions and spatial distribution of crystals, depends on the history of temperature and pressure during the solidification. In any case, the polymer will always contain a glassy fraction.

The interpretation of the facts (i) and (ii) above is very complicated and not completely understood: in any case it has to be reminded that polymeric chains of different length coexist in the material so that the polymer could hardly be considered as a pure substance. Moreover, during the growth of crystals, segregation of molecules of low molecular weight is frequently observed.

Thus the crystal fraction w at a given point x and at a given time t will be dependent on the past history of the polymer, and in any case it will never exceed a given maximum value w_m, depending on the temperature $\theta(x,t)$.

Since w_m drops steeply to zero at $\theta=\theta_m$, we will assume for simplicity

$$w_m(\theta) = a\ H(\theta_m-\theta)$$

where a is a constant, a<1, and H is the Heaviside graph.

2. Modelling isothermal solidification

In order to describe more clearly the model we will use, we start by considering the case of an isothermal process, i. e. we consider a sample of initially liquid polymer where temperature is mantained at $\bar{\theta}\in(\theta_g,\theta_m)$ throughout.

In this situation, crystals will appear in the melt. During their growth they tend to include glassy parts and will look like spheroids with an interior lamellar structure.

We will describe the process of nucleation and the

process of growth, separately.

a) Nucleation

In the case of heterogeneous nucleation it is assumed that a law of the type of the nuclear decay applies so that (see Avrami [1])

$$(2.1) \qquad \nu(t) = \nu_0 \, [1-w(t)] \quad e^{-pt}$$

where ν is the number of nuclei originating per unit time and per unit volume and ν_0, p are given positive constants. The term $[1-w]$ is introduced to take into account that spherulites can appear only in the fraction of the volume which is still free from crystals, i. e. $(1-w)$.

We will rather consider the case of homogeneous nucleation which is most interesting for industrial application. According to Tobin [13] the birth rate of new crystals is assumed to be proportional to the volume still occupied by the glassy material, and thus

$$(2.2) \qquad \nu(t) = \nu_0 \, (1-w(t)).$$

A drawback of (2.2) is that nucleation would continue also when $w > w_m$, a contradiction with the definition of w_m. For this reason, it has been proposed ([2], [7], [8]) to substitute (2.2) with

$$(2.3) \qquad \nu(t) = \nu_0 \, (1-w(t)/w_m).$$

This will correspond to the fact that, from the very beginning there is a volume fraction $1-w_m$ in which crystals can never appear.

Another way to interpret the fraction $(1-w(t)/w_m)$ —often referred to as the Mandelkern correction [8]— is to say that it takes into account that nucleation is absent not only in the volume fraction occupied by the crystals but also in the glassy parts "swallowed" by the growing spherulites.

More generally one could assume

$$(2.3') \qquad \nu(t) = \nu_0 \, f(1-\tfrac{w}{w_m})$$

with

(2.4) $f(0) = 0$, $f(1) = 1$, $f' \geq 0$.

b) <u>Growth</u>

    Let $v(t,\tau)$ be the volume at time t of a crystal born at time τ.

    For sake of simplicity, we assume that the nucleus is spherical of radius r_0 and that an <u>equivalent</u> <u>spherical</u> shape can be defined such that

(2.5) $v(t,\tau) = \frac{4}{3} \pi \left(r_0 + \int_\tau^t \rho(s) \ ds \right)^3$,

ρ being the crystal growth rate.

    Since

(2.6) $w(t) = \int_0^t \nu(\tau) \ v(t,\tau) \ d\tau$,

it is

(2.7) $\frac{dw}{dt} = \frac{4}{3} \pi \ r_0^3 \ \nu(t) + 4 \pi \ \rho(t) \int_0^t \nu(\tau) \left(r_0 + \int_\tau^t \rho(s) \ ds \right)^2 d\tau$.

    The starting point of all the proposed models is that ρ would be a constant (at constant pressure and temperature, as in the case we are considering) if the spherulites could grow independently of each other. The differences arise on how to incorporate the impingement effect among spherulites.

    According to Avrami, if one denotes by $\tilde{w}$ the (virtual) crystallinity obtained neglecting impingement, then

(2.8) $w(t) = 1 - e^{-\tilde{w}(t)}$,

so that in the case of homogeneous nucleation (2.2) and constant growth rate one obtains the following integral equation in terms of $y(t) \equiv 1-w(t)$

(2.9) $\ln y(t) = -\frac{4}{3} \pi \ \nu_0 \int_0^t y(\tau) \ [r_0 + \rho(t-\tau)]^3 \ d\tau$.

    Equation (2.9) can be easily reduced to an ordinary

differential equation by successive differentiations (see also [12]).

Moreover the same author claims that the correction introduced by (2.8) allows to neglect $1-w(t)$ in the nucleation term, so that (2.9) gives a simple law for the time evolution of $y(t)$. Further simplification leads to the "thumb rule" $y(t)=\exp[-k\ t^n]$ with n to be empirically determined.

A basic assumption in Avrami's model is that the ratio between nucleation and growth rate is constant (isokinetic processes).

In the case we are studying this assumption is not valid since as temperature grows from θ_g to θ_m the relative importance of the two mechanisms changes drastically.

We can add that Avrami's rule is largely used also in the study of solid-solid transitions, e. g. pearlite-austenite transformation in steel (see [4], [5], [14] and the literature quoted therein) and gives good results in isothermal processes.

Non-isothermal solid-solid transitions are usually modelled trying to relate them to isothermal processes rather than releasing the isokinetic assumption; this is done through some simplifying assumptions such as the so-called additivity rule [11].

Indeed the study of the dependence of ν_0 and ρ_0 on the temperature is experimentally much more difficult than in the case of polymers, when optical methods are avaiable.

In the model of Tobin (2.8) is substituted by

(2.10) $\quad w(t) = (1-w(t))\ \tilde{w}(t)$

so that instead of (2.9) one has

(2.11) $\quad \dfrac{1}{y(t)} = 1 + \dfrac{4}{3}\ \pi\ \nu_0 \displaystyle\int_0^t y(\tau)\ [r_0+\rho(t-\tau)]^3\ d\tau.$

We can note that (2.11) corresponds to say that

(2.12) $v(t,\tau) = \frac{4}{3} \pi (r_0+\rho(t-\tau))^3 (1-w(t))$

or, in the case of negligible r_0, to say that the speed of growth at time t is $\rho(1-w(t))^{1/3}$.

In Malkin's approach [7], besides the introduction of the correcting term $1-w/w_m$ instead of $1-w$, the second term in (2.7) is simply assumed to be proportional to $w(w_m-w)$.

Accordingly, (2.7) becomes

(2.13) $\dot{w} = k_0 (w_m-w) (1+c_0 w)$.

A slightly different overall crystallization kinetics has been proposed by Berger et al. [2] in the form

(2.14) $\dot{w} = k_1 w^{\frac{n-1}{n}} (1-w)^{\frac{n+1}{n}}$

where $n>1$ is an empirical coefficient and the term in brackets can be possibly substituted by (w_m-w). This model has been considered in [10] where the numerical calculations have been made also for the complete process with temperature and pressure variations. See also [3] for a crystallization model in a different context.

None of this models seems to be suitable to describe the process from a microscopic point of view and to predict in particular the distribution of the radii of the spherulites which are formed during a given process. Indeed —with the possible exception of Avrami's model which is consistent with experiments only in special cases— they are not based on physical assumptions but rather on artificial correcting terms.

For instance, the corrected growth speed in (2.12) is essentially $\rho(1-w(t))^{1/3}$, but in this case one should take into account this correction in the whole growth interval (τ,t) and not only at the final time t, and compute the integral in (2.5) accordingly.

In this paper we propose that the growth rate ρ is expressed as

(2.15) $\rho(t) = \rho_0 g(1-\frac{w}{w_m})$

where   g   is   a   given   function   with   the   same   qualitative
behaviour of   f, see (2.4).

Setting

(2.16) $\delta(t) = 1 - w(t)/w_m$

we have that the isothermal crystallization is governed by
the following integro-differential equation:

(2.17) $-w_m\dot{\delta} = \frac{4}{3}\pi r_0^3 \nu_0 f(\delta(t)) + 4\pi \rho_0 \nu_0 g(\delta(t)) \int_0^t f(\delta(t)) \left(r_0 + \rho_0 \int_\tau^t g(\delta(s))ds \right)^2 d\tau,$

$\delta(0) = 1.$

We will study equation (2.17) within the framework of
the general (non-isothermal) process, but, just to have an
idea of the behaviour of the solution, here we will consider
briefly the case in which

(2.18) $f(\delta) = g(\delta) = \delta,$

so that we have

(2.19) $\dot{\delta} = -k \, \delta \left\{ 1 + 3c \int_0^t \delta(\tau) \left(1 + c \int_\tau^t \delta(s) \, ds \right)^2 d\tau \right\},$

where

(2.20) $k = \frac{4}{3} \pi r_0^3 \nu_0/w_m,$

(2.21) $c = \rho_0/r_0.$

The determination of the numerical values of k and of c
is rather complicated, because the measurement of the involved quantities is not easy and largely different values are
found in the literature.

We have considered the case of Nylon-6 and (according to
[7]) we have used the value

$k = 1.7 \ 10^4 \ F(\theta) \ \text{sec.}^{-1}$

where

$$F(\theta) = \exp\left\{ -\frac{E}{R\theta} - \frac{\phi_1 \theta_m}{\theta(\theta_m - \theta)} \right\}, \qquad \theta \in (\theta_g, \theta_m).$$

The determination of c is even more complicated because of the difficulty in measuring (and sometimes in defining) the growth rate. Using values taken from [5] we set

$$c = 1.5 \ 10^7 \ F(\theta) \ \phi(\theta) \ sec.^{-1}$$

where $F(\theta)$ is as above and $\phi(\theta)$ cuts the growth rate for temperatures far from θ_m as it is experimentally evident.

In figure 1 we show typical isothermal crystallization curves for Nylon-6, which are in good agreement with the experimental data available in the literature.

It is worth noting that

$$\delta(0) = 1 \quad , \quad \dot{\delta}(0) = -k \ , \ \ddot{\delta}(0) = k(k-3c),$$

$$\lim_{t \to \infty} \delta = 0 \ , \quad \lim_{t \to \infty} \dot{\delta} = 0,$$

so that the condition $k<3c$ is sufficient to have at least one inflection point.

But, in the simplified case we are considering, equation (2.19) can be easily reduced to an ordinary differential equation, noting that

$$(2.22) \qquad 3 \ c \ \delta(\tau) \left(1 + c \int_\tau^t \delta(s) \ ds\right)^2 = -\frac{\partial}{\partial \tau}\left(1 + c \int_\tau^t \delta(s) \ ds\right)^3,$$

consequently (2.19) takes the form

$$(2.23) \qquad \dot{\delta} = -k \ \delta \left(1 + c \int_0^t \delta(s) \ ds\right)^3,$$

and setting

$$(2.24) \qquad Z(t) = 1 + c \int_0^t \delta(s) \ ds,$$

one reduces to

$$(2.25) \qquad \dot{Z} + \frac{k}{4} Z^4 = c + \frac{k}{4}.$$

This form is mainly useful because the qualitative analysis of the solution is easier. In particular we see that

$$\ddot{\delta} = -\frac{k}{4} Z^2 \dot{Z} [12 + 3\lambda - 7\lambda Z^4]$$

where $\lambda = k/c$.

Then, the square bracket is a decreasing function of time starting from $12-4\lambda$ and tending to $-16-4\alpha$. This means that if $k<3c$ we have one and only one inflection point. Otherwise $\ddot{\delta}$ is always positive.

The exact solution of (2.25) can be given in the implicit form

$$(\frac{k}{4} + c) t = G(Z) - G(1),$$

where

$$G(Z) = \frac{1}{2\gamma} \left\{\frac{1}{2} \ln \left|\frac{1+\gamma Z}{1-\gamma Z}\right| + tg^{-1}(\gamma Z)\right\},$$

$$\gamma = \left(\frac{k}{k+4c}\right)^{1/4}.$$

From a numerical point of view the form (2.25) is not particularly simpler than (2.19) since its solution is found in implicit form. In any case, this reduction is no longer possible when processes with varying temperature and/or pressure are considered.

3. The complete model

When temperature θ is no longer constant, recalling (2.3') and (2.15) the equation governing crystallization in $D\equiv\Omega\times\mathbb{R}^+$, $\Omega\in\mathbb{R}^3$ open, is

(3.1)

$$-w_m\frac{\partial\delta(x,t)}{\partial t}=\frac{4}{3}\pi r_0^3(x,t)\nu_0(x,t)\,f(\delta(x,t))+$$

$$+4\pi\rho_0(x,t)\,g(\delta(x,t))\int_0^t \nu_0(x,\tau)\,f(\delta((x,\tau))\left\{r_0(x,\tau)+\int_\tau^t \rho_0(x,s)\,g(\delta(x,s))ds\right\}^2\,d\tau,$$

where r_0, ν_0, ρ_0 are functions of x, t through θ (we assume the pressure to be constant).

The initial condition for (3.1) is

(3.2) $\delta(x,0) \equiv 1$, $x\in\Omega$.

The energy balance is described by the following equation

(3.3) $c(x,t) \frac{\partial\theta}{\partial t} = \text{div}(k(x,t) \nabla\theta) - L\, w_m \frac{\partial\delta}{\partial t}$,   in D,

c specific heat, k conductivity, L>0 (constant) latent heat of crystallization (per volume unit). Equation (3.3) is completed by the initial and boundary conditions

(3.4) $\theta(x,0) = \theta_0(x)$, $x \in \Omega$,

(3.5) $\theta(x,t) = \theta_1(x,t)$, $(x,t) \in \partial\Omega \times \mathbb{R}^+$.

To simplify notation we write (3.1) as

(3.1') $-\frac{\partial\delta}{\partial t}(\underline{x},t) = \mathcal{G}[\delta,\theta](\underline{x},t)$.

We assume that $\forall$ $(\underline{x},t) \in \bar{D}$, $\theta \in \mathbb{R}$,

(3.6) f, g, ν_0, r_0, $\rho_0 \in \text{Lip}(\mathbb{R}) \cap L^\infty(\mathbb{R})$; c, $k \in H^{1+\alpha,\frac{1+\alpha}{2}}(\bar{D})$; $\partial\Omega \in H^{2+\alpha}$;

(3.7) f', g'$\geq$0 and $f(s) \equiv g(s) \equiv 1$, $s \geq 1$, $f(s) \equiv g(s) \equiv 0$, $s \leq 0$;

(3.8) $\nu_0(\theta) \equiv \rho_0(\theta) \equiv 0$ if $\theta \geq \theta_m$;

   ν_0, ρ_0, $r_0 \geq 0$;

(3.9) $\theta_0 \in H^{2+\alpha}(\bar{\Omega})$, $\theta_1 \in H^{2+\alpha,1+\frac{\alpha}{2}}(\bar{D})$;

(3.10) A>1 s.t. $A^{-1} \leq c$, $k \leq A$ in D and A is a bound for all the norms involved in assumptions (3.6)-(3.9);

(3.11) $\theta_0(\underline{x}) \geq \theta_m$.

In (3.6), (3.9) $\alpha \in (0,1)$ is a given constant.

The notation of [6] has been used.

We note that assumptions (3.2) and (3.11), though not strictly necessary from a mathematical viewpoint, are made to keep the model consistent with the case of physical interest here, that is a polymer initially liquid at temperature above θ_m.

Some generalizations are possible e. g. one could let ν_0, r_0, ρ_0 depend explicitly on $(\underline{x}, t)$.

A solution to problem $(3.1')$-(3.5) is defined as a couple (θ, δ) such that $\theta \in C^{2,1}(D) \cap C^0(\bar{D})$, $\delta \in C^0(\bar{D})$, $\delta_t \in C^0(\bar{D})$ and $(3.1')$-(3.5) are fulfilled in a classical pointwise sense.

One can easily see that $\forall \, \theta, \, \bar{\theta}, \, \delta, \, \bar{\delta} \in C^0(\bar{D}_T)$, $D_T \equiv \Omega \times (0,T)$,

$$(3.12) \qquad 0 \le \mathcal{G}[\delta, \theta] \le C(A,T), \qquad \text{in} \quad D_T,$$

$$(3.13) \quad |\mathcal{G}[\delta, \theta](\underline{x}, t) - \mathcal{G}[\bar{\delta}, \bar{\theta}](\bar{x}, \bar{t})| \le C(A,T) \Big\{ |t - \bar{t}| + \|\theta(\underline{x}, \cdot) - \bar{\theta}(\bar{x}, \cdot)\|_t$$
$$+ |\bar{\theta}(\bar{x}, \bar{t}) - \bar{\theta}(\bar{x}, t)| + \|\delta(\underline{x}, \cdot) - \bar{\delta}(\bar{x}, \cdot)\|_t + |\bar{\delta}(\bar{x}, \bar{t}) - \bar{\delta}(\bar{x}, t)| \Big\},$$
$$\forall \, (\underline{x}, t), \, (\bar{x}, \bar{t}) \in D_T, \quad t \le \bar{t}, \quad \text{where} \quad \|\cdot\|_t = \max_{[0,t]} |\cdot|.$$

We need the following preliminary result

Lemma 3.1 *Let* $\theta \in C^0(\bar{D})$ *be a given function. Then* $(3.1')$-(3.2) *has a unique solution* δ *in* D, *satisfying* $\forall \, (\underline{x}, t), (\bar{x}, \bar{t}) \in D_T \equiv \Omega \times (0,T)$, $\forall \, T > 0$,

$$(3.14) \qquad 0 < \delta \le 1,$$

$$(3.15) \qquad |\delta(\underline{x}, t) - \delta(\bar{x}, \bar{t})| \le C(A,T) \Big\{ \|\theta(\underline{x}, \cdot) - \theta(\bar{x}, \cdot)\|_t + |t - \bar{t}| \Big\}.$$

Proof

Let us introduce, for any fixed $\underline{x} \in \Omega$, the unique solution $\mathcal{H}_{\underline{x}}[z]$ of

$$-v_t = \mathcal{G}[z, \theta](\underline{x}, \cdot), \qquad \text{in} \quad (0,T),$$

$$v(0) = 1,$$

where z is an arbitrary function in the class

$$\mathcal{B}_T = \{ z \in C^0([0,T]) \; : \; \|z\|_T \le A \}.$$

It is clear from (3.12) that $\mathcal{H}_{\underline{x}}[z] \in \mathcal{B}_t$ if $t \le t_A$, t_A depending on A only.

In view of (3.13) we have for $t \le t_A$, z_1, $z_2 \in \mathcal{B}_t$,

$$(3.16) \qquad \|\mathcal{H}_{\underline{x}}[z_1] - \mathcal{H}_{\underline{x}}[z_2]\|_t \le C(A,T) \, t \, \|z_1 - z_2\|_t.$$

Hence $\forall \, \underline{x} \in \Omega$, $\mathcal{H}_{\underline{x}}$ is a contraction of $\mathcal{B}_t$ in itself for a small t', $t' = t'(A)$, and it has a unique fixed point δ in $\mathcal{B}_{t'}$, i. e. a solution to $(3.1')$-(3.2).

Then (3.14) follows immediately from (3.1'), (3.2), (3.7), (3.8), noticing that

$$0 \leq \mathcal{G}[\delta,\theta] \leq C(A,T)\, \delta, \quad \text{in} \quad D_{t'}.$$

Existence in the whole D_T, $T>0$, follows iterating the step above with the same t'.

Finally we notice that, owing to (3.13)

$$(3.17)\ |\delta_t(\underset{\sim}{x},t)-\delta_t(\bar{x},t)| \leq C(A,T)\Big\{\|\theta(\underset{\sim}{x},\cdot)-\theta(\bar{x},\cdot)\|_t+\|\delta(\underset{\sim}{x},\cdot)-\delta(\bar{x},\cdot)\|_t\Big\}$$

$\underset{\sim}{x}$, $\bar{x}\in\Omega$, $t\in(0,T)$. Now (3.15) is a consequence of Gronwall's lemma and (3.17). $\square$

We prove now the main result of this section

<u>Theorem 3.1</u> *If (3.6)-(3.11) are fulfilled, and standard compatibility conditions are satisfied on $\partial\Omega\times\{0\}$, problem (3.1')-(3.5) has a unique solution.*

<u>Proof</u>

Fix $T>0$ and consider the class

$$(3.18)\ \mathcal{S}_T=\{(\theta,\delta)\in(H^{\beta,\frac{\beta}{2}}(\bar{D}_T))^2\ |\,\theta\equiv\theta_1\ \text{on}\ \partial\Omega\times(0,T),\ \theta\equiv\theta_0\ \text{on}\ \Omega\times\{0\},$$
$$\delta\equiv 1\ \text{on}\ \Omega\times\{0\},\ 0\leq\delta\leq 1\ \text{in}\ D_T\ \},$$

where $\beta\in(0,\alpha)$ is to be chosen.

Choose $(\theta^1,\delta^1)\in\mathcal{S}_T$, and define θ^2 as the solution to (3.3)-(3.5), where (3.3) has now the form

$$c\,\theta_t^2 - \text{div}(k\theta^2) = Lw_m\, \mathcal{G}[\delta^1,\theta^1].$$

Such a solution exists and is unique in the class $H^{2+\beta,1+\frac{\beta}{2}}(\bar{D}_T)$, because of (3.6)-(3.13).

Moreover [6 p. 419] $\exists\ \bar{\beta}\in(0,\alpha)$, $\bar{\beta}$ depending on A and α only, such that

$$(3.19)\quad |\theta^2|^{(\bar{\beta})}_{\bar{D}_T} \leq C_1 = C_1(A,T).$$

Let us define $\beta=\bar{\beta}$ and δ^2 as the solution to (3.1')-(3.2), where, in the right hand side of (3.1'), θ is replaced by the known function θ^2.

Estimate (3.15) implies $\delta^2 \in H^{\beta, \frac{\beta}{2}}(\bar{D}_T)$, and

$$(3.20) \qquad |\delta^2|_{\bar{D}_T}^{(\beta)} \le C_2 = C_2(A, T, C_1).$$

Thus the transformation

$$\mathfrak{F} : (\theta^1, \delta^1) \to (\theta^2, \delta^2)$$

takes the set

$$\mathfrak{s}_T^0 = \mathfrak{s}_T \cap \{|\theta|_{\bar{D}_T}^{(\beta)} \le C_1, \quad |\delta|_{\bar{D}_T}^{(\beta)} \le C_2\},$$

into itself.

Notice that $\mathfrak{s}_T^0$ is a compact subset of $(C_0(\bar{D}_T))^2$.

Define $\mathfrak{F}(\theta^1, \delta^1) = (\theta^2, \delta^2)$, $\mathfrak{F}(\theta^3, \delta^3) = (\theta^4, \delta^4)$. Using (3.13) and (3.16) we find

$$(3.21) \qquad |\theta^2 - \theta^4|_{D_t}^{(0)} + |\delta^2 - \delta^4|_{D_t}^{(0)} \le C(A, T) t \left\{ |\theta^1 - \theta^3|_{D_t}^{(0)} + |\delta^1 - \delta^3|_{D_t}^{(0)} \right\}.$$

Therefore $\mathfrak{F}$ is continuous in the max norm and Schauder's theorem implies the existence of a fixed point of $\mathfrak{F}$ in $\mathfrak{s}_T^0$, i.e. of a solution to (3.1')-(3.5) in the required class.

Uniqueness follows from noticing that $\mathfrak{F}$ is contractive for small t (see (3.21)).   □

Notice that $\delta_t \ge 0$ through D. Then (3.1)-(3.5) is a suitable model for processes where solidification only takes place.

Acknowledgement   We wish to thank Dr. S Mazzullo and Profesor S. Paveri-Fontana for many interesting discussions, and Dr. M. Bianchini for her help in performing numerical work.

REFERENCES

[1] M.AVRAMI, *Kinetics of Phase Change, I, II and III*, J. Chem. Phys. $\underline{7}$, 1103-1112 (1939); $\underline{8}$, 212-224 (1940); $\underline{9}$, 117-184 (1941).

[2] J.BERGER, W.SCHNEIDER, *A zone model of rate controlled solidification*, Plastics and Rubber Processing and Applications $\underline{6}$, 127-133 (1986).

[3] G.BRANDEIS et al., *Nucleation, Crystal Growth and the Thermal Regime of Cooling Magmas*, J. of Geophysical Res. $\underline{89}$, 10161-10177 (1984).

[4] J.W.CHRISTIAN, *The theory of transformations in metals and alloys*, Pergamon Press, Oxford 1965.

[5] W.J.HAYES, *Mathematical models in material sciences*, M. Sc. Thesis, Oxford 1985.

[6] O.LADYZENSKAJA et al., *Linear and Quasilinear Equations of Parabolic Type*, Translations of Mathematical Monographs, Vol. 23, American Mathematical Society, Providence R. I. 1968.

[7] A.YA.MALKIN et al., *General treatment of polymer crystallization kinetics, Parts 1 and 2*, Polym. Eng. Sci., Poym. Phys. Ed. $\underline{24}$, 1396-1401, 1402-1408 (1984).

[8] L.MANDELKERN, *Crystallization of Polymers*, Mac Graw-Hill, New York 1964.

[9] S.MAZZULLO, *Crystallization of polymers processing and the associated free and moving boundary problems*, Himont Italia 1987.

[10] S.MAZZULLO, M.PAOLINI, C.VERDI, *Polymer crystallization and processing: free boundary problems and their numerical approximation*, ECMI 88, Glasgow.

[11] E.SCHEIL, *Anlaufzeit den Austenitumwandung*, Arch. für Eisenhütten-wesen $\underline{8}$ (1935) 565-579.

[12] W.SCHNEIDER et al., *Non-isothermal Crystallization of Polymers*, Intern. Polymer Processing II 314, 151-154 (1988).

[13] M.C.TOBIN, *Theory of phase transition kinetics with growth site impingement, I and II*, J. Polym. Sci., Polym. Phys. Ed. $\underline{12}$, 394-406 (1974) and $\underline{14}$, 2253-2257 (1976).

[14] A.VISINTIN, *Mathematical models of solid-solid phase transitions in steel*, IMA J. Appl. Math. $\underline{39}$ (1987), 143-157.

REMARKS ON NEW MATHEMATICAL PROBLEMS
ARISING IN THE CONTEXT OF INFORMATION TECHNOLOGY

A. Bensoussan

General comments

Interaction between Computer Science and Mathematics has evolved considerably. We discuss here some features which have been emerging, among the main ones.

The plan is as follows :

    I.   SCIENTIFIC COMPUTING

    II. CONTROL, IDENTIFICATION, ESTIMATION

    III. DISCRETE EVENT SYSTEMS

    IV. NEW AREAS OF INFORMATION TECHNOLOGY

I. SCIENTIFIC COMPUTING

The traditional applications of mathematics arise in Physics, Mechanics,... Supercomputers have permitted :
- to study completely new areas of physical sciences.
- to consider new numerical techniques
- to investigate new approaches

I.1. New Areas of physical sciences

Among important developments in several fields, we mention the *Numerical Simulation of Reactive flow*. It applies to *combustion, aeronomy, partially ionized plasmas, aerodynamics, gas dynamic lasers, astrophysics, general multiphase and magnetohydrodynamic flows,....*

The model takes into account the *coupling* between *fluid dynamics* and *chemical reactions*.

The traditional model of an *homogeneous, viscous, incompressible flow* with *no chemical reactions* and *no external forces* consists of Navier Stokes equations :

$$\rho_0 \left(\frac{\partial u}{\partial t} + (u \cdot D)u \right) - \mu \Delta u + Dp = 0$$

$$\text{div } u = 0$$

17

Hj. Wacker and W. Zulehner (eds.),
Proceedings of the Fourth European Conference on Mathematics in Industry, 17–33.
© 1991 *B.G. Teubner Stuttgart and Kluwer Academic Publishers.*

If the fluid has a constant *specific heat* c and there are no external heat sources, then the temperature of the fluid is the solution of :

$$\rho_0 c(\frac{\partial T}{\partial t} + u \cdot DT) - \lambda \Delta T == 2tr\epsilon^2(u)$$

where $\epsilon = \frac{1}{2}(Du + (Du)^T)$ is the *velocity tensor*. The internal energy density is cT.

In general, all variables are coupled and appear as the solution of a complex system of P.D.E.

The main unknown are the mass density ρ, the velocity of the flow u, the number densities n^i of the individual chemical species and the Energy density E.

The system of equations is the following :

$$\frac{\partial \rho}{\partial t} + D \cdot (\rho u) = 0$$

$$\frac{\partial (\rho u)}{\partial t} + D(\rho uu) + D\sigma = \sum_i \rho^i a^i$$

$$\frac{\partial n^i}{\partial t} + D(n^i(u + u^i)) = Q^i - L^i n^i$$

$$\frac{\partial E}{\partial t} + D(Eu) + D(q + q_r) = \sum_i (u + u^i) \cdot m^i a^i$$

where σ is the *pressure tensor*, q the *heat* flux, q_r the radiative heat flux, a^i represent external forces, and Q^i, L^i represent the chemical production rates and losses, u^i is the diffusion velocity of species i. They are highly nonlinear expressions of the unknowns.

In view of the complexity, a *modular* approach is useful. Each physical process is calculated accurately and calibrated separately .

The physical properties should be incorporated in the numerical algorithms and a mathematical analysis of the behaviour of the algorithms should be performed. For more details, see [15].

I.2. Numerical methods

We shall illustrate the general idea of *decoupling* the difficulties in the case of Navier Stokes equations :

$$\frac{\partial u}{\partial t} + (u \cdot D)u - \mu \Delta u + Dp = f$$

$$\text{div } u = 0$$

$$u(x, 0) = u_o(x) \qquad (\text{div } u_0 = 0)$$

$$u = g \text{ on } \Gamma \qquad \int_\Gamma \nu \cdot g \, d\Gamma = 0$$

The two main difficulties are non linearities and incompressibility condition. *Operator splitting* will realize the decoupling.

Let θ be a parameter in $(0, \frac{1}{2})$ and α, β with $\alpha + \beta = 1$.

Knowing u^n, we compute $\{u^{n+\theta}, p^{n+\theta}\}$, $u^{n+1-\theta}$ and $\{u^{n+1}, p^{n+1}\}$ by the iteration :

(1)
$$\frac{u^{n+\theta} - u^n}{\theta \Delta t} - \alpha \mu \Delta u^{n+\theta} + D p^{n+\theta} = \beta \mu \Delta u^n$$
$$- (u^n \cdot D) u^n + f^{n+\theta}$$

$$\operatorname{div} u^{n+\theta} = 0$$

$$u^{n+\theta} = g^{n+\theta} \qquad \text{on } \Gamma$$

(2)
$$\frac{u^{n+1-\theta} - u^{n+\theta}}{(1 - 2\theta) \Delta t} - \beta \mu \Delta u^{n+1-\theta} + (u^{n+1-\theta} \cdot D) u^{n+1-\theta} = \alpha \mu \Delta u^{n+\theta}$$
$$- D p^{n+\theta} + f^{n+1-\theta}$$
$$u^{n+1-\theta} = g^{n+1-\theta} \qquad \text{on } \Gamma$$

(3)
$$\frac{u^{n+1} - u^{n+1-\theta}}{\theta \Delta t} - \alpha \mu \Delta u^{n+1} + D p^{n+1} = \beta \mu \Delta u^{n+1-\theta}$$
$$- (u^{n+1-\theta} \cdot D) u^{n+1-\theta} + f^{n+1}$$

$$\operatorname{div} u^{n+1} = 0$$

$$u^{n+1} = g^{n+1} \qquad \text{on } \Gamma$$

(2) is nonlinear and solved by a *least square technique* , and *conjugate gradient* minimization. (1) is linear and can be reformulated as a variational problem for the pressure p.

Various possibilities of finite element approximation, multigrid methods and domain decomposition can then be used at the discretization stage.

Efficient software packages result in the combination of all these techniques. For more details, see [8].

I.3. New approaches

We present two new directions :

I.3.1. Wavelets

An alternative to Fourier analysis has been developed in recent years, with applications to signal and image processing, sound analysis and numerical analysis. It has foundations in quantum field theory, statistical mechanics and pure mathematics (geometry of Banach spaces). This is the *Wavelet analysis*.

It combines advantages of the Haar system and of the trigonometrical system. The Haar system is defined by :

$$\psi(x) = \begin{cases} 1, & 0 \leq x < \dfrac{1}{2} \\ -1 & \dfrac{1}{2} \leq x < 1 \\ 0 & \text{otherwise} \end{cases}$$

$$\psi_{m,n}(x) = 2^{-\frac{m}{2}} \psi(2^{-m} x - n), \qquad m, n \in Z.$$

The $\psi_{m,n}$ form an orthonormal basis of $L^2(R)$, (and even L^p) but not for Sobolev spaces (unlike trigonometric series for periodic Sobolev spaces). On the other hand, the $\psi_{m,n}$ have good localization properties unlike trigonometric functions (the reverse being true for their Fourier transforms).

A wavelet system is defined by a function $\psi(x)$ and

$$\psi_{m,n}(x) = 2^{-\frac{m}{2}} \psi(2^{-m} x - n)$$

with the property

$$L^2(R) = \oplus_{m \in Z} W_m$$
$$W_m = \overline{\text{span } \{\psi_{m,n}\}}, \qquad \text{orthogonal spaces}$$

$\{\psi_{m,n}, n \in Z\}$ is an orthonormal basis for W_m.

Y. Meyer has constructed a wavelet system with ψ, C^∞ with rapide decay (faster that any power). Later one has constructed a wavelet system with ψ, C^k with exponential decay, and finally I. Daubechies has shown the existence of wavelet systems with compact support and arbitrary regularity. They will be very useful for all kinds of applications.

They are obtained from sequences h_n, with compact support, satisfying *additional assumptions* by the following procedure :

$$\phi(x) = \lim_{k \to \infty} \eta_k(x)$$
with
$$\eta_k(x) = \sqrt{2} \sum_n h_n \eta_{k-1}(2x - n)$$
$$\eta_0 = \mathbb{1}_{[-\frac{1}{2}, \frac{1}{2}[}$$
then
$$\psi(x) = \sqrt{2} \sum_n (-1)^n h_{-n+1} \phi(2x - n)$$

The *most* compact support corresponds to the two possible choices :

$$h_0 = \frac{1 \mp \sqrt{3}}{4\sqrt{2}}, \quad h_1 = \frac{3 \mp \sqrt{3}}{4\sqrt{2}}, \quad h_2 = \frac{3 \pm \sqrt{3}}{4\sqrt{2}}, \quad h_3 = \frac{1 \pm \sqrt{3}}{4\sqrt{2}},$$

For more details, see [12], [7].

I.3.2. Cellular Automata.

The availability of massively parallel computers, like the connection machine, has motivated the use of cellular automata on large lattices for obtaining solutions to P.D.E., in particular the incompressible Navier Stokes equations. A lot of work is necessary to justify this approach.

We describe here a model due to B.M. BOGHOSIAN, C.D. LEVERMORE,[3]. See also U. FRISCH, B. HASSLACHER, Y. POMEAU, [10].

Consider Burgers' equation :

$$\frac{\partial u}{\partial t} + c\frac{\partial}{\partial x}(u - \frac{u^2}{2}) = \nu\frac{\partial^2 u}{\partial x^2}$$

Replacing $\dfrac{\partial u}{\partial t}$ by $\dfrac{u(x,t+\Delta t) - u(x,t)}{\Delta t}$

Replacing $\dfrac{\partial}{\partial x}$ by $\dfrac{1}{2}\dfrac{u(x+\Delta x,t) - u(x-\Delta x,t)}{\Delta x}$

Replacing $\dfrac{\partial^2 u}{\partial x^2}$ by $\dfrac{1}{\Delta x^2}(u(x+\Delta x,t) + u(x-\Delta x,t) - 2u(x,t))$

and choosing $\Delta t, \Delta x$ such that $2\nu = \dfrac{\Delta x^2}{\Delta t}$, we obtain the discretization scheme

$$u(x,t+\Delta t) = \frac{1 - \frac{c}{2\nu}\Delta x}{2}u(x+\Delta x,t) + \frac{1 + \frac{c}{2\nu}\Delta x}{2}u(x-\Delta x,t)$$
$$+ \frac{c\Delta x}{8\nu}(u^2(x+\Delta x,t) - u^2(x-\Delta x,t))$$

This can be simulated "approximately" by the stochastic process :

$$\xi_1(x+\Delta x,t+\Delta t) = \frac{1+w(x,t)}{2}(\xi_1(x,t) + \xi_2(x,t))$$
$$- w(x,t)\xi_1(x,t)\xi_2(x,t)$$

$$\xi_2(x-\Delta x,t+\Delta t) = \frac{1-w(x,t)}{2}(\xi_1(x,t) + \xi_2(x,t))$$
$$+ w(x,t)\xi_1(x,t)\xi_2(x,t)$$

where ξ_1, ξ_2 take the values $0, 1$, w is random and takes the values -1 or 1. The random variables are independent and :

$$Ew = \frac{c}{2\nu}\Delta x$$

It can be proved that :

$$u(x,t) \sim E(\xi_1 + \xi_2)$$

The process ξ_1, ξ_2 is a cellular automata which can simulated on a *connection machine*. Reseach on similar types of stochastic processes is important in the context of solving nonlinear P.D.E. on massively parallel machines.

II. CONTROL, IDENTIFICATION, ESTIMATION.

The applications of these techniques are extremely diversified and come from physical sciences as well as from economic or even social sciences.

We describe some :
- new areas of applications
- new algorithms
- new approaches

II.1. New areas of application.

II.1.1. Environmental studies. The program "Global Change".

In view of the growing importance of environmental issues, a worldwide program of research has been developing in recent years, under the name of "Global Change". It connects specialists of climate dynamics, oceanography, Planetary Physics,... It seems that this direction is a source of important mathematical problems, of somewhat new nature.

The basic problem deals with the *prediction* of physical quantities, solutions of a set of nonlinear evolution P.D.E., with *unknown* parameters and unknown initial state. Nonlinearity creates an important sensitivity with respect to initial data and unknown quantities, resulting in a lack of predictability beyond some length of time. A fundamental question is to identify the *important regimes* of the physical variables, those which contain the main futures of interest and are *persistent*. There are several ways to give a mathematical meaning to this question. The interesting feature is that they result in a *mixture of statistical and dynamical methods*. A lot of work is needed in that direction, even for simple nonlinear systems.

The point of view of dynamical systems is to obtain the stationary solutions of the nonlinear P.D.E. (or system of P.D.E.) and the long-time behaviour of solutions. This is the theory of *attractors*.

A complementary statistical theory has been developed, for which we describe only two ideas, that of *persistent anomalies* and that of *EOF analysis* (Empirical orthogonal functions).

Consider a vector representing physical variables (typically a flow) which is computable through a model, which is not in general completely known (this is an important difficulty, which we leave aside). We represent it by $\psi_k(t), k = 1 \ldots N$ where k may represent a point x_k on a grid, or a component if the solution is obtained by an expansion.

We set $< \psi_k >=$ average of $\psi_k(t)$ over some record of data.

The instantaneous *anomaly* is defined by :

$$\widetilde{\psi}_k(t) = \psi_k(t) - < \psi_k >$$

The *pattern correlation* between an anomaly at time t and at a later time $t + \tau$ is defined by :

$$p(t, \tau) = \frac{\sum_k \widetilde{\psi}_k(t)\widetilde{\psi}_k(t + \tau) - (\sum_k \widetilde{\psi}_k(t))(\sum_k \widetilde{\psi}_k(t + \tau))}{\sigma(t)\sigma(t + \tau)}$$

where

$$\sigma(t)^2 = \sum_k \widetilde{\psi}_k(t)^2 - (\sum_k \widetilde{\psi}_k(t))^2.$$

We say that an anomaly $\widetilde{\psi}_k(t_0)$ *persists* from $t = t_0$, to $t = t_0 + J\tau$, if :

$$p(t_j, \tau) \geq p_0, \text{ where } t_j = t_0 + j\tau, j = 0 \ldots J - 1$$

and p_0 represents the persistence criterion. What is *expected* is that the anomalies which satisfy a reasonable persistence criterion fall into a *small number of easily identifiable patterns*, related to the attractors of dynamical system.

The EOF analysis goes as follows. Let :

$$\Gamma_{k\ell} = < \tilde{\psi}_k \tilde{\psi}_\ell >$$

Consider the eigenvalues of the matrix Γ $\lambda^1 \dots \lambda^N$, ranked in decreasing order and $e^1 \dots e^N$ are the corresponding eigenvectors, called the 1st EOF, the 2nd EOF, ...
Next expand the vector $\tilde{\psi}(t) = (\tilde{\psi}_k(t))$ on the basis $e^1 \cdots e^N$, hence :

$$\tilde{\psi}(t) = \sum_{i=1}^{N} \alpha_i(t)e^i$$

then one can easily check that :

$$< \alpha_i \alpha_j > = \lambda^i \delta_{ij}.$$

The coefficients $\alpha_i(t)$ are called the *principal components*. The EOF are interpreted as directions of variability of the anomaly, λ^i representing the part of the variance related to EOF e^i (the total variance being $\lambda^1 + \dots + \lambda^N$). The important *conjecture* is that the main OEF are related to the patterns associated to persistent anomalies.

In [13] theses connections are exhibited experimentally on some models.

Is there a general theory for these phenomenon, at least for some class of nonlinear dynamical systems ? This is an open question, which has a crucial importance for the understanding of the variability of atmospheric dynamics.

II.1.2. Computer vision

II.1.2.1. The segmentation problem

An image can be represented by a function $g(x, y)$ measuring the strength of the light signal striking a plane at point (x, y). Such a function is expected to have discontinuities reflecting edges of objects, and shadows. Outside such lines the function g is expected to behave more smoothly.

Having this in mind, one defines a segmentation of a region Ω, as a set of open connected subsets $\Omega_i, i = 1 \cdots n$, each one with a piecewise smooth boundary and Γ is the union of the parts of the boundaries of the Ω_i inside Ω.

An approximation of g is a function u which is differentiable on $\Omega - \Gamma$. One defines a cost function :

$$J(u, \Gamma) = \mu \int_\Omega (u - g)^2 dx + \int_{\Omega - \Gamma} |Du|^2 dx + \nu |\Gamma|$$

The *segmentation problem* consists in minimizing the functional J over the pair (u, Γ). Note that if $\nu = 0, \inf J = 0$.

This is a new class of problems in the calculus of variations, introduced in [14].

It has attracted a lot of interest and some progress has been made, concerning existence, and approximation.

It is interesting to consider the one dimensional problem, in which $\Omega = (0, 1), \Gamma = \{a_1; \cdots a_N, \text{ with } 0 < a_1 < a_2 \cdots < a_N < 1\}$ and $|\Gamma| = N$. One has :

$$J(u, a_1, \cdots, a_N) = \mu \int_0^1 (u - g)^2 dx + \sum_{i=0}^{N} \int_{a_i}^{a_{i+1}} u'^2 dx + \nu N$$

and we have defined $a_0 = 0, a_{N+1} = 1$.

Since we do not impose continuity at points a_i, we may write preferably :

$$J(u_1, \cdots, u_N; a_1, \cdots, a_N) = \mu \sum_{i=0}^{N} \int_{a_i}^{a_{i+1}} [(u_i - g)^2 + u'^2] dx + \nu N$$

There is a probabilistic interpretation of J. Consider in the segment (a_i, a_{i+1}) a process x_i such that :

$$x_i(t) = x_i(a_i) + w_i(t), \qquad t \in [a_i, a_{i+1}[$$

where $x_i(a_i)$ is not random, and $w_i(t)$ is a standard Wiener process. We observe on (a_i, a_{i+1}) the process $y_i(t)$ with :

$$dy_i(t) = x_i(t)dt + db_i \qquad y_i(a_i) = 0$$

where b_i is Wiener process independent from w_i.

The "a priori probability" of the trajectory $x_i(t)$ to coincide with a given function $u_i(t)$ which is $H^1(a_i, a_i + 1)$ is :

$$\exp -\frac{1}{2} \int_{a_i}^{a_{i+1}} [(u_i'^2 + u_i^2)dt - 2u_i dy_i]$$

For details see [17].

Considering independent processes in each interval, we obtain :

$$\exp -\frac{1}{2} \sum_{i=0}^{N} \int_{a_i}^{a_{i+1}} [(u_i'^2 + u_i^2)dt - 2u_i dy_i]$$

and the maximization of this probability results in minimizing J, up to the correspondence $dy_i \to g$ on (a_i, a_{i+1}). It would be extremely interesting to treat the 2 dimensional problem, which is the real one, by similar probabilistic methods. It is an open problem.

II.1.2.2. Mobile Robotics

Consider the problem of a mobile robot which tries to recover its environment, during its motion (the environment is assumed to be static). The robot is equipped with a camera, which takes images between time intervals. One way of approaching the problem is to extract tokens from the images in the sequence, match them from image to image and recover the motion and the structure of the environment.

Naturally, the tokens we compute in the images should be closely related to objects in the scene, if we want the matches to be meaningful. They are in general surface markings, shadows, depth discontinuities.

Let us explain the general ideas in the case of a point M, which is the object to be recognized by the mobile robot.

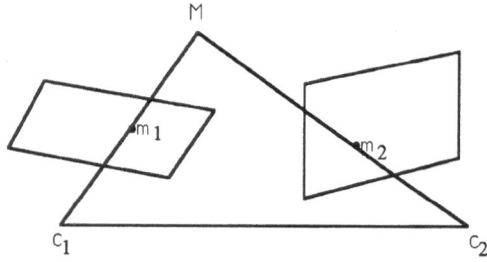

So M is the real point, C_1, C_2 represent the motion of the camera (installed on the robot), m_1, m_2 the images of M. The motion is decomposed into a rotation R with a rotation axis going through C_1, and a translation $t = C_1 C_2$.

If we consider a coordinate system attached to the camera, then we can measure $C_1 m_1$ and $C_2 m_2$ with the local coordinate system. The coordinates with respect to a common coordinate system, that related to C_1 are $C_1 m_1$ and $\mathcal{R} C_1 m_1$. Then one expresses the planarity constraint, namely that $C_1 m_1$, $C_2 m_2$ and t are coplanar ; it amounts to :

$$C_1 m_1 \cdot (t \wedge \mathcal{R} C_2 m_2) = 0.$$

The vector t has coordinate t_x, t_y, t_z but from the linearity, we can assume that $\|t\| = 1$, hence 2 parameters are enough. The matrix $\mathcal{R}$ depends of 3 parameters which characterize the unit rotation axis (2 parameters) and the rotation angle.

Conceptually, what is important is to recognize that the previous relations amounts to :

$$f(x, a) = 0$$

where a is a vector of parameters $\in \mathbb{R}^n$, and x is a vector of measurement $\in \mathbb{R}^n$ and f is a nonlinear relation.

Each successive image leads to a relation :

$$f(x_k, a) = 0$$

However the observation is not exact and rather described by the model

$$z_k = x_k + \nu_k$$

where ν_k is a white noise of covariance Γ. Considering that

$$a_{k+1} = a_k = a$$

we are in the framework on nonlinear filtering if we can express x_k as a function of a_k. It is of course natural to linearize around a given estimate of a, and to use extended Kalman filtering. Once t, R is obtained, one can recover M by expressing the relations :

$$\lambda C_1 m_1 = t + \mu R C_2 m_2$$

where λ, μ are unknown scalars. In this relation again t, R are known random variables, as well as $C_1 m_1, C_2 m_2$. Thus we are in a situation similar to the above and can use again a Kalman filter.

These techniques have been extensively used in the context of mobile robotics by O. FAUGERAS and his team, see for instance [9].

II.2. New algorithms

II.2.1. Parallel algorithms

The development of multiprocessors has generated a substantial interest in the obtaining of *parallel algorithms*. A thorough analysis is needed, since surprises can arise in comparison with the sequential approach.

Take for instance Jacobi and Gauss Seidel iterations for obtaining a fixed point of :

$$x = f(x) \qquad x \in \mathbb{R}^n$$

A Jacobi iteration is the following :

$$x_i^{k+1} = f_i(x^k), \qquad i = 1 \cdots n$$

and a Gauss Seidel is :

$$x_i^{k+1} = f_i(x_1^{k+1}, \cdots, x_{i-1}^{k+1}, x_i^k, \cdots, x_n^k)$$

The advantage of Gauss Seidel iteration is that it converges more frequently that Jacobi, and sequentially it performs much better (the convergence rate of Gauss Seidel iteration is better).

Parallel implementation will change the situation considerably.

Consider the case when there are n processors, and the sequence x^k such that :

$$x^{k+1} = f(x^k)$$

denoted by $x^{k,J}$ (Jacobi sequence) converges towards the fixed point. Suppose also f monotone, i.e. $f(x) \leq f(y) \; \forall x, y$ with $x \leq y$. Then take a sequence $x^{k,U}$ defined by :

$$x_i^{k+1,U} = f_i(x^{k,U}), \qquad \forall i \in U_k$$
$$x_i^{k+1,U} = x_i^{k,U}, \qquad \forall i \notin U_k$$

U_k is a subset of $\{1, \cdots, n\}$.

One can prove (T.N. TSITSIKLIS) that if one starts with the same initial value x^0 and $f(x^0) \leq x^0$ or $x^0 \leq f(x^0)$, then :

$$x^* \leq x^{k,J} \leq x^{k,U}, \qquad \forall k$$

where x^* is the limit fixed point. Hence Jacobi iteration performs better than any parallel version of Gauss Seidel iteration. When they are less than n processors available, or the assumption of monotonicity is not satisfied, no general statement can be made (see [2])

II.2.2. Simulated Annealing

This type of algorithm has been developed in the recent years in order to obtain a *global minimum* for a function $U(x)$, over $x \in B$, B compact, in the case when U is smooth. It is clear that such a problem occurs in many applications. Simulated annealing has first been used in the context of image processing.

The algorithm consists in a discrete version of the following stochastic differential equation:

$$dx_t = -DU(x_t)dt + c_t\sigma(x_t)dw_t, \quad x(0) = x$$

where the following assumptions are made

- U is C^2 from B to $[0, \infty)$ and

$$Min_{x \in B} U(x) = 0, \qquad DU(x) \cdot x > 0, \quad \forall x \in B - \overset{\circ}{B_1}$$

where B is a ball in $\mathbb{R}^n$, centered at the origin, and B_1 is an other ball, also centered at the origin and strictly included in B.

- σ is Lipschitz continuous from B to $[0, 1]$, with $\sigma = 1$, for $x \in B_1, \sigma = 0$ for $x \in \partial B, \sigma > 0$ on $\overset{\circ}{B}$
- $c_t = \dfrac{c}{Logt}$, for t large, $c > 0$.
- w_t standard Wiener process in $\mathbb{R}^n$
- $\pi^\epsilon(x) = \dfrac{1}{Z^\epsilon}(exp - \dfrac{2U(x)}{\epsilon^2}) \mathbb{1}_B$ with $\int \pi^\epsilon(x)dx = 1$ converges weakly to a probability π as $\epsilon \to 0$.

Note that π is a probability concentrated on the set of global minima of U(.).

Then the following result can be proved :

$$Ef(x_t) \to \pi(f)$$

$\forall f$ bounded, continuous, as $t \to \infty$, uniformly for x (the initial value) in B.
(For more details see [5]).

II.3. New approaches

Let us just mention the developments related to H_∞ theory and which permit to obtain protection of dynamic systems from disturbances via *feedback control* . We just mention some recent results concerning linear systems.

Let us consider the linear system

$$\dot{x} = Ax + Bu + Dw,$$
$$y = Cx$$

where w represents a disturbance, and u a control. We consider *feedback* controls, $u = Kx$. The *transfer* matrix $T_K(s)$ is given by

$$T_K(s) = C[sI - (A + BK)]^{-1}D$$

and we consider those K for which $A + BK$ is stable. The H_2 norm is defined by :

$$\|T_K\|_2 = \left(\dfrac{1}{2\pi} \int_{-\infty}^{+\infty} \text{tr } T_K(-j\omega)^* T_K(j\omega)d\omega\right)^{\frac{1}{2}}$$

and the H_∞ norm is defined by :

$$\|T_K\|_\infty = \sup_{\omega \in R}(\text{tr } T_K(-j\omega)^* T_K(j\omega))^{\frac{1}{2}}$$

which are finite since $A + BK$ is stable.

The problem of H_∞ or H_2 control *consists* in minimizing the above norms with respect to K.

Note that

$$\|T_K\|_\infty = \sup_w \{(\int_0^\infty |y(t)|^2 dt)^{\frac{1}{2}} | (\int_0^\infty |w(t)|^2 dt)^{\frac{1}{2}} \leq 1\}$$

and thus this norm expresses the *sensitivity of the system* with respect to external disturbances.

Among the important results obtained recently, it has been proven that *we can chose* a K such that $\|T_K\|_\infty \leq \gamma, \forall \gamma$ given, *if* there exists ϵ such that one can solve the Riccati equation

$$PA + A^*P - \frac{1}{\epsilon}PBB^*P + \frac{1}{\gamma}PDD^*P + \frac{1}{\gamma}CC^* + \epsilon I = 0$$

In fact $K = -\dfrac{B^*P}{2\epsilon}$ will serve for this purpose (for more details, see [11]).

III. DISCRETE EVENT SYSTEMS

New applications strongly related to information technology have created the need to develop a theory of DEDS, *discrete event dynamic systems*. Such applications are production or assembly lines, computer/communication networks, traffic systems,... A special issue of IEEE, Jan. 1989 is devoted to dynamics of discrete event systems.

Many new mathematical techniques have been developed in this context. We describe here one of them, the use of an algebraic structure, called *dioid*, in the modelling of *timed event graphs*.

Let us just recall the basic definition of a dioid. It is a set $\mathcal{D}$ provided with two inner operations $\oplus$ and $\otimes$ (addition and multiplication) such that
- they are both associative
- addition is commutative
- multiplication is right distributive with respect to addition
- there exists a null and identity elements

$$\exists \varepsilon \in \mathcal{D} : \forall a \in \mathcal{D}, \quad a \oplus \varepsilon = a$$
$$\exists e \in \mathcal{D} : \forall a \in \mathcal{D}, \quad a \otimes e = e \otimes a = a$$

- the null element is absorbing

$$\forall a \in \mathcal{D}, \quad a \otimes \varepsilon = \varepsilon \otimes a = \varepsilon$$

- the addition is idem potent

$$\forall a \in \mathcal{D}, \quad a \oplus a = a.$$

When addition is commutative, the dioid is called commutative. As an example take $\mathcal{D} = Z \cup \{-\infty\} \cup \{+\infty\}$ and

$$\oplus = \max, \quad \otimes = +$$
$$\varepsilon = -\infty, \quad e = 0$$

(note that we impose the rule $(-\infty) \otimes (+\infty) = (-\infty)$).

We can also consider

$$\oplus = \min, \quad \otimes = +$$
$$\varepsilon = +\infty, \quad e = 0$$

(in which case $(-\infty) \otimes (+\infty) = +\infty$).

A dioid is a structure somewhere between linear algebra and lattices. One can define a partial order relation

$$a \geq b \Leftrightarrow a = a \oplus b$$

and a *pseudo left inverse* denoted $a\backslash c$ which is the greatest subsolution of

$$a \otimes x = c.$$

Starting with these premises affine equations can be solved, as well as matrices defined and a matrix calculus is available. Matrix equations can also be solved. Let us see briefly how these concepts apply to timed event graphs.

Times event graphs are a special kind of Petri nets. They are directed graphs with two types of edges, *places* and *transitions*

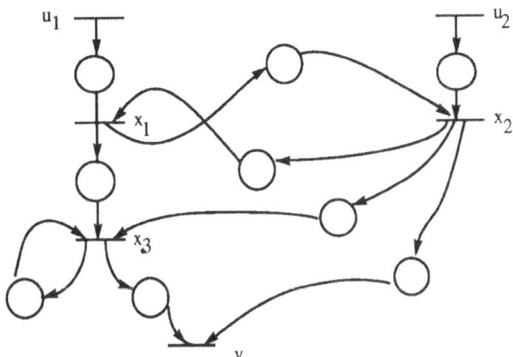

In the preceding figure, the transitions are $u_1, u_2, x_1, x_2, x_3, y$ and the places are denoted by $x_1|u_1, x_2|u_2, x_3|x_1, x_3|x_2, x_3|x_3, y|x_3, y|x_2, x_2|u_2, x_2|x_1$.

There is a single transition upstream and downstream, at each place.
In places, these are tokens or not. Tokens are created or consumed when transitions are fired, more precisely when a transition t is fired one token is consumed at each place which precedes t and one is created at each place which succeeds it.

Let us assume that transitions are immediate, but a token must stay at a place an amount of time called the holding time, which depends on the place. The following symbols are used

no token   1 token   1 token
holding time =3   holding time 0   holding time =2

For instance consider the places which precede x_1, we complete the information as follows

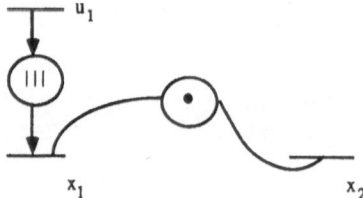

Let for a transition x, x_n be date at which transition x has been fired for the nth time. We can write the relation

$$(x_1)_n = \max[(x_2)_{n-1}, (u_1)_n + 3]$$

and of course similar relations for other transitions.

If the dates take values in $Z \cup \{+\infty\} \cup \{-\infty\}$, then we can work with the dioid considered above $\mathcal{D}$, with the operations $\oplus = \max$, $\otimes = +$.

The preceding relation writes

$$(x_1)_n = (x_2)_{n-1} \oplus 3(u_1)_n$$

where $3(u_1)_n = 3 \otimes (u_1)_n$ to simplify the notation.

One of the objectives of research in these directions is to obtain a theory similar to that of linear dynamic systems. In particular a theory of *stability* is being developed. This is important to obtaining an evaluation of performances for the real system which is modelled by the event graph. (See [6]).

IV. NEW AREAS OF INFORMATION TECHNOLOGY

Let us mention only two fields, artificial intelligence and neural networks and some recent mathematical problems motivated by them (again this is by no means exhaustive).

IV.1. Artificial intelligence

Since artificial intelligence needs to deal with *qualitative* aspects, more than with quantitative aspects (or in connection with them), this has motivated the development of *qualitative simulation* (or *qualitative physics*) in particular at Xerox Parc. Note that the economists needed much before similar techniques, in the context of the theory of *comparative economics* (P.A. SAMUELSON).

Our presentation here relies on some recent work of J.P. AUBIN.

We pose the problem of the *qualitative evolution* of solutions to a differential equation

$$\dot{x} = f(x_t) \qquad x \in \mathbb{R}^n$$

and more precisely to the *qualitative evolution* of a set of functionals

$$V_1(x_t), \ldots, V_m(x_t)$$

which are of importance (energy, entropy, indicators,...).

The qualitative behavior is expressed by the evolution of the functions sign $(\frac{d}{dt} V_j(x_t))$ with values in $\mathcal{R}^m = \{-1, 0, +1\}^m$.

This is the problem of interest. But we want to obtain this evolution, without solving the equation, since some independence should be obtained with respect to the *initial condition*.

Since sign $(\frac{d}{dt} V_j(x_t)) = $ sign $(DV_j(x_t)f(x_t))$ it is convenient to introduce in the closed subspace K of $\mathbb{R}^n$, where lives x_t, the *qualitative cells*

$$K_a = \{x \in K| \text{ sign } (DV_j(x)f(x)) = a_j\}$$

where $a \in \mathcal{R}^m$, and their closure (*large qualitative cells*)

$$\bar{K}_a = \{x \in K| \text{ sign } (DV_j(x)f(x)) = a_j \text{ or } 0\}.$$

Let $\mathcal{D}(f, V)$ be the subset of qualitative states a such that K_a is not empty. Let also denote by $x(t; x_0)$ the solution of the differential equation corresponding to an initial date x_0. One is interested in the study of *transitions between qualitative cells*.

If $b \in \mathcal{D}(f, V)$, we say that $c \in \mathcal{D}(f, V)$ is a *successor* of b, if $\forall x_0 \in \bar{K}_b \cap \bar{K}_c$, there exists $\tau > 0$, such that $x(t; x_0) \in K_c$, for all $t \in]0, \tau[$.

A qualitative state a is a *qualitative equilibrium*, if it is its own successor. It is said to be a *qualitative repellor* if $\forall x_0 \in \bar{K}_a$, there exists $t > 0$ such that $x(t; x_0) \notin \bar{K}_a$.

The theory developed by J.P. Aubin permits to characterize the map of successors, the qualitative equilibria, and the qualitative repellors.

It has been applied to the so-called "replicator systems", a prototype of which is the differential system ([1])

$$\dot{x}_i = x_i(\alpha_i - \sum_{j=1}^{n} \alpha_j x_j)$$

IV.2. Neural networks.

The basic neural network can be viewed as an undirected graph with n nodes, to which are attached a pair (W, θ) where

W is an $n \times n$ symmetric matrix, W_{ij} is the weight attached to the edge (i, j), $W_{ii} = 0$
θ is an n vector, θ_i is the threshold attached to the node i.

Nodes are called *neurons*. Each neuron has two possible states $(1, -1)$. Let v be the state of the neural network, v_i being the state of neuron i.

Let

$$E_i(v) = -\sum_{j=1}^{n} W_{ij} v_j + \theta_i$$

then the following calculation is performed by the network

$$v_i^{k+1} = \text{sign}\,(E_i(v^k)), \quad \text{for } i \in S^k$$
$$v_i^{k+1} = v_i^k \quad \text{for } i \notin S^k$$

where S^k is a subset of the neurons.

For instance if

$$k = hn + j \quad j = 0 \ldots n - 1$$

and $S^k = \{j + 1\}$, the *network operates in serial mode*.

Note that in our notation

$$\text{sign}\,(a) = \begin{cases} 1 & \text{if } a \geq 0 \\ -1 & \text{if } a < 0 \end{cases}$$

A *stable* state is a state such that

$$v^{k+1} = v^k = v.$$

A basic theorem of HOPFIELD is that if the network operates in serial mode, then it will converge to a *stable state*.

The applicability of neural networks in practice arises from the possibility of interpreting the stable states. For instance, in pattern recognition, the stable states are known patterns, and for a given input pattern, the network will converge to the known pattern which is the closest to the input. It is clear that the neural network realizes the following search problem

$$\min E(v) = -\frac{1}{2} \sum_{ij} W_{ij} v_i v_j + \sum_i \theta_i v_i \quad v_i = \{-1, +1\}$$

and attains a *local* minimum.

One can clearly consider many variants of the above problem. For instance consider the following model in continuous time

$$v_i(t) = g(u_i(t))$$
$$\frac{du_i}{dt} = -E_i(v(t))$$

where g is an increasing function from R to $[0, 1]$ and $E_i(v) = \dfrac{\partial}{\partial v_i} E(v)$, $E(v)$ energy function (for instance the above). It will converge towards a local minimum of $E(v)$. It can be realized as an analog integrated circuit.

In the spirit of *simulated annealing*, considered above, one can try to attain a *global* minimum of the Energy function, by considering a stochastic version of the preceding model. This has been done by E. WONG.

Consider the model

$$v_i(t) = g(u_i(t))$$

$$du_i = -E_i(v(t))dt + \sqrt{\frac{2T}{g'(u_i(t))}}\,dw_i$$

where T is a constant, and w_i are independent standard Wiener processes. The stationary probability density of the process $v(t)$ is

$$p(v) = \frac{1}{Z} \exp{-\frac{1}{T}E(v)}$$

where Z is the normalization factor.

The simulated annealing adaptation of the preceding algorithm (for instance take $T(t) \to 0$ as $t \to \infty$) remains to be done.For more details, you can see [4] and [16]).

REFERENCES

[1] J.P. AUBIN, *Mathematical Methods of Artificial Intelligence*, to be published.

[2] D.P. BERTSEKAS, J.N. TSITSIKLIS, *Parallel and Distributed Computation: Numerical Methods*, Prentice Hall, Englewood Cliffs, N.J. 1989.

[3] B.M. BOGHOSIAN, C.D. LEVERMORE, *Complex Systems* 1 (1987).

[4] J. BRUCK, J. SANZ, *A study on neural networks*, International Journal of intelligent systems, vol. 3, 59-75, (1988).

[5] CHIANG T.S., HWANG C.R., SHEU S.J., *Diffusion for global optimization in* $\mathbb{R}^n$, SIAM Control, 25, pp. 737-752, 1987.

[6] G. COHEN, P. MOLLER, J.P. QUADRAT, M. VIOT, *Algebraic tools for the performance evaluation of discrete event system*, Proceedings IEEE, special issue of dynamics of discrete event systems, Jan. 1989.

[7] I. DAUBECHIES, *Orthonormal Bases of Compactly supported Wavelets*, CPAM, 1988.

[8] E. DEAN, R. GLOWINSKI, C.H. LI : *Supercomputer solutions of P.D.E. problems in computational fluid dynamics and in control*, University of Minnesota, Supercomputer Institute.

[9] O. FAUGERAS, *A few steps towards artificial 3D Vision*, INRIA, Technical Report series, Fev. 88, N790.

[10] U. FRISCH, B. HASSLACHER, Y. POMEAU, *Lattice Gas Automata for the Navier Stokes Equation*, Physical Review Letters, 1986.

[11] P.P. KHARGONEKAR, I.R. PETERSEN, M. ROTEA,H_∞ *Optimal Control with State Feedback* IEEE Trans. Automatic Control, 1988.

[12] Y.MEYER, *Wavelets and Operators*, Book to appear.

[13] K.C. MO and M. GHIL, *Statistics and Dynamics of Persistent Anomalies*, Journal of the Atmospheric Sciences, March 1987, .

[14] D. MUNFORD, J. SHAH, *Optimal Approximation by piecewise smooth functions and associated variational problems*, Communications on Pure and Applied Mathematics, 1988.

[15] E.S. ORAN, J.P. BORIS, *Numerical Simulation of Reactive Flow* - Elsevier 1987.

[16] E. WONG, *Stochastic neural networks*, ERL, Berkeley, Feb. 89.

[17] O. ZEITOUNI, A. DEMBO, *A maximum a Posteriori Estimator for Trajectories of Diffusion Processes*, Stochastics, 1987, Vol. 20.

TIME DIVISION MULTIPLE ACCESS SYSTEMS AND MATRIX DECOMPOSITION

Rainer E.Burkard, TU Graz

Abstract

We discuss the time slot assignment problem in satellite communication. First we derive an optimal solution of the corresponding mathematical problem which, however, will turn out to be impracticable. Then we consider a modified problem , but this is NP-hard. We describe some heuristics for this modified problem. Then we treat the problem that onboard the satellite the class of switch-modes is fixed. We state an $O(n^3)$ method for solving this problem. Finally we comment on some extensions of switch-modes and on other nonlinear partitioning problems.

1. The Practical Background

In connection with communication via satellites a new technique has been introduced recently. Whereas for many years the frequency band has been divided under the participating ground stations (FDMA-method) the new technique uses time splitting. In the ground station the data are binary encoded and buffered until they can be remitted in very short *data bursts*. The ground stations transmit their buffered data periodically within short *time slots* specially assigned to the ground stations according to their momentary traffic needs. Within an interval of two milliseconds - the so-called *frame* - all participating stations should get some time-slot. The frame contains moreover two very short *reference bursts* for controlling the system. These reference bursts, whose lengths lie in the range of nano-seconds, give a hint which technical difficulties this new system has to focus: all signals have to be received at exactly the same time by all participating stations. But even a geostationary satellite makes some slight movements around its ideal position which influences the running time of the signals. This was one of the most difficult problems which had to be resolved before this technique could be implemented. For an excellent survey on this new technique and its relevance in communication via satellites see Schladovsky [16].

We shall deal in this paper with the question, how suitable time slots can be assigned to the participating stations that they can remit their data in the shortest possible time. But before we address this question we must know more about the way in which the participating stations communicate with each other.

35

Hj. Wacker and W. Zulehner (eds.),
Proceedings of the Fourth European Conference on Mathematics in Industry, 35–46.
© 1991 *B.G. Teubner Stuttgart and Kluwer Academic Publishers.*

Onboard the satellite *transponders* receive, amplify and return the signals
to earth. A transponder connects just one sending with one receiving
station. To simplify the following discussion we assume that there are n
transponders onboard the satellite and exactly n sending and receiving earth
stations sharing these transponders. The *switch mode* at time t contains the
information which of the n participating stations are connected at time t
and communicating with each other. For technical reasons it is assumed that
the switch mode changes simultaneously for all stations. Therefore we can
assume that for a certain time interval – the time-slot – the switch mode is
fixed. A switch-mode corresponds mathematically to an (nxn) permutation
matrix $P = (p_{ij})$, where $p_{ij} = 1$ means that currently station i and station j
are connected.

The momentary traffic needs are stored in the *traffic matrix* $T=(t_{ij})$ where
t_{ij} is a nonnegative integer which measures the amount of data to be
transmitted from station i to station j. W.l.o.g. we can assume that T is a
matrix with n rows and columns.

The *time slot assignment problem* asks now:

*Which switch modes are to be used and how long each of them to remit all
data in the shortest possible time?*

Mathematically these questions can be formulated as follows:

Time-slot assignment problem

*Find a number r and r permutation matrices $P_1, P_2, \ldots, P_r$ together with
scalars $\lambda_1, \lambda_2, \ldots, \lambda_r$ such that*

$$t^* := \sum_{k=1}^{r} \lambda_k \longrightarrow min$$

(P1)

$$s.t. \sum_{k=1}^{r} \lambda_k P_k \geq T.$$

In the next section we shall derive an $O(n^4)$ method for solving this problem
optimally. It will turn out, however, that this mathematically optimal
solution can not be implemented in practice. Therefore we shall modify our
problem in Section 3, but unfortunately the modified problem becomes
NP-hard. The way out of this dilemma are heuristics. In Section 4 we treat
another possibility to come to a practicable solution, namely fixing
possible switch-modes onboard the satellite. Finally we deal with some
mathematical generalizations of this problem in the concluding Section 5.

2. Optimal decomposition

Let an (nxn) traffic matrix $T = (t_{ij})$ with nonnegative entries t_{ij} be given. First we note that the optimum value t^* of (1) is bounded from below by every row and column sum of T, since these data cannot be remitted simultaneously. Secondly we remark that it is always possible to "fill up" the matrix T such that all row and column sums equal the maximum row or column sum τ. This can be done for example by adding

$$\frac{(\tau - r_i)(\tau - c_j)}{g}$$

to every entry t_{ij}, where r_i is the sum of all elements in row i of T, c_j is the sum of all elements in column j and g is the difference between $n\tau$ and the sum of all elements of T. Let us call the new, filled-up matrix $\hat{T}$. Then, dividing all elements of $\hat{T}$ by τ we get a doubly stochastic matrix with constant row and column sums 1. Due to a famous theorem of Birkhoff (1944) any doubly stochastic matric can be written as convex combination of at most $(n-1)^2+1$ permutation matrices. This follows from the fact that permutation matrices correspond to vertices of the assignment polytope, formed by all doubly stochastic matrices, which has dimension $(n-1)^2$. Thus $\hat{T}$ can be decomposed in a weighted sum of at most $r = (n-1)^2+1$ switch modes.

Now the question arises, how such a decomposition can actually be found. But this can be done by the following simple procedure.

Algorithm 1. Optimal decomposition

Let matrix $\hat{T}$ be given.
Step 1. Choose any permutation matrix P with the property $p_{ij}>0 \Rightarrow \hat{t}_{ij}>0$.
Step 2. Determine $\lambda := \min \{ \hat{t}_{ij} \mid p_{ij} = 1 \}$.
Step 3. Replace $\hat{T}$ by $\hat{T} - \lambda P$.
Step 4. If $\hat{T} = 0$, stop, else return to Step 1.

The only remaining question is, how we should implement the "choose" in Step 1. But this can be achieved by determining a perfect matching in a bipartite graph. Construct this bipartite graph as follows: its vertices correspond to the rows and columns of $\hat{T}$. Draw an edge between row vertex i and column vertex j if and only if $\hat{t}_{ij}>0$. A permutation matrix with the properties required in Step 1 corresponds to a perfect matching in this graph. Hopcroft and Karp [8] showed that perfect matchings can be determined by an algorithm of complexity $O(n^{2.5})$. Any time when we execute Step 3 a new 0 element is generated in $\hat{T}$. But starting from a perfect matching and taking into account this new 0 element the next perfect matching can just be found by growing an augmenting path which needs only $O(n^2)$ operations. Thus we get an algorithm of complexity $O(n^4)$ for finding an optimal decomposition of the given traffic matrix T. Note also that due to Step 3 at most t^* different switch modes will be generated. Summarizing we have shown:

Theorem..Given an (nxn) traffic matrix T Problem (1) can be solved optimally by an algorithm of complexity $O(n^4)$. The optimal value t^* equals the maximum row or column sum of T. At most $r = min$ $(t^*, (n-1)^2+1)$ different switch modes are used in the optimal solution. This number r may be attained.

Let us illustrate this approach by a small example.
Let $\hat{T}$ be given as follows

$$\begin{pmatrix} 1 & 4 & 5 \\ 5 & 3 & 2 \\ 4 & 3 & 3 \end{pmatrix}$$

According to the Theorem above we can decompose this matrix in a weighted sum of at most r = 5 permutation matrices. Choosing at the beginning of the algorithm $P = P_1$ as identity matrix we get $\lambda_1 = min \{ 1,3,3 \} = 1$. The reduction in Step 3 yields as new matrix

$$\begin{pmatrix} 0 & 4 & 5 \\ 5 & 2 & 2 \\ 4 & 3 & 2 \end{pmatrix}$$

We choose now as new permutation matrix

$$P_2 = \begin{pmatrix} 0 & 0 & 1 \\ 1 & 0 & 0 \\ 0 & 1 & 0 \end{pmatrix}$$

and get for $\lambda_2 = min \{ 5,5,3 \} = 3$. A new reduction step yields

$$\begin{pmatrix} 0 & 4 & 2 \\ 2 & 2 & 2 \\ 4 & 0 & 2 \end{pmatrix}.$$

In the next step we choose as permutation matrix

$$P_3 = \begin{pmatrix} 0 & 1 & 0 \\ 0 & 0 & 1 \\ 1 & 0 & 0 \end{pmatrix}$$

which yields $\lambda_3 = 2$ and we get after reduction

$$\begin{pmatrix} 0 & 2 & 2 \\ 2 & 2 & 0 \\ 2 & 0 & 2 \end{pmatrix}.$$

But it is easy to see that this matrix can be decomposed as above in

$$2 \cdot \begin{pmatrix} 0 & 0 & 1 \\ 0 & 1 & 0 \\ 1 & 0 & 0 \end{pmatrix} + 2 \cdot \begin{pmatrix} 0 & 1 & 0 \\ 1 & 0 & 0 \\ 0 & 0 & 1 \end{pmatrix}.$$

So we get the decomposition

$$\hat{T} = 1 \cdot P_1 + 3 \cdot P_2 + 2 \cdot P_3 + 2 \cdot P_4 + 2 \cdot P_5 \quad \text{with } t^* = 10.$$

Now let us discuss optimal solutions. When we assume that onboard the satellite there are 41 transponders, then we would have about 1600 different time-slots. But then the length of each of these is so small that they cannot be realized technically. Therefore the optimal solution of the problem is unsuited for the underlying practical problem. Thus we have to modify our question. This will be discussed in the next section.

3. The modified problem

As we have seen in the last section, an optimal solution of (1) is impractical because too many different switch-modes will be used. Therefore we modify the problem requiring a decomposition in at most n different switch-modes.

Modified decomposition problem

Find at most n switch-modes P_i and scalars λ_i such that

$$\sum_{i=1}^{n} \lambda_i \longrightarrow min$$

(P2)

s.t. $$\sum_{i=1}^{n} \lambda_i P_i \geq T$$

The formulation of the problem implies that every entry of T is now covered by just one 1-entry of a permutation matrix. Therefore the entries of T are not any more split in a solution. Rendl [13] showed that this modified problem (2) is NP-hard. Therefore no efficient algorithm is known for solving it. Two exact solution methods for this problem stem from Minoux [11] and Ribeiro, Minoux and Penna [14]. These authors transform Problem (2) into a set partitioning problem and use column generation techniques for solving it. In particular an $(n^2 \times n!)$ matrix A is defined by

$$a_{ik} = \begin{cases} 1, & \text{if the k-th permutation matrix has as i-th entry 1} \\ 0, & \text{otherwise.} \end{cases}$$

Moreover, a cost vector (c_k) is defined by

$$c_k = max \{ t_i | a_{ik} = 1 \}.$$

Then Problem (2) can be stated as

Find a Boolean vector x with n^2 components such that $\sum c_k x_k$ is minimized subject to $Ax = e$, where the vector e has n^2 entries equal to 1.

The authors report the following running times for this problem on an IBM
4381:

n	10	11	12	13	14	15
sec	≈50	≈160	≈700	≈3900	≈4200	≈34000

One can see from this table that the running times grow exponentially and
are prohibitively large even for small sized problems. Therefore there is a
special need for good heuristics. Camerini, Maffioli and Tartara [6]
proposed to choose the switch-modes $P = (p_{ij})$ successively such that

$$\sum \{ t_{ij} \mid p_{ij} = 1\} \longrightarrow \max$$

This amounts in solving n perfect maximum matching problems. Therefore the
complexity of this heuristic is $O(n^4)$. A small example shall illustrate this
approach:

Let $\hat{T}$ be given as follows

$$\begin{pmatrix} 1 & 4 & 5 \\ 5 & 3 & 2 \\ 4 & 3 & 3 \end{pmatrix}$$

Then

$$\hat{T} \leq 5 \cdot \begin{pmatrix} 0 & 0 & 1 \\ 1 & 0 & 0 \\ 0 & 1 & 0 \end{pmatrix} + 4 \cdot \begin{pmatrix} 0 & 1 & 0 \\ 0 & 0 & 1 \\ 1 & 0 & 0 \end{pmatrix} + 3 \cdot \begin{pmatrix} 1 & 0 & 0 \\ 0 & 1 & 0 \\ 0 & 0 & 1 \end{pmatrix}$$

which yields a value of $\Sigma\lambda_i = 12$.

Balas and Landweer [1] improved this approach by noting that it might be
better to solve successively perfect matching problems with a bottleneck
objective instead of sum objective:

$$\text{minimize} \quad \max \{ t_{ij} \mid p_{ij} = 1\}$$

In this case we get the following decomposition for the example above:

$$\hat{T} \leq 3 \cdot \begin{pmatrix} 1 & 0 & 0 \\ 0 & 1 & 0 \\ 0 & 0 & 1 \end{pmatrix} + 4 \cdot \begin{pmatrix} 0 & 1 & 0 \\ 0 & 0 & 1 \\ 1 & 0 & 0 \end{pmatrix} + 5 \cdot \begin{pmatrix} 0 & 0 & 1 \\ 1 & 0 & 0 \\ 0 & 1 & 0 \end{pmatrix}$$

which yields the same value of $\Sigma\lambda_i = 12$. The complexity of this heuristic is
$O(n^{3.5}\log n)$. The computational results reported by Balas and Landweer are
quite good. It is still possible to refine this last approach by solving
first perfect matching problems with bottleneck objective as above, but then
using as a second criterion *maximizing the sum* of all t_{ij} for which $p_{ij} = 1$
and $t_{ij} = \hat{t}$ holds, where $\hat{t}$ is the optimum value of the bottleneck
problem. This two-criteria problem can be solved effectively using the theory
of time-cost problems, but our computational experiments with randomly

generated test problems showed only a very small improvement compared with
the results found by the Balas-Landweer approach.

We have seen above that Problem (2) is *NP*-hard and difficult to solve
exactly. On the other hand we don't want to decompose the traffic matrix
optimally in $O(n^2)$ switch-modes. A compromise would be to decompose the
traffic matrix in m switch-modes with $n < m \ll n^2$. Can we do this
efficiently? Orlin [12] showed that the decision problem: *Given a doubly
stochastic nxn matrix with nonnegative rational coefficients and an integer
m with n≤m<r* is *NP*-complete by reducing it to the 3 partition problem. Thus
the decomposition in m switch modes is hard again and we ask for heuristics
yielding a suboptimal solution for this decomposition problem. Rote [15]
developed one such approach: Rote's algorithm makes use of the fact that the
traffic matrix can be decomposed in t^* switch-modes. He scales the traffic
matrix such that in the new, scaled matrix t^* equals m and decomposes this
scaled matrix according to Algorithm 1. Then the scaling is revoked and the
solution is rounded down to yield a decompositon of the originally given
traffic matrix. Again a small example shall illustrate this approach:

Let $\hat{T}$ be given as follows

$$\begin{bmatrix} 1 & 4 & 5 \\ 5 & 3 & 2 \\ 4 & 3 & 3 \end{bmatrix} \quad \text{and choose m = 4.}$$

By choosing the scaling factor F = 4 the scaled and rounded-up matrix U
becomes:

$$U = \left[\left\lceil \frac{t_{ij}}{F} \right\rceil \right] = \begin{bmatrix} 1 & 1 & 2 \\ 2 & 1 & 1 \\ 1 & 1 & 1 \end{bmatrix}$$

Decomposition of U according to the algorithm of Section 2 yields

$$U \leq 1 \cdot \begin{bmatrix} 1 & 0 & 0 \\ 0 & 1 & 0 \\ 0 & 0 & 1 \end{bmatrix} + 1 \cdot \begin{bmatrix} 0 & 1 & 0 \\ 0 & 0 & 1 \\ 1 & 0 & 0 \end{bmatrix} + 2 \cdot \begin{bmatrix} 0 & 0 & 1 \\ 1 & 0 & 0 \\ 0 & 1 & 0 \end{bmatrix}$$

and therefore

$$T \leq 4 \, U \leq 4 \cdot \begin{bmatrix} 1 & 0 & 0 \\ 0 & 1 & 0 \\ 0 & 0 & 1 \end{bmatrix} + 4 \cdot \begin{bmatrix} 0 & 1 & 0 \\ 0 & 0 & 1 \\ 1 & 0 & 0 \end{bmatrix} + 8 \cdot \begin{bmatrix} 0 & 0 & 1 \\ 1 & 0 & 0 \\ 0 & 1 & 0 \end{bmatrix}$$

By rounding down we get as solution:

$$T \leq 3 \cdot \begin{bmatrix} 1 & 0 & 0 \\ 0 & 1 & 0 \\ 0 & 0 & 1 \end{bmatrix} + 4 \cdot \begin{bmatrix} 0 & 1 & 0 \\ 0 & 0 & 1 \\ 1 & 0 & 0 \end{bmatrix} + 5 \cdot \begin{bmatrix} 0 & 0 & 1 \\ 1 & 0 & 0 \\ 0 & 1 & 0 \end{bmatrix}$$

with objective value 12.

Rote has shown that within $O(m \cdot n^{2.5})$ steps a solution can be found whose
value is not worse than $\frac{m}{m-n+1} \, t^*$.

The following table shows some preliminary numerical results for Rote's heuristic:

n	CPU sec (VAX 11/780)	$(d-t^*)/t^*$	Balas/Landweer
10	0.67	1.82%	7.83%
20	3.6	1.95%	5.13%
40	21.8	1.82%	4.09%
60	65.8	1.59%	2.37%
100	268.7	1.23%	1.78%

This table shows the average values of 100 test problems with randomly generated costs in (1,99). The third column gives the relative error of Rote's heuristic, where m=2n, and the last column that of Balas and Landweer's heuristic.

Note that with increasing size the solution quality improves for both heuristics. Since the running times stem from an experimental code they can surely be reduced using more sophisticated implementation techniques.

4. Fixed switch-modes

Lewandowski, Liu and Liu [10] made the proposal to consider the following *restricted decomposition problem* (P3):
Fix a class of switch-modes onboard the satellite and decompose the traffic matrix only in switch modes of this class.

They propose to consider classes of 2n switch-modes $P_1, P_2, \ldots, P_n$, $Q_1, Q_2, \ldots, Q_n$ with the following properties

(1) $\displaystyle\sum_{k=1}^{n} P_k = \sum_{l=1}^{n} Q_l = \underline{1}$, where $\underline{1}$ is a matrix with 1-entries only.

and

(2)   for every P_k and Q_l there is exactly one position where these matrices have a 1-entry in common.

Example:

$$P_1 = \begin{pmatrix} 1 & 0 & 0 \\ 0 & 1 & 0 \\ 0 & 0 & 1 \end{pmatrix} \quad P_2 = \begin{pmatrix} 0 & 1 & 0 \\ 0 & 0 & 1 \\ 1 & 0 & 0 \end{pmatrix} \quad P_3 = \begin{pmatrix} 0 & 0 & 1 \\ 1 & 0 & 0 \\ 0 & 1 & 0 \end{pmatrix}$$

$$Q_1 = \begin{pmatrix} 0 & 0 & 1 \\ 0 & 1 & 0 \\ 1 & 0 & 0 \end{pmatrix} \quad Q_2 = \begin{pmatrix} 1 & 0 & 0 \\ 0 & 0 & 1 \\ 0 & 1 & 0 \end{pmatrix} \quad Q_3 = \begin{pmatrix} 0 & 1 & 0 \\ 1 & 0 & 0 \\ 0 & 0 & 1 \end{pmatrix}.$$

According to a famous result of Euler such systems of orthogonal permutation matrices always exist unless n = 6.

The restricted decomposition problem can now be formulated as follows:

Given $2n$ permutation matrices which fulfill (1) and (2), find $2n$ numbers $\lambda_1, \lambda_2, \ldots, \lambda_n, \mu_1, \mu_2, \ldots \mu_n$ with

$$\sum_{k=1}^{n} (\lambda_k P_k + \mu_k Q_k) \geq T$$

(P3)

$$\sum_{k=1}^{n} (\lambda_k + \mu_k) \longrightarrow \quad min$$

Problem (P3) is in principle an LP. Lewandowski et al. presented a pseudo-polynomial algorithm of complexity $O(n^{3.5}t^*)$ where t^* is the maximum entry of the traffic matrix T. The following algorithm (Burkard [4]) transforms the problem to an assignment problem and solves it with $O(n^3)$ operations.

Algorithm 2: Restricted decomposition algorithm

1. Let $a \geq \max_{i,j} t_{ij}$

2. Define $u_{kl} := a - t_{ij}$, where (i,j) is determined as unique element with

   $p_{ij}^{(k)} = q_{ij}^{(l)} = 1$.

3. Solve the linear assignment problem with cost matrix $U = (u_{kl})$ and let (a_k, b_l) be an optimal dual solution.

4. Define δ such that

   $\lambda_k := a - a_k - \delta \geq 0$

   $\mu_l := -b_l + \delta \geq 0$.

   Then λ_k and μ_l are the coefficients of an optimal solution for (P3).

Example.

Let the traffic matrix T be given by

$$T = \begin{pmatrix} 1 & 4 & 5 \\ 5 & 3 & 2 \\ 4 & 3 & 2 \end{pmatrix}$$

and let $P_1, P_2, P_3, Q_1, Q_2, Q_3$ be given as in the example at the beginning of this section. We get $a = \max t_{ij} = 5$ and the following matrix U with

$u_{kl} := a - \{ t_{ij} | \quad p_{ij}^{(k)} = q_{ij}^{(l)} = 1 \}$:

$$U = \begin{pmatrix} 2 & 4 & 3 \\ 1 & 3 & 1 \\ 0 & 2 & 0 \end{pmatrix}.$$

An optimal dual solution of the linear assignment problem yields

$$a_1 = 2, \ a_2 = 1, \ a_3 = 0, \ b_1 = 0, \ b_2 = 2, \ b_3 = 0.$$

We choose δ such that min $(a - a_k) \geq \delta \geq$ max b_l. The value $\delta = 2$ yields

$$T = 1 \cdot P_1 + 2 \cdot P_2 + 3 \cdot P_3 + 2 \cdot Q_1 + 2 \cdot Q_3.$$

For $\delta = 3$ we get another solution.

5. Generalized switch modes

The ideas of the last section can be generalized to solve the following decomposition problem involving arbitrary Boolean (mxn) matrices:
Let $A_1, A_2, \ldots, A_K$, $B_1, B_2, \ldots, B_L$ be any 0-1 matrices with m rows and n columns with

$$\sum_{k=1}^{K} A_k = \sum_{l=1}^{L} B_l = \underline{1},$$ where $\underline{1}$ is a matrix with 1- entries only.

The <u>Boolean decomposition problem</u> is:

Let a nonnegative (nxm) matrix be given. Determine $\lambda_1, \lambda_2, \ldots, \lambda_K, \mu_1, \mu_2, \ldots \mu_L$
such that for positive coefficients c_k, d_l

$$\sum_{k=1}^{K} c_k \lambda_k + \sum_{l=1}^{L} d_l \mu_l \longrightarrow \quad min$$

s.t. $\quad \sum_{k=1}^{n} \lambda_k P_k + \sum_{l=1}^{L} \mu_l Q_l \geq T,$ $\quad \lambda_k \geq 0, \; \mu_l \geq 0 \quad$ *for all* $k, l.$

By using an analogue transformation as before this problem can be converted into a min cost flow problem in a network.

Another generalisation of switch-modes was suggested by Lewandowski and Liu [9] who introduced r-permutation matrices into the decomposition. An r-permutation matrix has exactly r 1-entries in each row and column. Thus the coincide for r=1 with usual permutation matrices. The authors show that an optimal decomposition in the sense of problem (P1) can be obtained in $O(n^5)$ steps. De Werra [7] related this problem to polyhedral decomposition and obtained a simpler and faster decomposition algorithm.

Problem (P2) is related to constrained partitioning problems. It can be stated in the following form:

Find a partition of $E = \{ (i, j) \mid 1 \leq i \leq n, \; 1 \leq j \leq n \}$ *into n sets* $P_1, P_2, \ldots, P_n$
where
(3) each of these sets P corresponds to a permutation matrix and
(4) $\sum max \{ t_{ij} \mid p_{ij} = 1 \} \longrightarrow min.$

In a recent paper Burkard and Yao [5] investigated partitioning problems of a similar form. They showed in particular that certain balancing and variance minimization problems can be solved efficiently, if condition (3) is relaxed to

(3′) each set P corresponds to a matrix which has just one 1-entry in each column.

In this case so-called *balancing problems* with the objective function

$$\sum_k \left[max \ \{ \ t_{ij} \mid \ p_{ij}^{(k)} = 1 \ \} \ - \ min \ \{ \ t_{ij} \mid \ p_{ij}^{(k)} = 1 \ \} \right] \longrightarrow min$$

as well as problems which minimize the sum of variances of the coefficients t_{ij} for the specific classes P can be solved by iteratively applying a Greedy-type algorithm. Thus these problems can be solved in an efficient way.

Acknowledgement

Partial support by Austrian Science Foundation, Project S32/01 is gratefully acknowledged.

References

[1] E.Balas and P.R.Landweer: Traffic assignment in communication satellites. Operations Res. Letters 2 (1983), 141-147

[2] G.Birkhoff: Tres observaciones sobre el algebra lineal. Rev. univ. nac. Tucumán (Ser.A) 5 (1940), 147-151

[3] R.A.Brualdi: Some applications of doubly stochastic matrices. Linear Algebra and its Applications 107 (1988), 77-100

[4] R.E.Burkard: Time-slot assignment for TDMA-systems. Computing 35 (1985), 99-112

[5] R.E.Burkard and E.Y.Yao: Constrained partitioning problems. Report 132, Institute of Maths, Graz University of Technology, Dec. 1988, to appear in Discrete Applied Maths.

[6] P.Camerini, F.Maffioli and P.Tartara: Some scheduling algorithms for SS/TDMA systems. Proceedings Fifth Intern. Conf. on Digital Satellite Communications, Genoa 1981, 405-409

[7] D.de Werra: A note on SS/TDMA satellite communication. To appear in J.Linear Algebra and its Applications

[8] J.Hopcroft and R.M.Karp: An $O(n^{2.5})$ algorithm for maximum matching in bipartite graphs. SIAM J. on Computing 2 (1973), 223-231

[9] J.L.Lewandowski and C.L.Liu: SS/TDMA satellite communications with k-permutation switching modes. SIAM J. on Algebraic and Discrete Methods 8 (1987) 519-534.

[10] J.L.Lewandowski, J.W.S.Liu and C.L.Liu: SS/TDMA time slot assignment with restricted switching modes. IEEE Transactions on Communications COM-31 (1983) 149-154

[11] M.Minoux: Optimal traffic assignment in an SS/TDMA frame: a new approach by set covering and column generation. RAIRO Recherche Opérationelle 20 (1986),273-286

[12] J.Orlin: Private communication 1985

[13] F.Rendl: On the complexity of decomposing matrices arising in satellite
     communication. Operations Res. Letters 4 (1985), 5-8

[14] C.C.Ribeiro, M.Minoux and M.C.Penna: An optimal
     column-generation-with-rank algorithm for very large scale set
     partitioning problems in traffic assignment. European J.Operational
     Res. 41 (1989), 232-239

[15] G.Rote: Eine Heuristik für ein Matrizenzerlegungsproblem, das in
     der Telekommunikation via Satelliten auftritt (Kurzfassung).
     ZAMM 69 (1989), 29-T31

[16] W.Schladofsky: TDMA/DSI Ein neues Übertragungsverfahren in der
     Satellitentechnik. Österr. Postrundschau, Heft 2 (1984), 24-29

Address of the author:

Rainer E.Burkard, Institut für Mathematik, Technische Universität Graz
Kopernikusgasse 24, A-8010 Graz (Austria)

**Fresh breeze around the ivory tower
- current trends in Austrian higher education**

by S. Höllinger

Presently Austrian universities are engaged in a period
of far-reaching change and transition. Economy and
society are expecting a new image of the universities and
more and better contribution to economic growth and the
solution of societal problems.

On the one side enterprises and firms are interested in
technological innovation and knowhow transfer. On the
other side society and decision makers look for expertise
when facing an accelerated degrading of environment.
Some further pressure on the performance of higher
education comes on behalf of the qualification profiles
of Austrian graduates. Critics find fault with the
duration of study courses and the quality of vocational
training. The structures of academic studies and the
labor market have to be reconsidered.

I am summarizing my point of view regarding current
trends in Austrian higher education in seven major
statements:

I. The Humboldtian heritage of the university system
   dissolves. High student enrolment and new demands
   define contemporary universities by way of analogy to
   a modern service industry.

Similar to the Federal Republic of Germany the university
system in Austria is rooted in the Humboldtian concept of
higher education and stresses an educational model which
is built around the humanities.
Therefore universities are not so much institutions for
vocational training but rather elitist institutions to

47

Hj. Wacker and W. Zulehner (eds.),
Proceedings of the Fourth European Conference on Mathematics in Industry, 47–56.
© 1991 B.G. Teubner Stuttgart and Kluwer Academic Publishers.

promote the advancement of higher learning and the
progress of pure science. This causes certain problems.

Within the Austrian university system, which discreetly
enforces an elitist concept of higher learning, there
always has existed on the side of the graduates a gap
between fiction and reality. Only a few graduates were
recruited for the sake of science (academic staff), but a
considerable number were recruited for the guilds of
medical doctors and higher civil servants. Others,
although trained for scientific research and the
advancement of culture, had to reduce their professional
expectations and accept posts in schools, administration
or even fields for which they had not been educated.

For a long time, this gap between fiction and reality
could survive because only two or three per cent of an
age group applied to study at the university. However,
the system ran into trouble with the rapidly growing
enrolment of students in the seventies and eighties.

Hence the process of expansion of student numbers now
stimulates a development within the university system
which can be described as a shift from a manufacture
paradigm - a small group of professors educating a small
number of students - to a new one, defining a modern
university by way of analogy to contemporary service
industry. This means that the higher education system has
to provide large numbers of students with high level
scientific and vocational training. This process of
change has not proved easy. The traditional
self-awareness of professors as the educational elite
collides with the reality of modern university life
characterized by a diversified student population and new
demands of the labor market.
The traditional image of the university as the privileged

institution of reflection and pure science collides with
manifold needs of society and economy expecting expertise
and commercialization of scientific knowledge.

II. <u>To a large extend traditional research work as an
    individual activity does not fit current demands any
    more. More planning and more co-operation within
    research are put on the agenda.</u>

Not only high student enrolment challenged the
Humboldtian tradition of Austrian universities in recent
years. Also the organization of modern research in pure
and applied sciences urges changes in research work. Most
of the Austrian university institutes are one person
units (comprising the professor himself). As a result a
high proportion of research is fragmented, consisting of
varyingly isolated "chairs". Therefore there is little
engagement in co-operative research projects and in
planning co-ordinated research programmes. The lack of
planning and focusing of research activities diminishes
the efficiency of the national R&D system and its ability
to respond to the needs of society and the economy.

III. <u>Austria's efforts to link up with the international
     scientific development are too scattered and cause a
     dispersion of limited resources. There is a need of
     more money and planning instead of a nostalgia for
     "Einsamkeit und Freiheit".</u>

Austria obviously suffers from a feeling of isolation
within the broader international scientific community.
This fact seems to be largely due to historical
circumstances and to the fact, that having held an
essential role in the history of Europe and the world for
centuries, it suddenly lost that role without being able
to find a place in the new political groupings. But

Austria's situation is not different from that of some
other small countries, though not having the same
history, accept their situation in the world as it now
is, without suffering the same psychological problems.
Sweden and Switzerland cultivate a variety of scientific
and technological relations to other countries and
succeed in occupying an important place within the
industrialized nations.

Therefore there is a need to promote a rational discourse
on Austria's opening towards the international scientific
community. Individual efforts in building up
international linkages have to be harmonized and
additional resources to be mobilised to take advantage
from international co-operation. This includes all kinds
of exchange between academics in Austria and different
scientifically developed countries of the world (personal
relations, participation in international conferences,
meetings and seminars, joint training courses etc.) as
well as the participation in international programmes
which I am considering later on.

IV. <u>Co-operation between university and industry is
improving but there is still a lot to be done to
promote a continuous dialogue between both.</u>

For a long time Austrian universities were quite isolated
from economic growth and industrial entrepreneurship.
Only the technical universities made some contribution to
industrial R&D. Recently the situation has changed. The
co-operation between enterprises and university
departments has been improving. But there still exists a
variety of problems. The major problem seems to be the
promotion of a continuous dialogue between the
universities and industry. Only a few university people
understand the language of the market forces and only a
few industrial representatives understand certain

delicacies of university life. This might be caused due
to the fact that Austrian economy is mainly based on
small and medium firms with a very low percentage of
graduates employed. On the other side most of the
university teachers and researchers never left the
educational system and therefore have no working
experience in enterprises and business.

Universities at the turn of the millenium – the reconstruction of the ivory tower

The quantitative expansion of the Austrian universities
within the last two decades in terms of student
enrolment, financial expenditures, curricula reforms
revealed all the structural problems mentioned above.
Government and the universities, politics and economy
therefore are engaged in reconstructing the ivory tower
to prepare Austrian higher education system soon facing
21st century. This policy proceeds along following lines:

* Internationalization as means to strengthen the
  performance of research and training
* Promotion of a continuous dialogue between science and
  economy to counterbalance splendid isolation
* Modernization of university education to improve the
  efficiency of vocational as well as scientific
  training of students.

V. Contemporary higher education policy aims at
   performance oriented universities.
   Internationalization and quality assessment of
   training and research are indispensable requirements.

To intensify internationalization of the Austrian higher
education system is one way to improve the performance of

university departments. The Ministry of Science and
Higher Education attempts to rise the budget devoted to
short term visits of Austrian scholars abroad,
partizipation in international conferences and meetings
as well as joint research and study programmes. But
internationalization is not an aim in itself. It has to
be combined with a policy of improving the performance of
research and training. Like other European countries
Austria will engage in evaluation procedures to assess
the quality of university research in order to promote
those departments in international co-operation which
already show a comparable high level of performance and
therefore are especially enabled to participate in joint
research and study programmes with foreign universities.

Some of the research and training programmes of the
European Communities are of interest to Austrian
universities. Soon Austria will officially take part in
the SCIENCE-programme which promotes the multinational
co-operation within science and is designed as a major
step to build up a homogenous European scientific
community. The programme is based on a peer review system
and delivers financial support to those academics and
research teams who have worked out proposals of high
scientific quality. Another programme which is of
definite interest to Austria is the COMETT-Programme
stimulating training partnerships between universities
and enterprises. Professor Wacker is already engaged in a
COMETT-project which might be known to you. When he
started his COMETT-initiative it was very difficult for
Austrians to participate. He had to find partners abroad
and to detect financial resources not supported by the
community budget but exclusively based on national
resources.
On May 22nd 1989 the European Communities decided to open
up the COMETT II-Programme to EFTA-countries. Hence we

hope that Austria soon will be an equally entitled participant in the COMETT II-Programme starting 1990 and that it will be much easier for Austrian university departments to engage into COMETT initiatives.

But it is not only necessary that Austrian scholars and students go abroad. Equally it has to be ensured that more foreign graduate students and visiting professors study and teach at Austrian universities. The mutual transfer of students and personnel is one of the most important means of international co-operation. Foreign graduate students bring in new incentives in training on the doctoral level, visiting professors bring in foreign language capacity and new ideas.

VI. Improving the co-operation between university and economy means to promote the expert dialogue with the "real world" while protecting university autonomy.

As I mentioned above Austrian universities had not had sufficiant co-operation with economy for a long time. Over the last few years, as a part of an innovation - oriented science and industry policy, a number of measures have been taken to intensify science transfer in order to make better use of the research capacities of universities.
These measures shall help to promote a continuous dialogue between the universities nd the enterprises enabling university people to understand the problems of knowledge transfer an application of science and technology without damaging university autonomy and basic research.

A stimulation programme enables university assistants to gain practical experience of work in a company for a

specific period. The assistant goes on nonactive status
whilst absent from university and can return at the end
of his work experience without his career suffering.

Recently this programme was completed by the pilot
experiment "Scientists go into business" which encourages
young academic's entrepreneurial initiatives. The main
targets in this experiment are university assistants who
want to start a business. The Ministry of Science and
Higher Education gives advice and provides financial
support.

The transfer of personell between universities and the
economy is one means to ensure better co-operation and to
improve the mutual recognition of needs. But innovation
policy is not restricted to this.
Several universities founded university extensions to
improve the linkages to local enterprises and to
establish post graduate training in applied sciences,
economics and technology.
Outside the universities local governments and
communities supported the federal government in founding
science parks and technology transfer centers which
actively and strategically link together university
departments, extra-university research institutions and
enterprises to promote technological innovation and
application of scientific knowledge.

The annual national "Austrian Science Fair" is organized
as a showplace for the scientific and technical higher
education establishments and acts as a centre for direct
encounter between know-how producers and the potential
appliers and users of that know-how.
The "Innovation Agency" was set up in 1984 under Federal
law outside the formal government structures. It acts as
an information, service and co-ordination center for the

technology-oriented innovation efforts of Austrian
industry. Its tasks include continuous public relations
work to improve the climate favorable to innovation,
activities designed to speed up technology transfer and
the provision of a comprehensive range of services for
Austrian entrepreneurs.

VII. The modernization of university education is not
only a case of curricula reform but has to be
supported by structural innovation and staff
development.

The internationalization of university training and
education is only one step to modernize study courses.
This action has to be completed by a comprehensive
curricula reform.

Currently the technical study courses are subject to far-
reaching changes. A variety of weaknesses forced the
administration to start up with reform activities:
Education in engineering lasts too long. The duration of
the study courses is much longer than in comparable
European countries. The drop-out rate is close to 60
percent and the graduates lack foreign language skills
and knowledge in environmental and social context of
modern technology. The application of computer science in
different engineering sciences is not integrated in the
study courses.

Therefore the curricula reform will shorten the average
duration of study courses so that graduates are enabled
to enter the labor market earlier and hence can better
compete with graduates from other European universities.
The shortage of studies shall also give the opportunity
to integrate new study contents, for example,

environmental science, foreign language training and
computer science application like CAD and CAM. Presently
the number of examinations in the technical study courses
is very high and therefore prevents autonomous learning
of the students. This shall be changed to give more
opportunity to interdisciplinary methods of learning and
acquiring new knowledge.

But the reform activities will not be restricted to a
pure curricula reform. It will also entail structural
measures. At the moment the university bodies do not have
an obligation to modernize the contents of study courses
within a certain period of time. Henceforth bodies will
be forced to renew study plans every two years to ensure
that students will be confronted with the most recent
technological development. Additional to this staff
development will be promoted to ensure the quality of the
teaching staff.

Determination of the Transient Thermal Behaviour of Thyristors with Cooling Devices.

Herbert Niessner

Abstract: For practical purposes the termal behaviour of a thyristor plus cooling device may be studied on an electrical analogous two-terminal network of resistors and capacities [2] . Thereby the impedance as a function of frequency is rational and representable as a continued or partial fraction [4]. Numerical values for resistors and capacities can be identified approximating the transient rise in temperature due to switching on a constant heat source, i.e. the transient thermal resistance, by a sum of exponentials [5]. Identified values differ considerably from those naively calculated. Particular mathematical problems arise when analogous networks of thyristor and cooling device, whose elements are determined separately, are combined [1] or when transient thermal resistances are simply added [3].

[1] Anwander, E.; Lawatsch, H.: Thermische Messungen und thermische Ersatzschaltbilder von Halbleiterelementen und Kühlern für die rechnergestützte Bemessung und Simulierung von Stromrichtern. ETZ-A 96 (1975) 261-265

[2] Anwander, E.; Lawatsch, H; Neumeister, E.: Rechnergestützte Dimensionierung und Simulierung von Stromrichteranlagen. ETZ-A 96 (1975) 117-122

[3] Arbenz, P.; Niessner, H.: Zulässigkeit der Addition von transienten Wärmewiderständen. Techn. Rep. CTT ST 84/5, BBC Brown Boveri, Baden/AG 1984

[4] Cauer, W.: Theorie der Wechselstromschaltungen. Berlin: Akademie-Verlag 1954.

[5] Niessner, H.: Multiexponential fitting methods. Numerical Analysis (ed. Descloux, J.; Marti, J.) 63-76, Basel: Birkhäuser-Verlag 1977

Herbert Niessner, Dept.TECT-W, ABB Informatik AG, CH-5401 Baden/SWITZERLAND

Hj. Wacker and W. Zulehner (eds.),
Proceedings of the Fourth European Conference on Mathematics in Industry, 57.
© 1991 B.G. Teubner Stuttgart and Kluwer Academic Publishers.

Mathematics for Industry

A.B. Tayler

1. General Remarks

A Presidential address provides an opportunity to ask broad questions about ECMI, this lively consortium of many nationalities and diverse mathematical interests, and its annual meeting. I only have time to examine two such questions - why are we here and why am I giving this address?

There may be as many answers to the first question as conference participants but my guess is that almost all of us have come to improve our morale. The applied mathematician or mathematical modeller is often located in some isolation, whether in a university or industry, and may be beset by pure mathematicians, computer scientists, engineers or physicists, all of whom know they are superior because of the unique importance of their profession and in particular their own specialist knowledge or research. An applied mathematician, attempting to communicate with all these professions and requiring generalist knowledge, is clearly a different species with a different philosophy and goals. So we come together to be reassured that this philosophy is both intellectually respectable and valuable to the real world. On a more practical level we come here to pool our experience, to discover colleagues from other countries interested in problems similar to our own but often from a new angle, with a novel complication, or with a new insight. We may also discover new problems, or new applications for our mathematical skills, and

Hj. Wacker and W. Zulehner (eds.),
Proceedings of the Fourth European Conference on Mathematics in Industry, 59–69.
© 1991 *B.G. Teubner Stuttgart and Kluwer Academic Publishers.*

become more generally aware of new areas of mathematics ripe for application. We come to educate ourselves and renew and extend personal
relationships, which are an important stimulus in the growth of interest in
mathematics for industry.

There is no simple answer to the question why am I speaking, except a
mixture of unexpected opportunities taken, and persistence with an idea
over more than twenty years. Over this period the Oxford Study Groups with
Industry have organised annual workshops on industrial problems involving
mathematics. This activity has led to closer research collaboration with
industry, new research problems in applied mathematics, training courses
for postgraduate students in applications of mathematics, and a need to
link with expertise available outside Oxford University and outside the
United Kingdom. My expertise, and that of the newly formed Oxford Centre
for Industrial and Applied Mathematics, is in mathematical modelling,
mostly of continuous processes which may be described by differential
equations. I defend my profession with three quotations:

- Mathematical modelling is a logical thought process applicable to
   problems in many diverse areas and a defence against confusion and
   disorder (Melzak, 1983);

- Applied mathematics is the consistent pursuit of the understanding
   which the interplay of mathematics and applied science can give
   (Carrier, 1962);

- Mathematics provides the cutting edge for technology, needed in an
   increasingly competitive world (R. Reagan, U.S. President, 1986).

So what can Mathematics do for industry? In my experience it rarely
'solves' industrial problems, indeed most industrial problems are never
solved and compromise decisions have to be made. Mathematics can provide a
detailed analysis of a problem area, a framework for thinking about the

problems, a check on experimental evidence and computer simulation, and
a predictive tool. It should be an essential component of any decision
taking and the more complicated the problem the more valuable a mathematical analysis is likely to be. In particular when the scales (length,
temperature, etc.) of the problem are very large (or very small) experiments may be extremely difficult to carry out. I will give three examples
of this from the mining, semiconductor, and steel industries, giving some
details for the continuous casting of steel process because of its economic
importance and its relevance to several participants at this meeting.

2.   Two processes - large and small length scales

(a) Cliff blasting (Please, Wheeler & Wilmott, 1987)

If a mineral layer in rock is roughly horizontal and near the surface
one method of mining is to progressively reduce a surface layer containing
the mineral to a muck pile by blasting. Thus at each stage there is a
cliff, and a sequence of muckpiles as in Fig. 1. The problem is to
determine where the mineral layer will end up in the muck piles so that it
can be efficiently collected. Four stages may be observed in this process
and mathematical models are being developed for all of them.

Stage one - about 10 μs - surface vibrations only.

The mathematical model proposes a shock wave in a fluid-like region
due to the high pressure gas in the borehole. In front of the shock
is a transition region in which cracks are propagated, and in front of
this is an elastic region with a seismic wave reflected at the cliff
face.

Stage two - about 100 μs - no observations at all.

The transition region grows as the cracks propagate due to the high
pressure gas streaming into them.

Cliff blasting

Single borehole

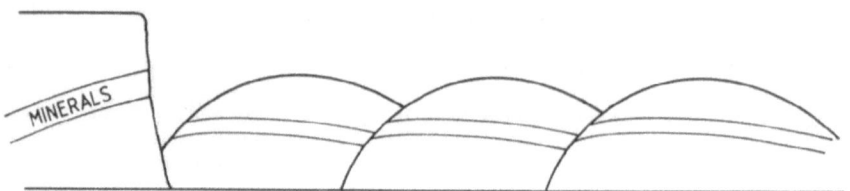

Periodic boreholes

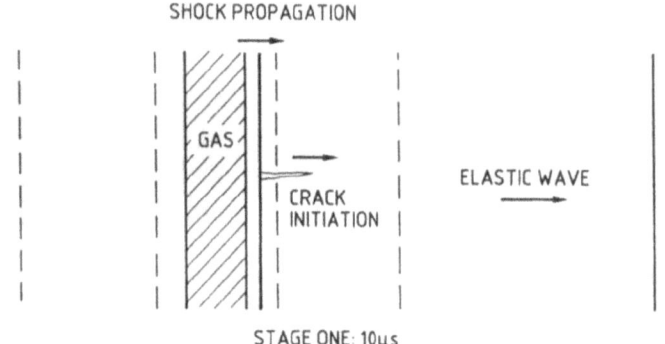

Fig. 1

Stage three - further 10 μs - cracks seen at the cliff face.

The fractured rock rapidly accelerates due to the high pressure

gradients as the cracks penetrate the surface.

Stage four - about 1 sec - cliff explodes.

Free fall under gravity for each rock particle is assumed.

The mathematical analysis allows the process to be separated into four

fairly distinct stages, which cannot be done by observation or experiment.

It does include the effects of varying strength explosive charges and their

location.

(b)   Semi-conductor process modelling (Tayler & King, 1989)

One of the many complicated processes which a silicon chip undergoes is

that of oxidation to provide a useful thin layer (μm) of non-conducting

silicon oxide. In general, erosion processes reduce the surface at a rate

q creating a moving boundary $G(\underline{x},t) = 0$ where $q|\nabla G| = \frac{\partial G}{\partial t}$. Thus if q is

prescribed, $q^2(\frac{\partial G}{\partial t})^2 = (\frac{\partial G}{\partial x})^2 + (\frac{\partial G}{\partial y})^2 + (\frac{\partial G}{\partial z})^2$, and G satisfies the geometrical

optics equation, with wave speed 1/q. The properties of this equation are

well known and a number of interesting phenomena may occur (Smith & Carter,

1986). However for the oxidation of silicon q is the result of a surface

chemical reaction of silicon with oxygen, which diffuses through the

silicon oxide surface layer. Thus q has to be found from boundary value

problems for the slow viscous flow of the oxide and the diffusion of the

oxygen through it. This leads to complicated equations which may be

simplified, however, if the oxide layer is thin (Tayler, 1988).

3.   The continuous casting of steel

This is another complicated process which has been simplified in Fig.
2, where both the overall vertical and horizontal scales have been reduced
relative to the metal sheet thickness. Molten steel is fed continuously
into the top of the mould under gravity, is cooled near the exit of the
mould to form a thin solid skin, runs through guide rollers while still
flexible, which feed it horizontally into squeeze rollers, which
progressively reduce the sheet thickness. Four parts of this process have
been studied in detail by the Oxford Study Groups and each one is a
challenging problem.

(i) Hot rolling

Under the squeeze rollers the metal is assumed to be visco-plastic so
that an appropriate constitutive law is

$$\underline{\dot{\epsilon}} = \lambda(T)\,|\underline{\sigma}|^{n-1}\underline{\sigma} \ ,$$

where $\underline{\sigma}$ is the stress tensor, $\underline{\dot{\epsilon}}$ the rate of strain tensor, and T the
temperature. The viscous model $n = 1$ may be analysed for a sheet, or thin
film asymptotics used to obtain a generalised Reynolds equation (Tayler,
1986). A major problem is the possibility of oncoming material backing up
upstream of the rollers to form a 'bolster', that is a region in which
there may be circulatory flow and lower temperatures. Neither model can
adequately predict such a phenomenon.

(ii) Solidification front

For a pure material there is a unique phase change boundary and the
liquid steel is assumed to be at the melting temperature. This leads to a
two dimensional one phase Stefan problem whose properties are well

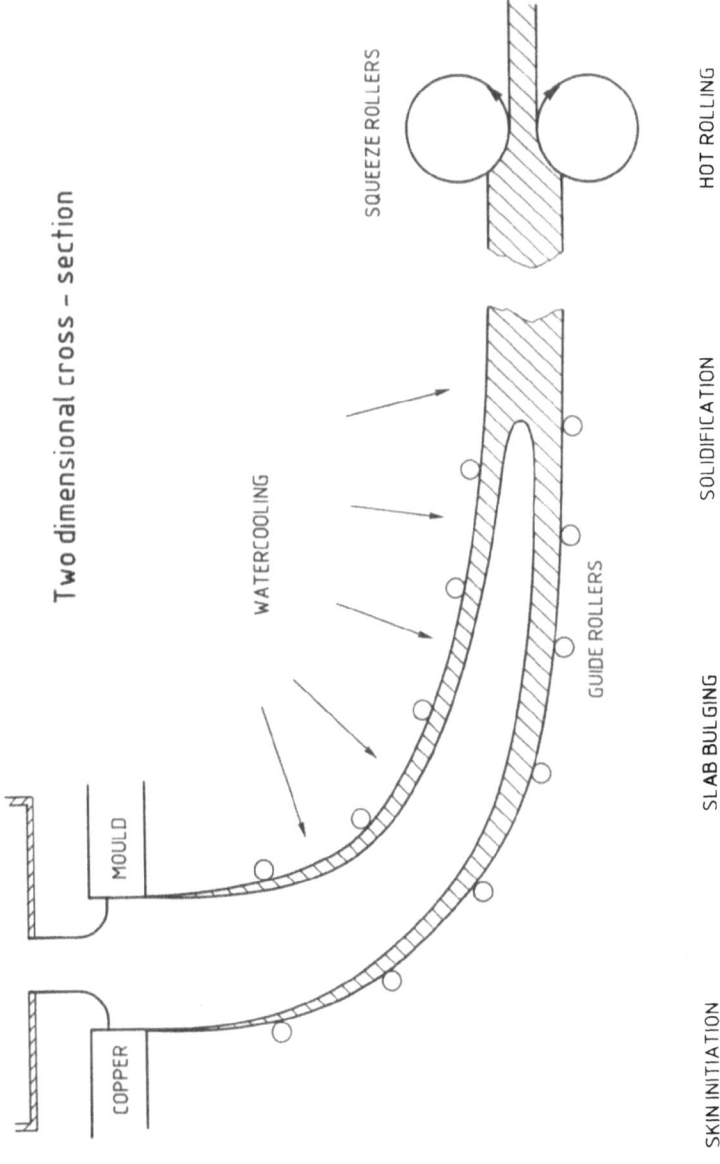

Fig. 2

understood (Tayler, 1986). However, for a binary alloy the liquidus and solidus boundaries are separated by a 'mushy' region in which the volume fraction solidified depends on the local concentration of the secondary material. The local structure in this mushy region may be heterogeneous, with 'grains' growing from nuclei, or 'dendritic', and the quality of the product depends on the particle histories in the mushy region. Models for the mushy region which can predict grain size adequately have not yet been developed despite considerable efforts (Ockendon & Crowley, 1979).

(iii) Skin initiation

The processes occurring in the mould region are more complicated than is shown in Fig. 2. The steel skin must be prevented from touching the mould wall and this is achieved by interposing a thin layer of flux powder which melts to form a lubricant layer on the wall. The flux is drawn through by oscillating the mould, creating a pumping action. However, this oscillatory motion is responsible for periodic notches on the surface of the solidified steel, which spoil its quality. Work is in progress on several possible models which analyse this situation, and further details are given in Bland (1984) and Dewynne, Howison & Ockendon (1990).

(iv) Slab bulging

The solidifying metal strand is passed through a double series of rigid rollers. The skin of solidified steel is supported by the ferrostatic pressure of the fluid core and one set of rollers. The deflection and thickness of the skin are small compared to the radius of the rollers and the total strand thickness in normal operation, but skin deflection may become comparable with strand thickness causing undesirable 'slab bulging'.

The proposed model is a visco-plastic beam for the skin with the liquid
steel at constant pressure moving with constant velocity, and the contact
area with the roller assumed to be small. Then in each span $r < z < r+1$ the
displacement δ is given in terms of the bending moment M by

$$\frac{d^3 \delta}{dz^3} = -|M|^n \text{sgn } M$$

where $\frac{d^2 M}{dz^2} = -P$, the ferrostatic pressure.

At the roller $z = r$, δ is prescribed as δ_r, and the slope $\frac{d\delta}{dz}$, the strain
$\frac{d^2 \delta}{dz^2}$ and the bending moment are continuous. The shear $\frac{dM}{dz}$ will be discontinuous, by an amount equal to the force applied by the roller. More
elaborate roller conditions can be applied if needed; the problem is how to
numerically iterate solutions through each span and what initial data to
use. In the simplest model we take $n = 1$, which is clearly not physically
realistic but may give us some mathematical insight, and examine the
transition from $z = r$ to $z = r+1$. In each section between rollers $\frac{d^5 \delta}{dz^5} = p$,
and

$$\frac{\delta - \delta_r}{P} = \frac{(x-r)^5}{120} + \frac{\delta_r^{iv}}{24} (x-r)^4 + \frac{\delta_r^{\prime\prime\prime}}{6} (z-r)^3 + \frac{\delta_r^{\prime\prime}}{2} (z-r)^2 + \delta_2^{\prime}(z-r) \ ,$$

where the coefficients contain the derivatives of δ evaluated at $z = r$.
Applying the continuity conditions on these derivatives at $z = r+1$, and
eliminating δ_r^{iv} and δ_{r+1}^{iv}, we obtain

$$\underline{u}_{r+1} = \begin{bmatrix} \delta_{r+1}^{\prime} \\ \delta_{r+1}^{\prime\prime} \\ \delta_{r+1}^{\prime\prime\prime} \end{bmatrix} = M \underline{u}_r + \underline{c}_r \ ,$$

where M is a constant matrix with real eigenvalues $\lambda_1 = -1$, $|\lambda_2| > 1$, $|\lambda_3| < 1$ and $\underline{c}_r$ is a given vector. Thus there is growing mode (which is not the case for the elastic beam $\frac{d^4 \delta}{dz^4} = P$) and we may expect to have to apply two initial conditions, in addition to $\delta_0 = 0$ and a 'downstream radiation' condition.

A solution, periodic in each span, is clearly possible if $\underline{c}_r = \underline{c}$ and $\underline{u}_r = (I-M)^{-1}\underline{c}$. A solution periodic every two spans is also possible using the eigensolution corresponding to the eigenvalue -1, but is not observed in practice. The difficulty in using a numerical scheme is how to damp out the growing mode by the use of some downstream condition, and a number of conditions have been tried. Such a condition will also be needed in the nonlinear problem $n \neq 1$, but I am unaware of any published work on this, despite a number of very interesting questions which can be asked.

4.   Concluding Remarks

These examples have been chosen to illustrate the breadth of application of some fairly simple mathematical ideas which have neverthe-less stimulated interesting new questions. It is this stimulus given by real problems to academic research which motivates me to interact with industry, and has sustained our effort in Oxford over twenty years. I see every reason to believe that real world problems will continue to stimulate us all in the future.

5. References

1. Bland D.R.; Flux and the Continuum Casting of Steel. IMA JAM 32
    (1984).

2. Carrier G.F.; Applied Mathematics: What is needed in Research and
    Education. SIAM Review 4 (1962).

3. Dewynne J.N., Howison S.D., Ockendon J.R.; The Numerical Solution of a
    Continuous Casting Problem. Proc. Oberwolfach Conference on Free
    Boundaries and Optimal Control (ed. J. Sprekels) Birkhauser (1990).

4. Melzak Z.A.; Bypasses. A Simple Approach to Complexity. Wiley
    (1983).

5. Ockendon J.R., Crowley A.B.; The Numerical Solution of an Alloy
    Solidification Problem. Int. J. Heat Mass Transfer 22 (1979).

6. Please C.P., Wheeler A.A., Wilmott P.; A Mathematical Model of Cliff
    Blasting. SIAM J. App. Math. 47 (1987).

7. Smith R., Carter G.; The Theory of Surface Erosion by Ion Bombardment.
    Proc. Roy. Soc. A 407 (1986).

8. Tayler A.B.; Mathematical Model in Applied Mechanics. OUP (1986).

9. Tayler, A.B.; Mathematical Models of Silicon Chip Fabrication. Proc.
    ECMI Strathclyde (1988).

10. Tayler A.B., King J.R.; Free Boundaries on Semiconductor Fabrication.
    Proc. Int. Conf. on Free Boundaries: Theory and Application (ed.
    Hoffmann) Pitmans (1989).

Mathematical Problems in Research at HILTI

K. Weiss, Schaan

Summary:

The essence of this paper are six examples of varying complexity and involving different kinds of mathematical techniques which all show the usefulness of mathematics in industry. They also show that this usefulness depends
strongly on the degree to which the problems are explicitely solved and on
how strongly the solutions contribute to the development and marketing of
new products.

0   Introduction

The bulk of this paper consists of six examples chosen to illustrate
how different kinds of mathematical methods of varying degrees of complexity
are useful to industry. As a matter of course it cannot be the intention to
show this in general but rather in relation to a specific industrial environment. To evaluate the relevance of the examples it is therefore essential
to give a brief description of the company where the work to be described
was done.

The HILTI group, founded in 1941 presently employs about 10'600 people
in over 80 countries realizing (in 1988) a turnover of about 1,700 Million
sFr.. The company is the market leader to cover the needs of selected professional customers in the construction industry and allied trades on a
worldwide basis.

Products are developed, manufactured and distributed by 4 product
divisions to be sold directly by the local market organisations. The main
products of the 4 divisions are

o   drilling & electric tools

o   powder-actuated fastening systems

o   anchors and

o   construction chemicals and cladding

Production plants are to be found in FL, FRG, GB, USA, and F while R & D
centers are situated in Schaan (FL), Munich and Kaufering (FRG), and Tulsa
(USA).

It will be clear from the above that all Hilti products have to meet
very thorough quality requirements when it comes to questions of durability,

Hj. Wacker and W. Zulehner (eds.),
Proceedings of the Fourth European Conference on Mathematics in Industry, 71–80.
© 1991 *B.G. Teubner Stuttgart and Kluwer Academic Publishers.*

reliability, and endurance of the most adverse conditions as encountered
worldwide on construction sites.

Having understood this it will be easier to appreciate the kind of
research which went into the examples as presented in section 2.

1   Tools Needed
Clearly, solving mathematical problems for industry nowadays asks
for a generous supply of electronic computing equipment. Our research can
rely on such a supply including on the software side codes for
o   Finite Elements (Marc/Mentat)
o   Mathematical manipulation (Macsyma)
o   Data handling and control of experiments (Assyst)
o   CAD (Anvil)
o   Statistics (Stratgraph), etc and
many general purpose programs and of course compilers as needed. It will
be seen in the next section how these tools are used.

2   Six Examples
The titels of the following subsections indicate that the six examples
are chosen two demonstrate two things, viz.
(i)   the various mathematical techniques of different complexity needed
to solve our problems and
(ii)   some typical physical or engineering problems encountered when it
comes to ensure the leading market position for our products.

2.1   Curve Fitting: Preload Relaxation of Anchors in Concrete
Anchors subjected to loads varying with time (e.g. fastening of vi-
brating machines, rails, wind or water exposed structures, etc.) are to be
applied with a preload imposed on the bolt such that the strains caused by
the variations of the load appear inside the bolt and not at the interface
between the anchor and the material where in the anchor is fastened. If the
material is concrete, the most frequently encountered case, one has to cope
with the well known phenomenon of creep under load. (The reason is mainly
that typically about 20 % of the concrete volume are pores.) This creep
tends to reduce the preload initially imposed on the anchor thereby endan-
gering the long term (typically 40 years) safety. It is therefore paramount
to extrapolate safely from relatively short term experiments (4-5 years) to
the long term influence of creep.

In our case we chose as a model creep mechanism Kohlrausch relaxation [1], originally introduced to describe elastic aftereffects of silk and glass fibres in electrometers and later applied to more recent problems such as the conductance relaxation in semiconductors [2] and thermal relaxation in a random magnet [3]. The model equation reads

$$[(\sigma(t)-\sigma_\infty)/\sigma_0]^n = -\tau(t)\dot{\sigma}(t)/\sigma_0 \qquad\qquad (1)$$

where $\sigma(t)\sigma_0$, σ_∞ are the preloads at time t, at time $t=0$ and asymptotically for $t\to\infty$, and where $\tau(t)$ is a time dependent relaxation time characteristic for the material (i.e. concrete). Solving for a constant relaxation time $\tau(t) = \tau_0$ one finds

$$n = 1: \quad \sigma(t)-\sigma_\infty = (\sigma_0-\sigma_\infty)\,\exp{-t/\tau_0} \qquad\qquad (2)$$

$$n > 1: \quad \sigma(t)-\sigma_\infty = (\sigma_0-\sigma_\infty)[1+(n-1)(1-\sigma_\infty/\sigma_0)^{n-1}t/\tau_0]^{1/(1-n)} \qquad (3)$$

Numerical curve fitting then yields the best estimates for the parameters n, τ_0, and σ_∞. In the Figure a typical result is shown. It eventually enabled us to make safe reliable long term predictions for anchors under preload.

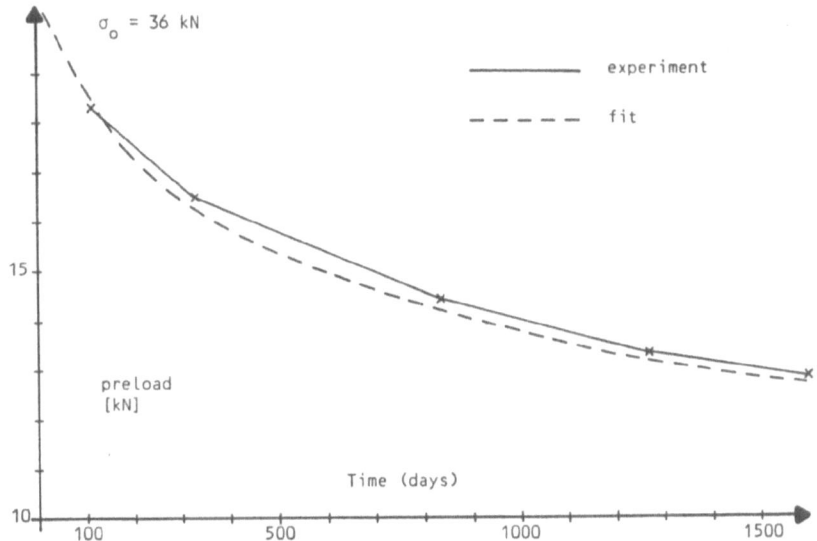

Of course, the problem described in this section can also be tackled from
a more fundamental point of view as was done in a an investigation on the
constitutive equations of concrete by two of my collegues [4].

2.2   Simple Models: Pull-Out Values of Adhesive Anchors

A very important property of anchors is their pull-out value which
of course depends on many parameters including design, anchoring mechanism,
and geometry of the anchor and the properties of the material the anchor
is fastened to. Here, on the basis of a model [5] for grouted anchors we
developed the following simple method to calculate the pull-out values of
anchors fastened to concrete by a two-component glue. Consider the following
arrangement.

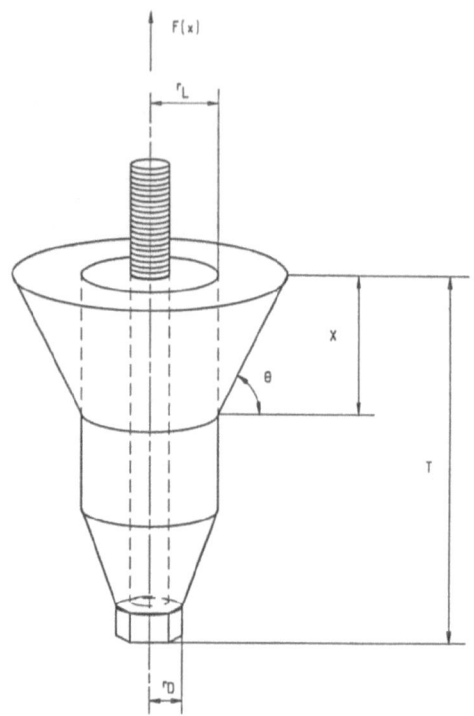

r_L :hole radius

r_D :effective anchor radius

T :anchoring depth

The force at which the anchor breaks loose as a function of the height of the failure cone x is given by

$$F(x) = [(x/\tan\theta+r_L)^2-r_L^2]\pi\sigma_B \qquad \text{concrete} \qquad\qquad (4)$$

$$+ (r_L^2-r_D^2)\pi\sigma_V \qquad\qquad \text{adhesive}$$

$$+ (T-x-r_L)2\pi r_L\sigma_{BV} \qquad \text{concrete/adhesive}$$

where σ_B, σ_V, and σ_{BV} are the failure stresses for concrete, the glue, and the interface glue/concrete, respectively. Differentiating with respect to x gives the height $x_o \cong 3r_L$ of the failure cone for which the anchors preferably fails. Evaluating $F_{min} = F(x_o)$ to an appropriate approximation yields in a typical case $F_{min} \cong 100$ kN which compares favourably with the experimental value of about 80 kN ± 10 %.

2.3 Sophisticated Models: Prediction of Tool Performance by Continuum Mechanical Methods

Drill hammers and drill bits are an apparently simple mechanical system to make holes into concrete by taking advantage of its weak point: its brittleness or, more technically speaking, its low tensile strength. Apparantly simple it is, but in order to increase the processes efficiency, to ensure a reliable durability, and to reduce the noise and vibration levels, quite sophisticated theoretical and experimental investigations are called for. This subsection well address - very superficially - some of the theoretical work we have done in this context. As a first step well known theories describing shock waves in homogeneous cylindrical bodies created by the impact of one such body on a second one [6] were studied. This represents a very crude model for the impact of the pneumatically driven piston inside the drill hammer onto the drill bit. This theory was then generalized to describe cylinders of smoothly varying diameters and to include not only longitudinal waves but also bending and torsional waves. Next a real drill bit with spiral was considered. A cleverly devised coordinate transformation [8] reduced this case to the previous one where the cylinder is now characterized by an effective speed of sound - different from the speed of sound of the actual material -, an effective length, and an effective diameter. Furthermore and very importantly, in the new corrodinate system the longitudinal waves were decoupled from the transverse and the longitudinal ones.

These and other theoretical results led to several important improvements. For instance, it was possible to devise a new geometrical design in
the region where the thickness of the drill bit changes and thereby to reduce the sensitivity to production tolerances of the efficiency at which the
shock wave travels through this region. Secondly resonances in the sound
emission could be eliminated by designing a spiral with varying pitch which
led to a considerable reduction of the sound emission.

2.4   Very Sophisticated Models: Finding Re-bars in Concrete

Re-bars in concrete are a well known nuisance when they are hit during
drilling with hammer drills. In many cases it is unsafe to cut them and in
all cases other drilling or cutting methods like for instance systems with
diamond equipped tools are necessary. A method, easy to handle, which would
allow to locate the re-bars with sufficient precision beforehand would therefore be very helpful. Quite a few physical effects could in principle be
exploited to achieve this goal including sound waves, microwaves, thermal
effects [9], X-rays, and electromagnetic fields. The last one, specifically
the re-bar induced distortions of magnetic fields is probably the most frequently investigated possibility. A thorough mathematical analysis of the
correspondig reconstruction problem [10], which is similar to the methods
used when reconstructing holographic [11] or tomographic information did,
as yet, not lead to a really satisfactory solution of the problem. Progress
will be possible if the measurement techniques will be more refined and more
easy to perform and if faster algorithms are found to do the calculations on
a reasonably small computer in a reasonable time.

2.5   Probability: Failure of Multiple Fastening Systems

If some object is fastened not by one but by many fastening elements,
the probability of failure for the whole system is in general a complicated
function of the failure probability of the individual elements. Depending on
the risk of damage to people or property in a given case it is mandatory to
calculate reliably the system's failure probability.

This calls first for a thorough knowledge of the failure probability
of the individual elements to be acquired through experiments and to be modeled by a distribution function based on these experiments. The figure
below shows how for a specific case a model distribution function is fitted
quite successfully to the experimental results.

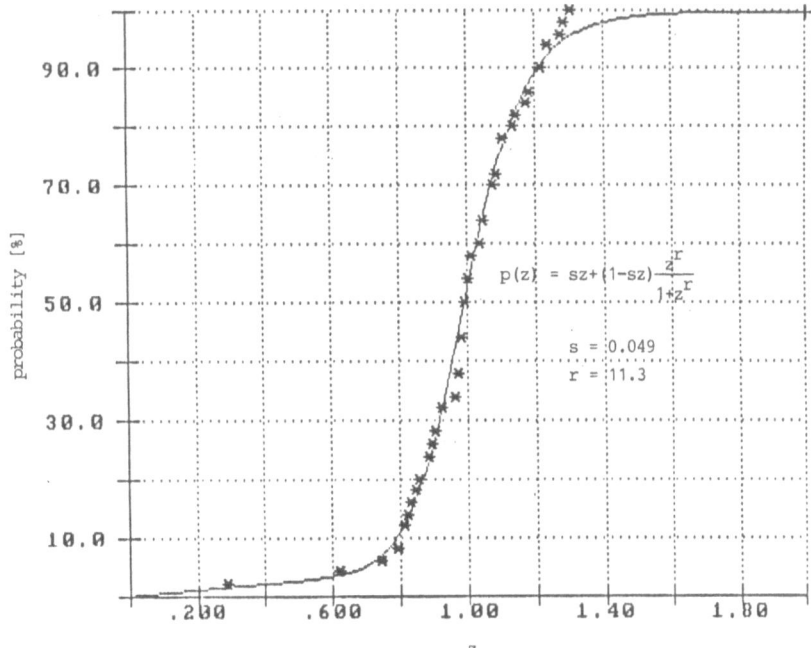

Secondly a probabilistic model has to be conceived describing the
collective failure mode whose probability one wants to calculate. In all
practical cases of our applications this mode can be described as a kind of
chain reaction: if one element fails its neighbours will in general be able
to still keep the system fastened by redistributing the additional load
among the intact elements. If a second element fails it is perfectly
possible that again the load can be redistributed, etc. But there will be a
point, where the additional load created by the failure of several individual elements exceeds the carrying capacity of the rest of the elements so
that the system eventually fails.

Thirdly a mechanical problem arises in that the load distribution of
the total load on the individual elements has to be calculated for the intact system and at all stages of the chain reaction as described above.

Preliminary calculations proved to be quite successful [12] and the
work is in full progress, also incorporating Monte-Carlo calculations which
are mathematically simple, but less flexible and calling for a much larger
numerical effort as is clear if one realises that failure probabilities of
the order of 10^{-6} and smaller have to be calculated.

2.6   Finite Elements: Fastening of Steel Slabs - Top-Hat Geometry

When fastening steel slabs to, say, concrete sometimes the difficulty arises that the head of the nail is pushed through the slab so that no safe fastening results. To avoid this problem a so called Top-Hat was developped as shown schematically in the Figure.

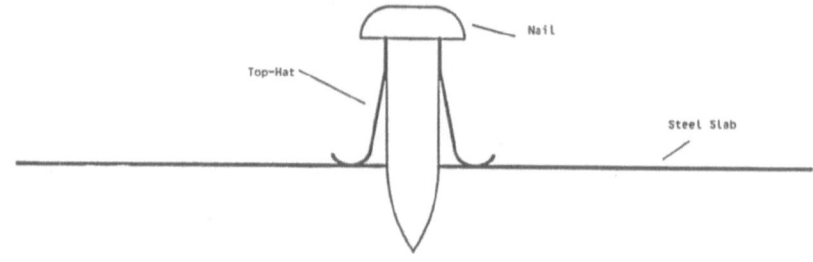

The design of this Top-Hat was determined by FE-methods as displayed in the following Figure for one successful case:

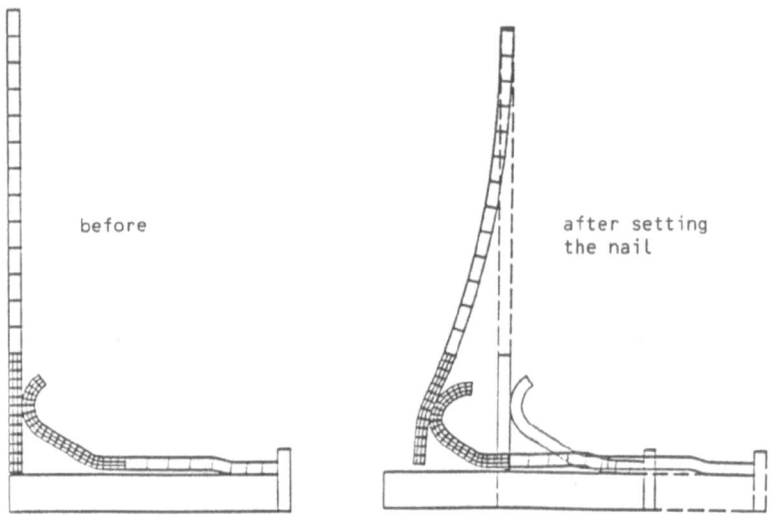

Clearly this is a fine example for the usefulness of FE-methods in cases where the underlying physics is relatively simple but where the geometrical constraints are complicated.

3.   Conclusions

We hope to have shown, that mathematics is an indispensable tool for Research & Innovation at Hilti. The methods used span the range from back of the envelope feasibility calculations through analytical models to complicated computer dependent detailed design problems. It should be clear from what was said that one needs a generous supply of soft- and hardware to be able to perform these tasks. The following table which is of course open to the right and to the bottom summarizes the problem aereas where Hilti needs mathematics and the methods used for the time being. Crosses stand for actual activities

Hilti needs	curve fit.	simple mod.	soph. mod.	very so. mod.	prob. stat.	FE BE	....
	Mathematical tools						
upstream research			X	X			
new products			X	X		X	
product modification			X			X	
feasibility		X					
reduced exp. effort		X	X		X	X	
customer software	X	X			X		
reliability	X				X	X	
safety	X	X			X		
quality control	X				X		
regulations	X				X		
.							
.							
.							

To conclude it should be emphasized that success (like everywhere else) depends on dedicated qualified people with, in this case, the desire to link abstract thinking with innovatives products to satisfy our very demanding customers.

Literature

[1] Kohlrausch, R.: Ann. Phys. (Leipzig) 72, 393 (1847), Sect. 15

[2] Queisser, H.J.: Phys. Rev. Lett. 54, 234 (1985)

[3] Coey, J.M.D.; Ryan, D.H.; Buder, R.: Phys. Rev. Lett. 58, 385 (1987)

[4] Gerber, W.H.K.; Magyari, E.: Engineering Fracture Mechanics, to be publ. (1989)

[5] Cones, M.A.: Paper presented at the 1982 annual convention ACI, Atlanta, Ga.

[6] Hecker, R.; Riederer H.: Forschungsbericht BMFT-FB-HA 85-009 (1985)

[7] Gerber, W.H.K.: Proceedings FASE/DAGA 1982, 715 (1982); ebenso 1984, Darmstadt, 335 (1984); Proceedings Nichtlineare Strukturdynamik (1986), Zürich, 1 (1986)

[8] Gerber, W.H.K.: Proceedings FASE/DAGA 1987, Aachen, 381, 1987

[9] Hillemeier B.; Müller-Run, U.: Beton und Stahlbau 4/1980, 83 (1980)

[10] Engl H.W.; Neubauer, A.: Proceedings of CTAC 89, Brisbane, (1989); A. Neubauer, Proceedings of the Int. Symposium and Team Workshop on 3-D Electromagnetic Field Analysis, Okayama (1989) to be publ.

[11] Dändliker, R.; Weiss, K.: Optics Comm. 1, 323 (1970)

[12] Weiss, K.: Proceedings of the 2nd Int. Conf. on Structural Failure, Product Liability and Technical Insurance, Vienna, 657 (1986); Mitteilungen aus dem Institut für Hochbau und Industriebau der Universität Innsbruck, Festschrift Prof. Henschker, 65 (1986)

Minisymposium: Steel Processing

WATER DROPLET COOLING OF GALVANISED IRON

N.G. Barton

1. Introduction

Consider the solidification of a layer of molten zinc on a steel substrate, a central step
in the manufacture of galvanised iron. Under present operations, the zinc layer cools in air
after it is passed through a bath of molten zinc and air jet strippers. It is proposed to cool
the zinc layer by a bank of nozzles which spray a cloud of microscopic water droplets at
high speed towards the surface (Figure 1).

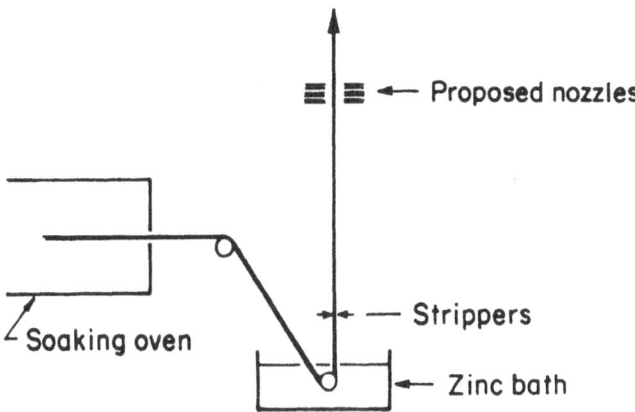

Figure 1: A sketch of the proposed process line

The following values are representative of the manufacture of galvanised iron by water

cooling:

speed of strip	1.5 m.s^{-1}
thickness of strip	10^{-3} m
thickness of zinc layer	$10 \times 10^{-6} \text{ m}$
temperature of zinc bath	$450° \text{ C}$
fusion temperature of pure zinc	$420° \text{ C}$
speed of water droplets at nozzles	200 m.s^{-1}
diameter of water droplets	$50 \times 10^{-6} \text{ m}$
distance of nozzles from zinc	0.1 m
thickness of spray region	0.1 m

83

Hj. Wacker and W. Zulehner (eds.),
Proceedings of the Fourth European Conference on Mathematics in Industry, 83–96.
© 1991 B.G. Teubner Stuttgart and Kluwer Academic Publishers.

Under normal manufacturing conditions, the layer of molten zinc on the steel strip solidifies about 3 m above the zinc bath. For about the last 0.5 m below the point where solidification is complete, the zinc surface takes on a dull finish associated with the breakthrough of zinc crystals to the surface. The vertical region where solidification takes place is called the mushy region in this paper.

The zinc solidifies from the inside to the outside leaving a pronounced pattern called 'spangles' on the zinc. The spangles are regions where solidification has taken place from a single nucleation site. Typically, the spangles have a diameter of about 5×10^{-5} m or less, and at the boundary between the spangles can be found a very small valley. The small variations in coating thickness from spangle to spangle and between spangles is called 'spangle relief'. Spangle relief does not cause a problem for many applications of galvanised iron, but can be unwanted for some applications (such as car panels) where a very smooth surface is required. Experiments have shown that spraying water onto the surface can achieve a spangle size of about 10^{-3} m; even smaller spangles with a diameter of about 10^{-4} m can be obtained when a proprietary nucleation agent is mixed in with the water.

A thorough investigation of this industrial process involves many features including solidification of the zinc layer, surface tension effects, wave propagation on a highly viscous thin layer, and the impact of high speed droplets with mushy zinc. These effects and their associated time scales are now examined using simple models.

2. Heat transfer calculations

First of all, consider the cooling of galvanised iron under normal conditions without the water spray operating. Two simple calculations are used to estimate the vertical extent of the zone where solidification of the molten zinc takes place. Before doing so, the physical behaviour of a solidifying material is briefly reviewed.

For most of this work, it is assumed that the molten zinc is pure. This is not likely to be the case, and, moreover, the solidification of a fluid containing impurities is a delicate phenomenon which is still imperfectly understood. In general, the presence of impurities leads to the situation depicted in Figure 2 in which solidification at a certain concentra-

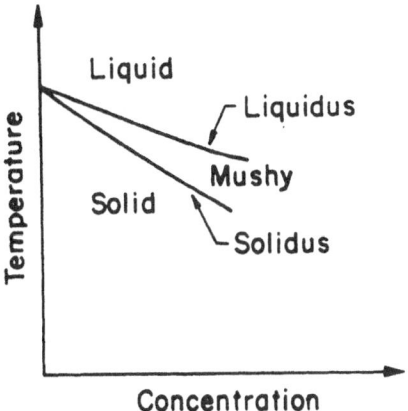

Figure 2: Typical solidification behaviour in the presence of impurities

tion of contaminant starts at the *liquidus* temperature and is completed at the *solidus* temperature. Impurities generally lower the solidus and liquidus temperatures below that of the pure material. There is another significant complication in that the impurities are transported by diffusion in the liquid ahead of the solidification front. Thus, solidification in the presence of impurities involves a consideration of diffusion processes for both heat and impurities, as well as the moving boundary aspects of the solidification front. It is also possible for liquid to be *supercooled*, or remaining in liquid state below the liquidus temperature. The solidification front is typically unstable when the liquid is supercooled, and consequently the front forms a dendritic structure. The dendrites and the molten material trapped between them form what is commonly called a "mushy zone" which is neither completely solid nor completely liquid.[5]

For the case of a very thin layer of zinc on steel strip, solidification usually is initiated at a nucleation site at the interface of the zinc and steel. Thereafter, crystals of solid zinc grow away from the nucleation site in all directions and eventually either collide with crystals from another nucleation site or reach the free surface. This is called "equiaxed growth" [9] and each nucleation site produces a spangle.

Constant heat transfer

During the time interval τ_1, suppose there is constant heat transfer Q per unit area at the free surface, and that the temperature is reduced from the bath temperature T_0 to the fusion temperature T_f. If the contribution of the zinc layer to the sensible heat is neglected and if thermal diffusion throughout the steel is so fast that the temperature T at a vertical location is almost independent of depth in the strip, then a simple heat balance gives

$$\frac{dT}{dt} = \frac{Q}{\rho_s c_s \delta_s}$$

[The notation is described at the end of this report.]

When the strip has reached the fusion temperature T_f of pure zinc, we therefore have the estimate

(1) $\tau_1 = (T_f - T_0)\rho_s c_s \delta_s / Q.$

Suppose that solidification of the zinc layer takes place during the time interval τ_2. This gives the simple estimate

(2) $\tau_2 = \rho_z \delta_z L_z / Q.$

It follows that

(3) $\tau_1 + \tau_2 = [(T_f - T_0)\rho_s c_s \delta_s + \delta_z \rho_z L_z]/Q$

and, inserting the values $\tau_1 + \tau_2 = 2$ s and the material properties given at the end, it follows that

(4) $Q = 8.7 \times 10^4 \text{ Wm}^{-2},$

(5) $\tau_1 = 1.25 \text{ s},$

(6) $\tau_2 = 0.75 \text{ s}.$

The above calculation indicates that, if the strip speed is 1.5 m.s^{-1}, the vertical extent of the zone where solidification takes place is about 1.1 m. This calculation is almost certainly an underestimate, since it is likely that more efficient heat transfer will take place near the bottom of the vertical strip, thereby leading to a smaller zone for sensible heat loss and a longer solidification zone.

Newton's law of cooling

This hypothesis just raised can be tested using a slightly more realistic heat transfer boundary condition at the free surface of the zinc. Specifically, the boundary condition at the edge of the zinc is taken to be Newton's law of cooling

$$k_z \frac{dT}{dy} = -H(T - T_a)$$

where y is a co-ordinate measured outwards from the steel-zinc interface. A simple heat balance for the stage when sensible heat is removed gives

$$\rho_s c_s \delta_s \frac{dT}{dt} = -H(T - T_a)$$

which may be integrated to give

$$T - T_a = (T_0 - T_a)e^{-Ht/(\rho_s c_s \delta_s)}.$$

The time interval τ_1 for the strip to reach the fusion temperature T_f is now

(7) $$\tau_1 = -\frac{\rho_s c_s \delta_s}{H} \log[(T_f - T_a)/(T_0 - T_a)].$$

The calculation to give the length of the zone where solidification takes place requires more care. We assume that the movement of the front is governed by approximately steady state diffusion processes[7] and write down the following system of equations for the moving boundary problem:

$$\frac{d^2 T}{dy^2} = 0, \quad Y(t) < y < \delta_z,$$

$$k_z \frac{dT}{dy}(\delta_z, t) = -H(T - T_a),$$

$$T(Y(t), t) = T_f,$$

$$-k_z \frac{dT}{dy}(Y(t), t) = \rho_z L_z \frac{dY}{dt}.$$

The solution satisfying the conditions at the strip edge δ_z and the moving front $Y(t)$ is

$$T(y, t) = A(t) + B(t)y$$

where

$$A(t) = [HT_a Y - (k_z + H\delta_z)T_f]/(HY - (k_z + H\delta_z))$$

$$B(t) = H(T_f - T_a)/(HY - (k_z + H\delta_z))$$

whilst the moving boundary condition gives the equation

$$-k_z B = \rho_z L_z \frac{dY}{dt}$$

which may be integrated to give

$$0.5 \, \rho_z L_z Y [HY - 2(k_z + H\delta_z)] = -k_z H (T_f - T_a)t.$$

Finally, the solidification front will have moved all the way through the zinc layer after an interval τ_2 when $Y(\tau_2) = \delta_z$, and this gives

(8)
$$\tau_2 = \frac{\rho_z L_z \delta_z^2}{2k_z (T_f - T_a)}\left(1 + \frac{2k_z}{H\delta_z}\right).$$

Again, it is known that $\tau_1 + \tau_2 = 2$ s. We assume that $2k_z/H\delta_z \gg 1$ and insert values (at end) for the other constants to obtain

$$H = 790 \text{ W}/(^\circ \text{Cm}^2),$$

$$\tau_1 = 0.35 \text{ s},$$

$$\tau_2 = 1.65 \text{ s}.$$

It may be now confirmed that $2k_z/H\delta_z \gg 1$.

The results of this calculation also confirm that the zone where solidification takes place is surprisingly large indeed if a slightly more realistic cooling law is used. Basically, therefore, solidification takes place shortly after the zinc emerges from the bath, and it is not until the last part of the process that the crystals of zinc growing from the inside to the outside are evident to the eye.

3. Surface tension effects

The enhanced solidification processes discussed here also involve very fine granules of a proprietary nucleation agent. These granules might cause important surface tension effects - for example, if they dissolved and spread across the surface almost instantly, they would cause a local reduction in surface tension. The consequence would be a reduction

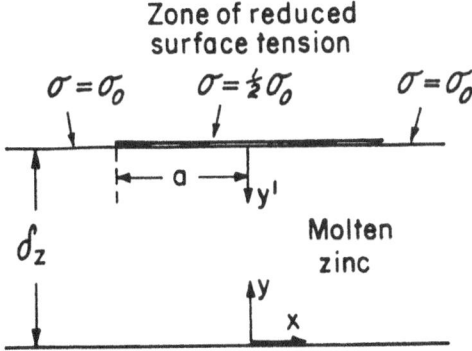

Figure 3: The physical situation investigated in Section 3

in the depth of the molten zinc layer in the region surrounding the granule. A two dimensional calculation is now presented to estimate how quickly surface tension could cause an appreciable reduction in the depth of the layer.

Consider the situation sketched in Figure 3 where the nucleation agent has caused the surface tension in a region of width $2a$ to be $\sigma_0/2$ in comparison to the undisturbed value of σ_0. A velocity profile in the y direction will develop as a result of the surface tension; the time for this profile to develop fully into Couette flow is of the order of

$$\tau_3 = \delta_z^2/(2\nu_z) \approx 100 \times 10^{-6} \text{ s}$$

The vorticity equation in the zinc layer is

$$\frac{\partial \omega}{\partial t} = \nu_z \frac{\partial^2 \omega}{\partial y'^2}$$

which has solution

$$\omega = \omega_0 \operatorname{erfc}(y'/2\sqrt{\nu_z t})$$

provided $t \ll \tau_3$. Here, ω_0 is determined by its value at the free surface as follows:

(9)
$$\omega_0 = -\frac{du}{dy} = \Sigma/\mu_z \approx \frac{\sigma_0}{2a\mu_z}$$

where Σ is the shear stress at the surface and the reduced value $\sigma_0/2$ of the surface tension applies in a region of width a.

The speed of the molten zinc as a response to the surface tension is therefore

$$u = 2\omega_0\sqrt{\nu_z t} \operatorname{ierfc}(y'/2\sqrt{\nu_z t});$$

and the volume flux out of a control surface of length $2a$ surrounding the original site of the granule is

$$2 \int_0^{\delta_z} u \, dy' = 4\omega_0 \nu_z t \, i^2 \text{erfc}(0).$$

Here, i^nerfc is the nth integral of the complementary error function and $i^2\text{erfc}(0) = 1/4$.

Hence, if h denotes the depth of the molten zinc, it follows that

$$\text{rate of change of volume} = 2a\frac{dh}{dt} \approx \omega_0 \nu t = -\frac{\sigma_0 \nu_z t}{2a\mu_z}$$

where equation (9) has been used for ω_0.

This equation can be simplified and integrated to give

$$h = h_0 - \frac{\sigma_0 t^2}{8a^2 \rho_z},$$

from which it follows that the time required for the zinc layer to halve its thickness under the action of surface tension is

(10) $$\tau_4 = 2a(\rho_z h_0/\sigma_0)^{1/2}.$$

This produces the typical value for τ_4 of 61×10^{-6} s when a is 100×10^{-6} m and the other physical values are as given at the end.

The calculation produced in this Section indicates that surface tension might have dramatic consequences in the extremely thin layers of molten zinc under consideration, although, in practice, drawing down of the surface by reduced surface tension would be affected by the growth of the solid crystals of zinc from below.

4. Waves on a shallow viscous fluid

In this Section, we consider another illustrative calculation which might be considered a precursor to full numerical solutions discussed in the next Section. Specifically, suppose that an impact by a water droplet has caused the surface of the zinc to deform, and it is desired to find out how long it would take for the surface to return to its original condition neglecting any thermal effects. The required theory is that of wave propagation in a very shallow viscous fluid with surface tension effects included.

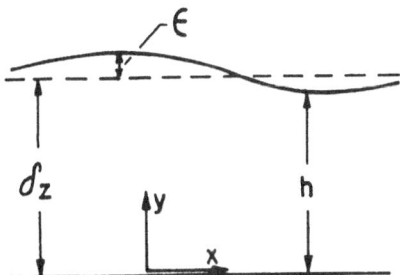

Figure 4: Definition sketch for waves on a shallow fluid

Consider the situation shown in Figure 4 in which the surface of the molten zinc is initially in a non-equilibrium state, and we seek to find out how long it would take to recover the equilibrium position if surface activity is ignored. The starting point is the set of equations

(11)
$$\rho_z\left[\frac{\partial u}{\partial t} + u\frac{\partial u}{\partial x}\right] = -\frac{\partial p}{\partial x} + \mu_z\frac{\partial^2 u}{\partial y^2},$$

(12)
$$0 = \frac{\partial p}{\partial y},$$

for viscous flow in a shallow region, together with the surface tension condition

(13)
$$p = -\frac{\sigma}{R} \approx -\sigma\frac{\partial^2 h}{\partial x^2},$$

the continuity equation

(14)
$$\frac{\partial h}{\partial t} + \frac{\partial}{\partial x}\int_0^h u\,dy = 0$$

and the approximate boundary conditions

$$u(x,0,t) = 0, \quad \partial u/\partial y(x,h,t) = 0.$$

In general, these equations require a numerical solution, but some simplification is possible in certain cases as is now shown. A solution satisfying the boundary conditions is sought in the form

$$u = W(t)\frac{y}{h}\left(\frac{y}{h} - 2\right).$$

This enables the integral in the continuity equation to be evaluated. Further, if the acceleration terms in equation (11) are neglected [see the end of this section for discussion of this point], the governing equations become

(15)
$$0 = \sigma \frac{\partial^3 h}{\partial x^3} + \frac{2\mu_z W}{h^2},$$

(16)
$$\frac{\partial h}{\partial t} = \frac{2}{3}\frac{\partial}{\partial x}(Wh).$$

The term W can now be eliminated from (15,16) to give

(17)
$$-\frac{\partial h}{dt} = \frac{\sigma}{3\mu_z}(3h^2\frac{\partial h}{\partial x}\frac{\partial^3 h}{\partial x^3} + h^3\frac{\partial^4 h}{\partial x^4}).$$

This equation also requires a numerical solution in general, but further simplification is possible when the wave amplitude ϵ is small in comparison with the depth h. In this case, the ratio of the terms on the RHS is ϵ/h, so that (17) reduces to the governing equation for the shallow viscous capillary waves

(18)
$$-\frac{\partial h}{dt} = \frac{\sigma \delta_z^3}{3\mu_z}\frac{\partial^4 h}{\partial x^4}$$

where the term h^3 has been approximated by the mean term δ_z^3.

Equation (18), which represents a gross simplification, allows a simple solution which gives some insight. It is straightforward to show that the damping rate of the Fourier coefficient in

$$h = \delta_z + \epsilon \cos kx \ e^{-\gamma t}$$

is

(19)
$$\gamma = \sigma \delta_z^3 k^4/3\mu_z.$$

Thus, in this approximation, the viscous waves are standing waves which are very highly damped at large k (corresponding to small wavelength λ).

The numerical values given at the end may be used to calculate the relaxation timescale $1/\gamma$ in the following table:

λ	$1/\gamma$
10^{-3} m	8.2×10^{-3} s
5×10^{-4} m	5.1×10^{-5} s
10^{-4} m	8.2×10^{-7} s

Thus the time required for these small wavelength waves (say of wavelength slightly smaller than 5×10^{-4} m) to decay is much faster than the thermal processes associated with the solidification of the zinc in the absence of water drops (Section 2), and approximately the same as that for the surface tension effects described in Section 3.

The above theory is not, however, applicable at very small wavelengths. If $1/\gamma$ is taken as an appropriate timescale, the acceleration terms in (11) are of order $\rho_z W \gamma$ compared to the order $\mu_z W/\delta_z^2$ of the viscous term. Acceleration terms should therefore be included when γ is of ν_z/δ_z^2, or for timescales less than about 2×10^{-4} s. Thus the last two lines of results in the above table are not strictly applicable because of the neglected acceleration terms. Detailed numerical work is required if equations (11,12,14) need to be solved.

5. Drop and particle impact studies

There is quite a large literature on drop and particle impact studies. Experimental studies on droplet splashing are available [4,8,10,11,15,16] and special mention is made of a paper by Araki & Moriyama [1] which provides a semi-empirical model for the deformation behaviour of a liquid droplet impinging on a hot surface. Many computational studies [3,6,12,14] have been made of the impact of drops on either solid or liquid surfaces. Recently, particle methods [13] have been shown to be particularly suitable for free surface problems.

The problem discussed in this paper involves a combination of impact of fluid drops onto a fluid surface *and* solidification. The physical scales are also rather special, involving a consideration of high speed microscopic drops incident on a very shallow fluid layer. The computational literature cited above would require substantial modification for application to the present problem.

6. Conclusions

This paper provides simple descriptions for some physical processes that occur in the manufacture of galvanised iron, either by present techniques or by a proposed new method. Each aspect of the study has potential for further work which would certainly be of academic interest and hopefully of industrial interest.

The paper is a modified form of the report [2] on a problem presented to the 1989 (Australian) Mathematics-in-Industry Study Group by BHP Coated Products Division.

Contributions by Rodney Carr, Paul Cleary, John Harper, David Jenkins, Mark McGuinness, Chee Ng and Cat Tu are gratefully acknowledged, as is the permission of CSIRO Division of Mathematics and Statistics to publish the article.

Nomenclature

a	diameter of region affected by drop (m)
A, B	functions of time ($^{\circ}$C, $^{\circ}$C m^{-1})
c	specific heat (Jkg^{-1} $^{\circ}$C^{-1})
C_D	drag coefficient
h	depth of molten zinc (m)
H	heat transfer coefficient (Wm$^{-2\circ}$C^{-1})
k	thermal conductivity of heat (Wm$^{-1\circ}$C^{-1})
k	wave number (m^{-1})
L	latent heat of fusion (Jkg^{-1})
p	pressure (Pa)
Q	heat transfer coefficient (Wm^{-2})
R	radius of curvature of surface (m)
t	time (s)
T	temperature ($^{\circ}$C)
u, v	velocity components (ms^{-1})
W	velocity component (ms^{-1})
x	coordinate parallel to the zinc layer (m)
y, y'	coordinates normal to the zinc layer (m)
Y	position of solidification front (m)
γ	damping rate (s^{-1})
δ	thickness (m)
ϵ	wave amplitude (m)
μ	viscosity (kgm^{-1}s^{-1})
ν	kinematic viscosity (m^{2}s^{-1})
ρ	density (kgm^{-3})
σ	surface tension (Nm^{-1})
Σ	shear stress (Nm^{-2})
τ	time interval (s)
ω	vorticity (s^{-1})

Subscripts

   s   steel

   z   zinc

Physical values

c_s	$450 \ \mathrm{Jkg^{-1}{}^{\circ}C^{-1}}$
k_z	$100 \ \mathrm{Wm^{-1}{}^{\circ}C^{-1}}$
L_z	$10^6 \ \mathrm{Jkg^{-1}}$
μ_z	$3 \times 10^{-3} \ \mathrm{kgm^{-1}s^{-1}}$
ν_z	$0.46 \times 10^{-6} \ \mathrm{m^2 \ s^{-1}}$
ρ_s	$8 \times 10^3 \ \mathrm{kg \ m^{-3}}$
ρ_z	$6.5 \times 10^3 \ \mathrm{kg \ m^{-3}}$
σ	$0.7 \ \mathrm{Nm^{-1}}$

References

[1] Araki, K.; Moriyama, A.: Theory on deformation behaviour of a liquid droplet impinging onto hot metal surface, *Trans ISIJ* 21 (1981), 583-590.

[2] Barton, N.G.: Droplet cooling of galvanised iron, *Proceedings of the 1989 Mathematics-in-Industry Study Group*, N.G. Barton and J. Ha (eds) (CSIRO Division of Mathematics and Statistics, 1989), 18-35.

[3] Daly, B.J.: Numerical study of the effect of surface tension on interface instability, *Phys. Fluids* 12 (1969), 1340-1354.

[4] Edgerton, H.E.; Killian, J.R.: *Flash* (Branford, Boston, 1954).

[5] Fowler, A.C.: Theories of mushy zones: application to alloy solidification, magma transport, frost heave and igneous intrusions, *Structure and Dynamics of Partially Solidified Systems*, D.E. Laper (ed.) (Martinus-Nijhoff, Dordrecht, 1987), 159-200

[6] Harlow, F.H.; Shannon, J.P.: The splash of a liquid drop, *J. Appl. Phys* 38 (1967), 3855-3866.

[7] Hill, J.M.; Dewynne, J.N.: *Heat Conduction* (Blackwell, 1987).

[8] Hobbs, P.V.; Kezweeny, A.J.: Splashing of water drop, *Science* 155 (1967), 1112-1114.

[9] Kurz, W.; Fisher, D.J.: *Fundamentals of Solidification* (Trans. Tech. Pub., Switzerland, 1984).

[10] Levin, Z.; Hobbs, P.V.: Splashing of water drops on solid and wetted surfaces: hydrodynamics and charge separation, *Phil. Trans A* 269 (1971), 555-585.

[11] Macklin, W.C.; Metaxas, G.J.: Splashing of drops on liquid layers, *J. Appl. Phys* 47 (1976), 3963-3970.

[12] Monaghan, J.J.: Application of the particle method SPH to hypersonic flow, *Computational Techniques and Applications CTAC-85*, J.Noye & R. May (eds) (Elsevier, 1986), 357-365.

[13] Monaghan, J.J.: An introduction to SPH, *Comp. Phys Comm.* 48 (1988), 89-96.

[14] Nishikawa, N.; Amatatu, S.; Suzuki, T.: Simulation of impact phenomenon by impinging droplet, *Computational Fluid Mechanics*, G. de Vahl Davis & C. Fletcher (eds) (Elsevier, 1988), 569-578.

[15] Sahay, B.K.: Rupture of water drops over liquid surfaces, *Indian J. Phys* 18 (1944), 306-310.

[16] Worthington, A.M.: *A Study of Splashes*, (MacMillan, N.Y., 1963).

Arbeitsgruppe Technomathematik   and   CSIRO Division of Mathematics & Statistics
Universität Kaiserslautern   P.O. Box 218 Lindfield, N.S.W. Australia 2070
Erwin-Schrödinger-Str
D-6750 Kaiserslautern, FRG

A MATHEMATICAL MODEL FOR THE CALCULATION OF COOLING CRITERIA FOR ROUND BLOOM

H.Holl, R.Scheidl, K.Schwaha

VOEST ALPINE Industrieanlagenbau Ges.m.b.H.

Abstract

In the continuous casting machine the solidification of the liquid steel is controlled by extracting the heat in a defined way. One of the most difficult problems in the continuous casting process is to find the proper cooling conditions. High complexity arises form the correlation of thermal effects on product quality.

The heat transfer coefficient in a continuous casting machine is described as a stepwise function. This may lead to thermal stresses and to defects in the continuously cast product. The formation of cracks and segregation as well as the stucture of the solidified steel can be influenced by a proper secondary cooling concept.

In particular round blooms are rather susceptible to center cracking and segregation. In this model a Stefan-like problem is treated with the thermal boundary conditions being the control variable. The objective is to achieve high cooling rates (to minimize segregation) but without getting center cracking (constraint condition). To analyse the cracking thermally induced stresses have to be evaluated. This is done with a non-linear thermo-visco-elastic material law. The calculation can be adapted so that additional metallurgical experience can be brought in. This leads to a cooling concept, which is based on theoretical concepts and practical experiences.

Hj. Wacker and W. Zulehner (eds.),
Proceedings of the Fourth European Conference on Mathematics in Industry, 97–108.
© 1991 *B.G. Teubner Stuttgart and Kluwer Academic Publishers.*

A MATHEMATICAL MODEL FOR THE CALCULATION OF
COOLING CRITERIA FOR ROUND BLOOM

Introduction

In the continuous casting process liquid steel is
continuously solidified. It comprises the so called
primary cooling in the mold. Here the cross sectional
shape of the strand is determined by forming an initial
shell. In the so called secondary cooling region further
shell growth up to total solidification is achieved by
spray cooling onto the strand. A typical machine design
for bloom casting is shown in fig. 1. Two main cast
sections are distinguished in continuous casting:
- **Slabs** have a width to thickness ratio »1
- **Blooms** and **Billets** have a width to thickness
  ratio ≈1

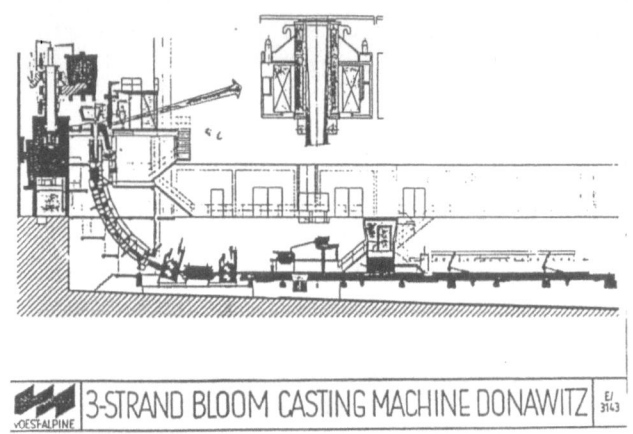

Fig. 1

The secondary cooling region is subdivided into several
zones, which are controlled separately. This division into
several controlled units results in a defined spray water
in each zone and therefore a stepwise function for the
heat transfer coefficient.

The continuous casting process is in a rapid progress.
As the increasing automation permits to influence several

quality parameters, adequate calculating tools have to be developed, to optimize the quality results. This requires good theoretical models for the relevant physical processes of strand solidification.

Secondary cooling is the main control variable to influence solidification and related physical processes. Besides operational requirements like machine protection, solidification within the machine length and energy saving, the quality features are to be optimized or have to fulfill constraint conditions.

In this paper segregation is considered, which widely influences the mechanical behaviour of the continuously cast product. Furthermore the occurance of cracks must be avoided which is considered by limiting the reheating temperature of the surface. In order to verify the maximum reheating temperature for cracking, a stress and strain calculation was carried out. The results confirm the importance of the reheating temperature, as a critical tension strain in the mushy zone will cause centre cracking.

These both targets i.e. low segregation and a crack free product lead to controversial requirements, see fig. 2, which must be considered in a practical cooling strategy.

<u>Controversial control variables</u>

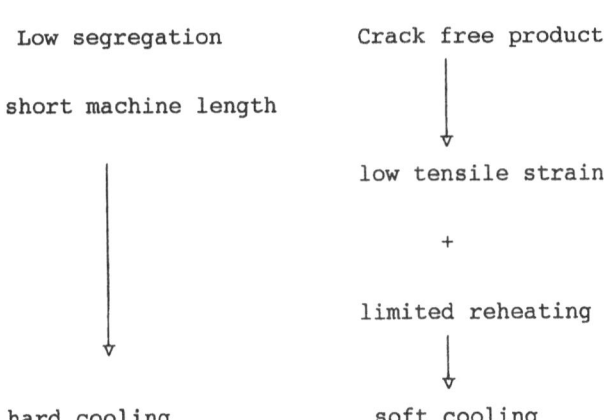

Fig. 2

Mathematical formulation of the heat transfer

The heat transfer in the continuous casting machine can be
described in a one dimensional formulation as follows [1]:

$$\frac{\partial}{\partial x}[k(T(x,t))\frac{\partial T(x,t)}{\partial x}] = \varrho\, c\ T(x,t)\frac{\partial T(x,t)}{\partial t} \quad \text{for } (x,t)\in[0,d/2]\text{x}[0,t] \tag{1}$$

$$k(T(0,t))\frac{\partial T(0,t)}{\partial x} = h(t)[T(0,t)-T_w] \quad \text{for } t \in [0,t_e] \tag{2}$$

$$\frac{\partial T}{\partial x}(d/2,t) = 0 \quad \text{for } t \in [0,t_e] \tag{3}$$

$$T(x,0) = T_o(x) \quad \text{for } x \in [0,d/2] \tag{4}$$

$$T(s(t),t) = T_1 \quad \text{for } t \in [0,t_e] \tag{5}$$

    x .... space variable
    c .... specific heat
    k .... heat conductivity
    h .... heat transfer coefficient
    T .... temperature
    T_1 ... initial temperature of the steel
    T_w ... temperature of the spray water
    t .... time; ϱ .... density

This problem looks like a Stefan problem [2]. But as
there are alloying elements in the steel, there is a
temperature range for the solidification and additionally
temperature depending material parameters have to be
considered, so this is not a classical Stefan problem.
    Solving the classical Stefan problem for the round
geometry, after some simplifications (constant material
parameters, constant heat transfer coefficient,
solidification temperature)

$$1 - 2\ \tau = \Gamma^2\ (1 - 2\ \ln\ \Gamma) \tag{6}$$

is derived, with
    $\tau = t/t_e$ ... dimensionless time
    $\Gamma = r/R$ ... dimensionless radius
    $t_e= R^2/(2\ k_e^2)$ ... time for completely solidification
    k_e ... solidification coefficient

as an implicit formula for the shell thickness depending
on time. In previous studies it was shown that such simple
analytical formulas do not represent the real
solidification behaviour /3/, so a numerical procedure
which can solve this problem very efficiently must be
found.

The calculation procedure of this direct problem which
is implemented as a finite difference model is shown in
fig. 3. The geometry is divided into finite volumes. For
all of these volumes the corresponding equations (1) to
(5) are formed and also the boundary conditons can be
formulated in an analogous way. Furthermore the material
parameters depending on the temperature are known. As can
be seen in fig. 3, the latent heat of solidification is
treated by appropriate values of the specific heat.

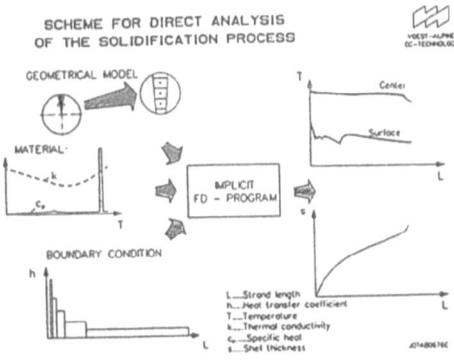

Fig. 3

For a given water flow in each cooling zone, the heat
transfer coefficient in the whole continuous casting
machine can be computed /4/. With the mentioned equations,
the temperature of the strand at each position in the
machine (in each segment) and the shell thickness curve is
calculated (fig. 3).

As a lot of variables (e.g. the water flow rate in each
zone) influence the temperature curve, it is nearly
impossible to get an adequate secondary cooling strategy
only by these direct calculations. The necessary parameter
study requires many different cases to be calculated and
is hence very time and cost consumptive.

The inverse solidification analysis

To solve the inverse problem means to compute the
suitable heat transfer coefficient, when all other
parameters are known. This is a suitable tool for finding
the proper cooling conditions that means to find the
stepwise heat transfer function which leads to optimal
results /1,5/. In this constrained optimization problem
constraints exist both for the
-   control variables i.e. the spray cooling intensity at
    each zone due to a limited operating range of the
    nozzles and
-   dependent variables, in our case the thermally induced
    interface strains.
Segregation minimization is the optimality criterion.
Minimization of segregation is equal to fast
solidification, hence we formulate the condition

$$\int \dot{s}^2 \, dt \quad \longrightarrow \max \qquad (7)$$

which reads for the discretized version:

$$g = \sum_i \frac{(\Delta s_i)^2}{\Delta t_i} \quad \longrightarrow \max \qquad (8)$$

$\dot{s}$ ... solidification rate

Δs_i ... shell thickness increase during time step $\_t_i$

This assures a good segregation profile due to a fast
shell growth rate /6/ and further the metallurgical length
of the continuous casting machine is shortened. This is
important when considering the investment costs of a
plant.

But the hard cooling gives a high reheating temperature
at the end of the secondary cooling which induces strains
and stresses in the cast product. So a condition for the
maximum permissible reheating temperature must be
formulated. This constraint is realized via a penalty
function method. The value of the penalty function
$f(T_{re})$ (with T_{re} as the maximal reheating temperature)
is added to the cost functional $-g$ to form a new cost
funcional $g^* = f - g$, which has to be minimized. The
resulting function f is defined as follows:

$$f(T) = \begin{cases} \dfrac{w}{T_{crit}-T_{re}} & T_{re} < T_K \\[3ex] 10 \ g + w \ \dfrac{(T_{re}-T_K)}{(T_{crit}-T_K)^2} & T_{re} \geq T_K \end{cases} \tag{9}$$

$$\text{with } T_K = T_{crit} - \frac{w}{10 \ g} \tag{10}$$

and w being a weighing factor (0-1).

For these basic considerations a reheating of 100 °C was taken as the critical value. Other authors state that 130°C reheating temperature is a critical temperature where the first cracks can be discovered /7,8/.

The so formulated optimizations problem is solved with a difference approximation of Rosen's projected gradient method as - only few function evaluations are required and - this modified algorithm does not need a differentiable funcional /9/.

The calculation steps of this procedure are schematically shown in fig. 4 for a minimization problem. We refer to /5/ for more information on the mathematical nature of this problem in particular its ill-posedness.

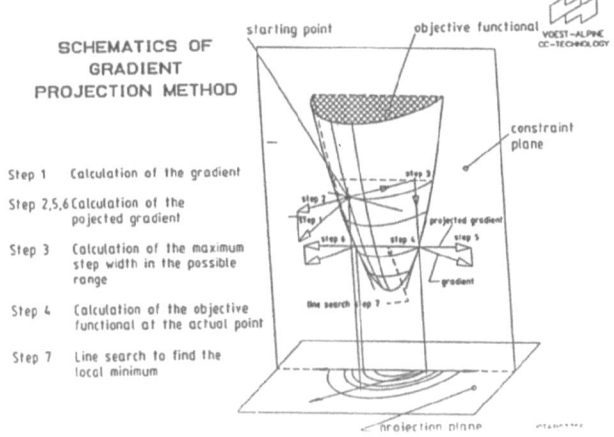

Fig. 4

With the mentioned algorithm we are able to calculate a suitable cooling depending on the steel grade. A maximum reheating temperature corresponds to a specific critical strain value.

The mechanical model

As round products are susceptible to centre cracking a model is to be formulated which allows the calculation of the tension strain in the solidifying region in the zone of maximal reheating. Here a FE-model is used for a sectional piece, as axial symmetry is asumed and a generalized plain strain condition, as the axial strain is considered to be constant over the section at a certain time. The formulation of the three sets of governing equations leads us to the following formulation of the mechanical model:

Geometrical equations

Axisymmetry: $\dfrac{\partial}{\partial \varphi} = 0$

Axial strain ϵ_{zz}^{*} (generalized plain strain)

Linearized strains: $\epsilon_{rr(r)} = \dfrac{\partial u_{(r)}}{\partial r}$; $\epsilon_{\varphi\varphi(r)} = \dfrac{u_{(r)}}{r}$; $\epsilon_{zz(r)} = \epsilon_{zz}^{*}$

Equilibrium (virtual displacement principle)

$$\int_{0}^{R} (\sigma_{rr}\, \delta\epsilon_{rr} + \sigma_{\varphi\varphi}\, \delta\epsilon_{\varphi\varphi} + \sigma_{zz}\, \delta\epsilon_{zz})\, r\, dr = 0$$

Material behaviour

$$\dot{\epsilon}_{ii} = \dot{\epsilon}_{ii}^{el} + \dot{\epsilon}_{ii}^{ne} + \dot{\epsilon}_{ii}^{th}$$

$$\dot{\epsilon}_{ii}^{th} = \mathcal{L}(T) \cdot \dot{T}$$

$$\dot{\epsilon}_{ii}^{el} = \frac{1}{E(T)} \left[\dot{\sigma}_{ii}(1+\mu) - \mu \left(\sum_{k} \dot{\sigma}_{KK} \right) \right]$$

$$\dot{\epsilon}_{ii}^{ne} = \frac{S_{ii}}{S_{v}} \, \dot{\epsilon}_{v}^{ne} \quad \text{with} \quad S_{v} = \sqrt{\frac{3}{2} \sum_{k} S_{KK} \cdot S_{KK}}$$

$$S_{ii} = \sigma_{ii} - \frac{1}{3} \sum_{K} \sigma_{KK}$$

$$\dot{\epsilon}_{v}^{ne} = \frac{A(T) \cdot \left(\frac{S_{v}}{T} \right)^{m}}{\epsilon_{v}^{ne} + d}$$

Steel at continuous casting temperature is described by a thermo-visco-elastic material law /10/.

The discretization is done in coordination with the FD-model for the thermal analysis. The notation corresponding to an integration point is shown in fig. 5.

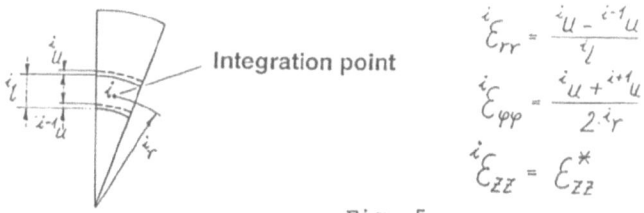

$${}^{i}\varepsilon_{rr} = \frac{{}^{i}u - {}^{i-1}u}{{}^{i}l}$$

$${}^{i}\varepsilon_{\varphi\varphi} = \frac{{}^{i}u + {}^{i+1}u}{2 \cdot {}^{i}r}$$

$${}^{i}\varepsilon_{zz} = \varepsilon_{zz}^{*}$$

Fig. 5

The relevant strain is the strain of a particle, which passes the temperature range between the zero strenght temperature (usually above the solidification temperature) and the zero ductility temperature (mostly below the solidification temperature), where the material behaviour is very brittle. As can be seen in fig. 6 one can define a corresponding critical phase in the passline of a particle. The strain accumulated in this phase (the so called interface strain) must be below a defined value to avoid centre cracking.

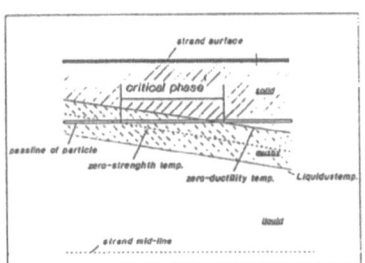

Critical phase for internal crack creation

Fig. 6

Applications

The introduced FD-model and FE-model were applied to the casting of round blooms with 219 mm diameter. The temperature profiles for the conventional cooling strategy are shown in fig. 7a for the surface and for several points in the bloom. The reheating temperature of the iteratively found cooling, which consideres cracking avoidance and segregation minimization (fig. 7b) is much higher and the length of the liquid pool L_{LP} is shortened.

Temperature profiles

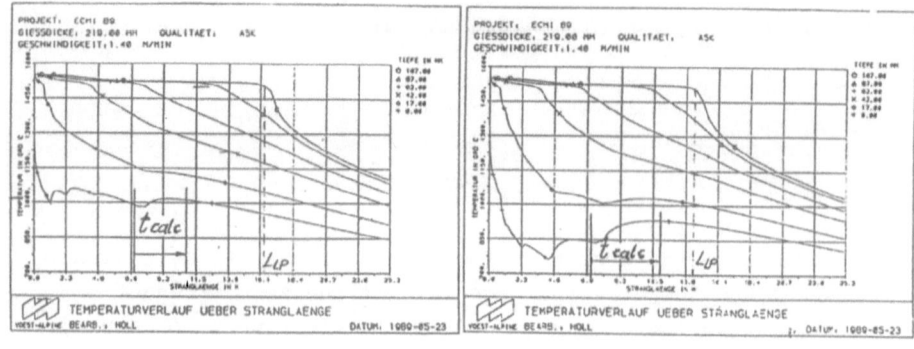

conventional   iteratively found

a   fig. 7   b

Stresses and strains based on these temperature values
were calculated with the FE-model. The strains and not the
stress values are relevant for cracking. As steel in this
zero-strength/zero-ductility temperature region is brittle
only perpendicular to the dendrites just the
circumferential and the axial strains are relevant. Fig. 8
shows that the circumferential strains are very little for
the conventional cooling case and they are considerably
higher with the new cooling strategy. The axial strain has
an analogous qualititative curve but the values are lower,
so only the circumferential strain is the important one.

Circumferential strains

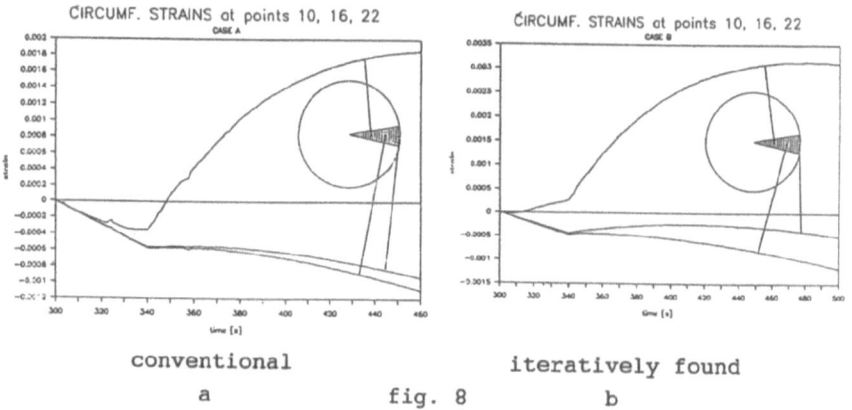

conventional   iteratively found

a   fig. 8   b

The maximum interface strain for both cases was
- conventional cooling :   0.08 %
- iteratively found cooling: 0.16 %

The significant higher strain value for the new cooling
is still below the critical value, which was determined
experimentally. Other autors /7,8,11/ found a permissible
reheating temperature of up to 150 °C. A computation with
such a limit value leads to a strain value of 0.25%, which
is within the scatter range of the critical values of the
investigated steel grade. It has to be emphasized, that
strain values are tremendously higher, if reheating
coincides with total solidification (1.4% for 150 °C
reheating).

So it can be expected that the modified cooling with
100 °C maximum reheating improves
- segregation profile and
- the solidification length
without creation of centre cracks.

Summary

The introduced procedure calculates an optimized
secondary cooling for weak segregation and avoidance of
cracks due to reheating. Methods applied were FD- and
FE-analysis for temperature and stress/strain calculations
respectively and Rosen's projected gradient method for
optimization.

Acknowledgement

The present work is based on a former analysis,
frequent discussions and support by T. Langthaler and
Prof. H.Engl.

References:

1. H.W. Engl, Th. Langthaler, 'Control of the
   solidification front by secondary cooling in continuous
   casting of steel' in H.W. Engl, H. Wacker, W. Zulehner
   (eds.), Case Studies in Industrial Mathematics
   (Teubner, Stuttgart 1988) p. 51-77

2. H.S. Carslaw, J.C. Jäger, <u>Conduction of Heat in Solids</u> (Oxford at the Clarendon Press 1959)

3. K.L. Schwaha, H. Holl, H.W. Engl, Th. Langthaler, 'Inverse solidification Analysis - A valuable tool for computer aided process engeneering' in H.Y. Sohn, E.S. Geskin (eds.), <u>Metallurgical processes for the Year 2000 and Beyond</u> (TMS, Las Vegas 1988),p.132-149

4. U. Reiners, R. Jeschar, R. Scholz, 'Wärmeübergang bei der Stranggußkühlung durch Spritzwasser', <u>Steel research</u> 60 (1989), p.442-450

5. H.W. Engl, Th. Langthaler, 'Numerical solution of an inverse problem concerning with continuous casting of steel', <u>Zeitschrift für Operation Research</u> 29, 1985, B185-B199

6. D.H.Kirkwood, 'Microsegregation' <u>Materials Science and Engineering</u> 65 (1984), p.101-109

7. A.Palmaers, A.Etienne, J.Mignon, 'Calculation of the Mechanical and Thermal Stresses in Continuous Cast Strands' <u>Stahl und Eisen</u> 99 (1979), p.1030-1050

8. A.Grill, J.K.Brimacombe, F.Weinberg, 'Mathematical Analysis of Stresses in Continuous Casting of Steel' <u>Ironmaking and Steelmaking</u> (1976) No.1, p.38-47

9. J.B. Rosen, 'The gradient projection method for nonlinear programming (Part I: linear constraints)' <u>SIAM Journal Appl. Math.</u> 8 (1960), p.181-217

10. O.M.Pühringer, 'Strand Mechanics for continuous slab casting plants' <u>Stahl und Eisen</u> 96 (1979), p.279-284

11. G.Van Drunen, J.K.Brimacombe, F.Weinberg, 'Internal cracks in strand-cast billets' <u>Ironmaking and Steelmaking</u> (1975) No.2, p.125-133

Automatic Secondary Cooling Control for the Continuous Casting Process of Steel

LAITINEN, E.,[1] LOUHENKILPI, S.,[2] MÄNNIKKÖ, T.[1] AND NEITTAANMÄKI, P.[1]

Abstract. Two simulation models for finding the optimal secondary cooling strategy in continuous casting process are presented. The first one is a two-dimensional off-line model for optimizing steady state casting operation. The second one is a one-dimensional model which can be used for simulating casting operation in real-time during varying casting conditions.

INTRODUCTION

The essential features of a continuous casting machine and the real-time control system are shown in Fig. 1. Molten steel is poured down from the tundish into the water cooled mold. The mold must be capable of extracting sufficient heat from the incoming metal in order to form a solid shell which can support the liquid pool after the mold exit (point z_1 in Fig. 1; the distances z_i are measured from the meniscus level z_0). After the mold the strand is supported by rollers and cooled down by water sprays. In this secondary cooling region more heat is extracted from the steel so that eventually the solidification is completed (point z_3). The secondary cooling region is further divided into several cooling zones. After the water sprays (point z_2) the strand is cooled down only by radiation. The strand is straightened at the unbending point z_4 and in the cutting point z_5 it is cut up. The control of the strand temperature field and the shell thickness along the casting machine are of central importance in continuous casting operation. Both have a considerable influence on defects which can be formed in cast material. When using direct hot charging/rolling the slabs with maximum heat content should be obtained at the machine exit. Accurate knowledge of the liquid pool end location is important especially when using soft reduction near the pool end.

In this paper two simulation models for optimizing and controlling the casting conditions are presented. These methods permit the use of temperature dependent material properties, complex boundary conditions and moving boundaries at the solidification front. Numerical calculations are performed by using the the finite element (FEM) method in space discretization and finite difference (FD) in time discretization.

---

[1] University of Jyväskylä, Department of Mathematics
[2] Helsinki University of Technology, Institution of Process Metallurgy

Hj. Wacker and W. Zulehner (eds.),
Proceedings of the Fourth European Conference on Mathematics in Industry, 109–121.
© 1991 B.G. Teubner Stuttgart and Kluwer Academic Publishers.

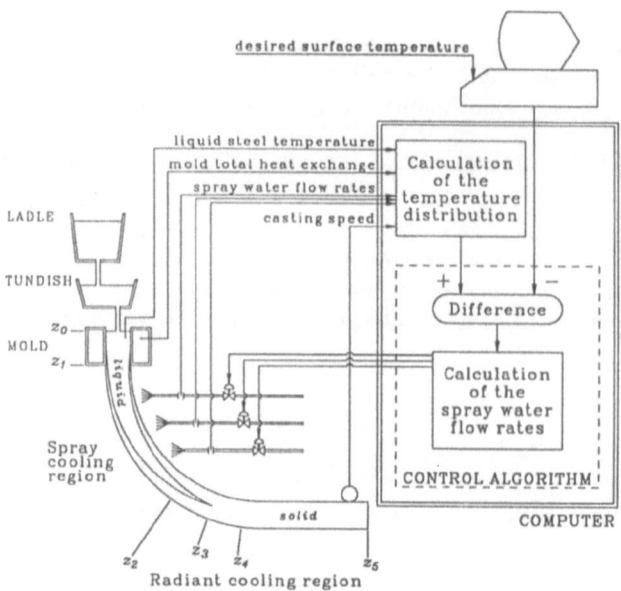

Fig. 1. Schematic representation of the continuous casting process.

The first method is based on the optimal control theory and it can be used for optimization of steady-state casting condition. By this method the heat transfer coefficients along and across the strand surface can be controlled in order to meet metallurgical cooling criteria determined beforehand. When several different cooling criteria are used we are dealing in fact with the optimization of multicriterion function with constraints in state. When optimizing the obtained multicriterion function one must be carefull when setting the state constraints which define the quality of the cast steel. Without exact information of the thermal conditions which lead to crack formation the determination of the state constraints is difficult. In spite of these difficulties the optimization model gives an improved basis for the design of the secondary cooling system and to define the maximum casting speed to produce defect-free products.

The second method is used to control the secondary cooling water sprays of the continuous casting machine during unsteady-state casting conditions. The aim of this control method is to keep the surface temperature of the strand constant with respect to time in spite of casting speed variations, and in this way minimize the formation of cracks in the final product. The temperature distribution of the strand is calculated by using a mathematical model based on a heat conduction formula with phase changes. The optimal spray water flow rates for each spray cooling zone are calculated by minimizing the deviation between the target surface temperature and the calculated actual surface temperature.

TEMPERATURE SIMULATION MODEL AT STEADY-STATE CASTING CONDITIONS

The heat transfer from the interior of the strand cross-section, $\Omega \subset \mathbb{R}^2$, to the external surface, Γ, is assumed to be described by means of the nonlinear heat conduction equation which can be written in the form:

$$(1) \qquad \rho(T)c(T)\frac{\partial}{\partial t}T = \mathrm{div}(k(T)\nabla T) + q,$$

where $T = T(x,t)$ is the temperature in a point $x \in \Omega$ at a time t; $q = q(T)$ describes the latent heat of the material. Furthermore, $c(T)$ denotes the specific heat, $\rho(T)$ the density and $k(T)$ the thermal conductivity.

By applying an expression $f_s = f_s(T)$ which describes the solid fraction, q can be expressed as:

$$(2) \qquad q(T) = \rho L \frac{\partial f_s}{\partial T}\frac{\partial T}{\partial t} \ .$$

By defining a smooth enthalpy function $H(T)$, which takes into account both the latent heat and the specific heat, and by applying the Kirchhoff transformation $K(T)$, equation (1) reduces to the form

$$(3) \qquad \frac{\partial}{\partial t}H(T(x,t)) = \Delta K(T(x,t)),$$

where

$$(4) \qquad H(T(x,t)) = \int_0^{T(x,t)} \left[\rho(\xi)c(\xi) - \rho(\xi)L\frac{\partial f_s}{\partial T}\right] d\xi$$

and

$$(5) \qquad K(T(x,t)) = \int_0^{T(x,t)} k(\xi)d\xi.$$

In order to solve equation (3) for the strand temperature distribution, the boundary and initial conditions must be specified. The following assumptions considering the boundary and initial conditions, as well as the other continuous casting characteristics were assumed in developing the model:

(i) Heat conduction in the transverse direction of the strand alone is considered. Heat conduction in the longitudinal direction can be neglected, due to the relative high withdrawal rate and low thermal conductivity of the steel [16].

(ii) In the liquid pool the heat flows by conduction and forced convection. It is difficult to exactly simulate the convective heat transfer, because it depends strongly on flow velocity which is not very well known. However, the convective heat transfer has only a minor effect on the thickness and temperature of the solid shell because the superheat of the steel is small. In this model convection was taken into account by using an effective thermal conductivity for the liquid steel.

(iii) The solid fraction f_s is assumed to vary piecewise linearly between solidus and liquidus temperatures. The latent heat emission is proportional to the change of the solid fraction.

(iv) Specific heat, density and thermal conductivity of steel were defined separately for several disjoint temperature intervals, inside which they were assumed to be constant.

(v) Heat transfer from the strand surface to the mould cooling water is mainly governed by the air gap formation between the solid shell and the mould face. The air gap has a very strong influence on the heat transfer coefficient between the strand surface and the mould cooling water. For the approximation of the heat transfer coefficient we refer to [3, 7]. Hence, the heat flux through the mould wall can be expressed as:

(6)
$$-k(T)\frac{\partial}{\partial n}T = h_{mould}(T - T_{mould}),$$

where T_{mould} is the ambient mould temperature and h_{mould} is the heat transfer coefficient in the mould.

(vi) In the secondary cooling region the relationship between the heat transfer coefficients h and the spray conditions like spray water flow rates, spray pressure, nozzle type and strand surface temperature must be determined experimentally. For such relations we refer to [13]. Hence, the heat flux across the boundary can be expressed as:

(7)
$$-k(T)\frac{\partial}{\partial n}T = h(x,t)(T - T_{H_2O}) + \sigma\varepsilon(T^4 - T_{ext}^4),$$

where $h(x,t)$ is the heat transfer coefficient in the secondary cooling region. T_{H_2O} is the spray water temperature and T_{ext} is the air temperature.

(vii) In the air cooling region it is assumed that heat is extracted by radiation only. Hence, the heat flux across the boundary can be expressed as:

(8)
$$-k(T)\frac{\partial}{\partial n}T = \sigma\varepsilon(T^4 - T_{ext}^4).$$

The emissivity value of 0.8–0.9 is normally used for the oxidized strand surface. In cases, where insulating covers are used to protect the strand against heat loss, the value of emissivity must be adjusted respectively.

(viii) Initial conditions (at meniscus level): the melt is assumed to have a temperature equal to the incoming metal temperature at time t=0:

(9) $$T(x,0) = T_0(x)$$

OPTIMIZATION OF THE STEADY STATE CASTING CONDITIONS

The aim is to define the optimum cooling conditions, i.e. the heat transfer coefficients on the strand surface and the casting speed to achieve a good productivity and to cast defect-free products. The basis for the model is to minimize a criterion function which is formulated by means of metallurgical cooling criteria. Thence our control problem reads

(10) $$\min_{h \in U_{ad}} \{J(T(h); h) = J_1(T(h); h) + J_2(T(h); h) + J_3(T(h); h) + J_4(T(h); h)\},$$

where the temperature $T(h)$ satisfies the equations (3)–(9).

Here U_{ad} denotes the set of admissible controls (heat transfer coefficients):

(11) $$U_{ad} = \{h(x,t) | h_{min}^i \leq h(x,t) \leq h_{max}^i; \quad i = 1, ..., 6\},$$

where h_{min}^i and h_{max}^i are the minimum and maximum possible heat transfer coefficients in the cooling zone i.

The metallurgical cooling criteria are formulated as a mathematical criterion function which expresses if some cooling criteria are met or not. The criterion function is defined as a sum of partial costs J_1, J_2, J_3, J_4 for which we apply the penalty technique with the nonmeeting of the constraints.

The cost of nonmeeting the first criterion is expressed as

(12) $$J_1(T(h); h) = \frac{c_1}{2} \int_{t_3}^{t_5} \int_{\Omega} \left[(T(x,t) - T_s)^+ \right]^2 d\Omega dt,$$

where t_3 is the maximum allowable length for the liquid pool and t_5 is the cut-off point for the strand.

The second criterion is expressed as

(13) $$J_2(T(h); h) = \frac{c_2}{2} \int_{t_1}^{t_5 - \Delta t} \int_{\Gamma} \left\{ \left[\left(\frac{T(x, t + \Delta t) - T(x,t)}{\Delta t} - CP \right)^+ \right]^2 \right.$$
$$\left. + \left[\left(CN - \frac{T(x, t + \Delta t) - T(x,t)}{\Delta t} \right)^+ \right]^2 \right\} d\Gamma dt,$$

where $CP > 0$ and $CN < 0$ are the allowable upper and lower bounds of the variation rate of the surface temperature and Δt is the time interval during which this parameter has to be controlled.

The third criterion can be expressed as

$$(14) \qquad J_3(T(h); h) = \frac{c_3}{2} \int_\Gamma \left[(T_2 - T(x, t_4)^+ \right]^2 d\Gamma,$$

The temperature T_2 is the upper bound of the temperature interval where the steel exhibits a low ductility and z_4 is the location along the caster, where a deformation is given to the product (the unbending point on a curved mold caster, for instance, in Fig. 1 the point t_4).

The fourth criterion is expressed as

$$(15) \quad J_4(T(h); h) = \frac{c_4}{2} \int_{t_1}^{t_5} \int_\Gamma \left\{ \left[(T_{min} - T(x, t))^+ \right]^2 + \left[(T(x, t) - T_{max})^+ \right]^2 \right\} d\Gamma dt,$$

where T_{min} and T_{max} are the minimum and maximum surface temperatures.

Above c_i, $i = 1, 4$ represents the relative importance of meeting one criterion with respect to the other, $a^+ = a$, if $a > 0$ and $a^+ = 0$, if $a \leq 0$ For the integration limits we refer to Figure 1, where $z_i = V * t_i$; V is the casting speed. For further details for dealing with the control problem we refer to [**6, 7**]. For earlier models see [**2, 9, 15**].

NUMERICAL EXAMPLE

The numerical values of material and casting data are shown in Table 1.

Carbon content	0.5 %		ρ	$7200 \frac{kg}{m^3}$
$T_l = T_0$	1485 °C		L	$272 \frac{kJ}{kg}$
T_s	1377 °C		σ	$5.67 \times 10^{-8} \frac{W}{m^2 K^4}$
T_{mold}	80 °C		ϵ	0.8
T_{H_2O}	27 °C		Cross-section: 100×100 mm^2	
T_{ext} spray z.	97 °C		Casting speed 2.70 $\frac{m}{min}$	
T_{ext} rad. z.	437 °C		Mold lenght 0.70 m	

Table 1 Material and casting data used.

The optimal adjustment of the spray system is doing with help of the control model (10). The desired values for the metallurgical cooling criteria which the temperature distribution must meet is presented in Table 2.

Criterion	Aimed value	Criterion
Length of the liquid pool	≤ 8.3 m	J_1
Maximum reheating rate on the surface	$+1\frac{^\circ C}{sec}$	J_2
Minimum cooling rate on the surface	$-1\frac{^\circ C}{sec}$	J_2
Surface temperature at the unbending point	$> 1100\ ^\circ C$	J_3
Minimum surface temperature along the secondary cooling zones	$900\ ^\circ C$	J_4

Table 2 Metallurgical cooling criteria.

Solving the optimization problem we obtain the heat transfer coefficients presented in Figure 2. The corresponding surface temperature in three points of cast cross-section is presented in Figure 3. It can be seen that the temperature field meets quite well all the criteria set. The value of the criterion function decreased 93 % from its initial value during optimization. The upper and lower bounds for the heat transfer coefficients was $h_{min} = 0.1$ and $h_{max} = 2.0$. For details see [7].

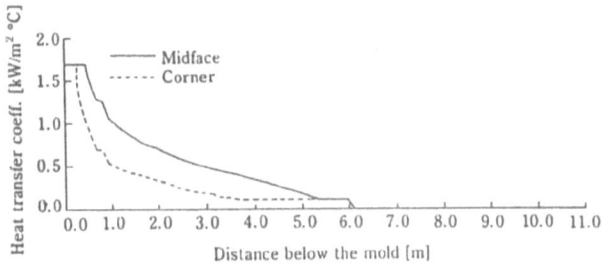

Figure 2 Heat transfer coefficients as a function of distance.

TEMPERATURE SIMULATION AT UNSTEADY-STATE
CASTING CONDITIONS IN REAL-TIME

Since in dynamic, real-time simulation a high calculation speed is required, only a one-dimensional formula of the heat conduction equation is applied. The strand

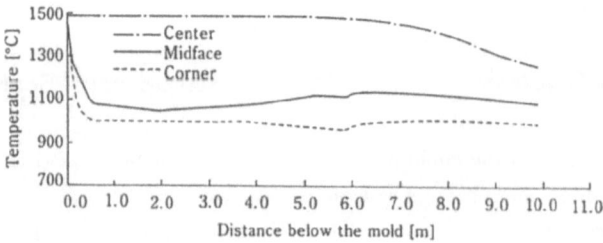

Figure 3 The calculated temperature distribution in corner, midface and centre of billet's cross-section as a function of distance.

temperature field and shell thickness profile are calculated in real-time using actual measured input data such as casting speed, mould heat flux (kW), liquid steel temperature, water flow rates, etc. The assumptions are the same as in the two-dimensional case with two exceptions:

1. Because the model is one-dimensional, heat transfer through the thickness of the wide side alone is considered. It means that the model can be applied only to the cross-sections which have a sufficiently large wide side/narrow side ratio.

2. The heat transfer across the mold boundary is defined as:

(16)
$$-k(T)\frac{\partial}{\partial n}T = \frac{Q}{A_m},$$

where $Q = Q(x, z; \tau)$ denotes the heat exchange in the mold and A_m is the area of the mold faces.

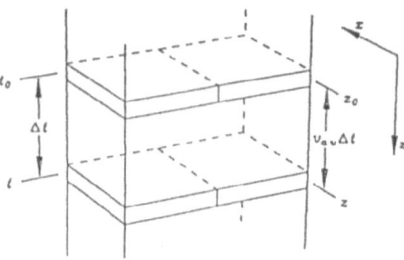

Figure 4 Movement of a strand cross-section during the time step

The calculation procedure can be described as follows:
The strand is divided into tracking planes (see Fig. 5) and the temperatures at

mesh points through the thickness of the plane are calculated. The positions of
the tracking planes are changed after every temperature calculation, depending on
the actual casting speed. The temperatures of the tracking planes are calculated
starting from the temperature values calculated at the previous time event and
using the actual cooling conditions. The temperature of the initial tracking plane
is set equal to the liquidus temperature.

The mathematical formulation of the model reads as follows:

At a present time $t \in \mathbf{R}$, the temperature distribution $T(z, x; t)$ at a point x along
the thickness of the strand cross-section and at a distance z from the meniscus
can be calculated starting from the slab temperature distribution $T(z_0, x; t_0)$ at
previous time event $t_0 = t - \Delta t$, where $z_0 = z - v_{av}\Delta t$:

$$
(17) \quad
\begin{cases}
v_{av}\dfrac{\partial}{\partial z}H(T(z, x; t)) = \dfrac{\partial^2}{\partial x^2}K(T(z, x; t)) \\[2mm]
-k(T)\dfrac{\partial}{\partial x}T(z, x; t) = \begin{cases} \dfrac{Q}{A_m} \\ h(T - T_{H_2O}) + \sigma\epsilon(T^4 - T_{ext}^4) \\ \sigma\epsilon(T^4 - T_{ext}^4) \end{cases} \\[2mm]
T(z_0, x; t_0) = T(z - v_{av}\Delta t, x; t - \Delta t)
\end{cases}
$$

The functions of enthalpy, $H(T)$, and Kirchhoff transformation, $K(T)$, are defined
by the equations (4) and (5).

REAL-TIME CONTROL ALGORITHMS

Control methods based on minimizing some multicriterion function would be too
time consuming for the real-time control using microcomputers. Therefore, we
have to be content with fewer criteria in order to be able to use simpler and faster
control algorithms.

In order to have defect-free final product it is necessary that the surface temperature of the strand behaves rather smoothly. Thus, one must define on the surface
a metallurgically good temperature distribution $T_d = T_d(x, z)$ which is a target
temperature in our control algorithms. In the control algorithms the new water
flow rates is calculated so, that the difference between target temperature and
actual temperature (calculated by the simulation model) will be minimized.

The first control algorithm we shall present is called PID-control. Let Θ^n denote
the average difference between the calculated and target surface temperature at
the time event t^n, and let W^n denote the water flow rate at time event t^n. Then
the new flow rate is obtained from

$$
(18) \quad W^n = W^{n-1} + K_P\left[\Theta^n - \Theta^{n-1} + \frac{\Delta t}{T_I}\Theta^n + \frac{T_D}{\Delta t}(\Theta^n - 2\Theta^{n-1} + \Theta^{n-2})\right],
$$

where K_P, T_I and T_D are tuning parameters.

The second control algorithm is based on the solution of the heat equation with a Dirichlet-type boundary condition, and thus, we shall call this method Dirichlet-control. At each time event t^n the algorithm consists of three steps: First we solve the system

(19)
$$\begin{cases} v_{av}\dfrac{\partial}{\partial z}H(T(z,x;t)) = \dfrac{\partial^2}{\partial x^2}K(T(z,x;t)), & \\[2mm] -\dfrac{\partial}{\partial n}K(T(z,x,;t)) = Q/A_m & \text{on } \Gamma\times]0,z_1], \\[2mm] T(z,x;t) = T_d & \text{on } \Gamma\times]z_1,z_2], \\[2mm] T(z_0,x;t^0) = T(z - v_{av}\Delta t, x; t - \Delta t) & \end{cases}$$

Secondly, we determine the heat transfer coefficient h from the equation

(20) $$-\frac{\partial}{\partial n}K(T) = h(T - T_{H_2O}) + \sigma\varepsilon(T^4 - T_{ext}^4) \quad \text{on } \Gamma\times]z_1,z_2].$$

And thirdly, we calculate the average heat transfer coefficient h_j for each cooling zone j and determine the water flow rates W_j from some empirical relation.

Off-line simulation results for a slab caster (with slab dimensions 1030×210 mm^2) are shown in Fig. 5. In Fig. 5 (a) is demonstrated how the surface temperature of the strand changes when there is a considerable drop in the casting speed and the secondary cooling is not controlled; at the distance 4 m from the meniscus the temperature decreases ca. 120 °C, at the distance 9 m ca. 70 °C and at the distance 16 m ca. 40 °C. Moreover, the change affects on the surface temperature still several minutes after the casting speed has returned to normal. In Figs. 5 (b)--(c) we see the effects of the same drop in casting speed, but the secondary cooling is controlled, in Fig. 5 (b) with the PID-control and in Fig. 5 (c) with the Dirichlet-control (the water flow rates are represented by the dashed lines). As we notice, the variation in the surface temperature is now only 10 °C at the distance 4 m, and even less further away.

CONCLUSIONS

Two simulation models for optimizing and controlling the casting conditions in the continuous casting process have been described. The developed two-dimensional model can be used for optimizing the casting conditions of the continuous casting process under steady-state operation. The one-dimensional dynamic model can be used in real-time operation and thus gives a good basis to develope a modern control system for secondary cooling.

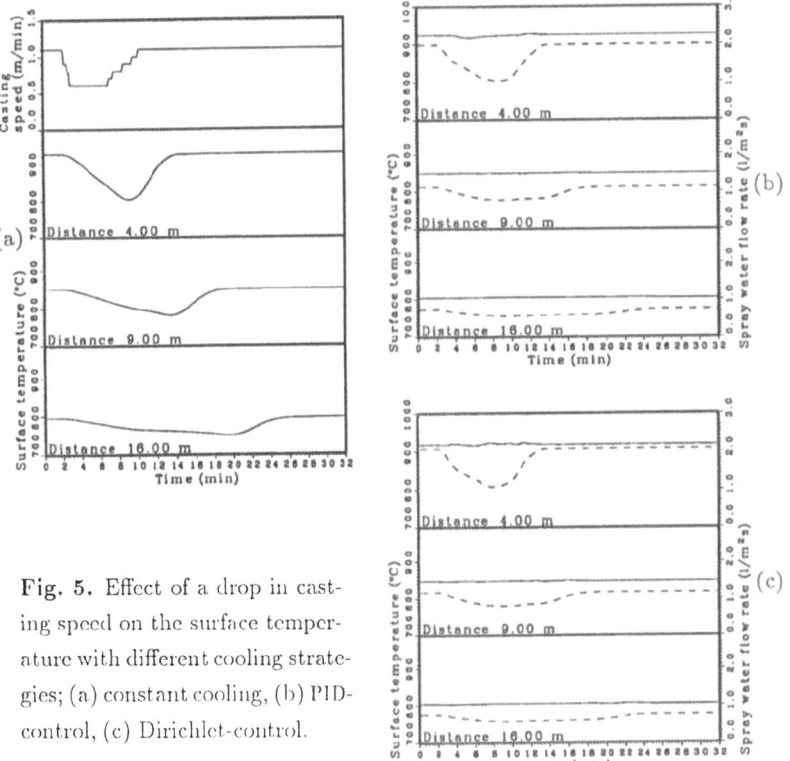

Fig. 5. Effect of a drop in casting speed on the surface temperature with different cooling strategies; (a) constant cooling, (b) PID-control, (c) Dirichlet-control.

The relevant casting conditions, steel properties, machine layout and water flow rates are as input parameters in these models. The strand temperature field and the shell thickness along the caster are calculated.

Before the models can be employed in casting machines, the correlation of the heat transfer coefficient along the strand surface, h, versus cooling parameters must be carefully determined. In the secondary cooling zones, h depends on spray conditions (water flux, spray pressure, nozzle type, spray distance) and on the state of the strand surface and the support rolls (temperature, scale). This correlation should be determined separately for each cooling zone. In the mould, h depends among other things on the air gap formation. For such relations see, for example [13].

The real-time control system is implemented as a microcomputer program and is under the industrial tests on steel works in Finland. The system provides means to the manufacture of defect-free slabs: The operator sees all the time the status of the cast, and in case of any disturbance can try to correct it at early stage. The system can also be used in testing new caster designs and new cooling strategies, instead of full-scale runs on production machines.

REFERENCES

[1]   Baptista, L.A., *Control of spray cooling in the continuous casting of steel*, Theses of University of British Columbia (1979).

[2]   Birat, J.P., Larrecq, M., Le Bon, A., Jeanneau, M., Poupon, M. and Senaneuch, D., *Control of secondary cooling in continuous casting of plate grades*, IRSID, ACI 83 RE 1009, Août 1983 (1983).

[3]   Brimacombe, J.K., Samarasekera, I.V. and Lait, J.E., *Continuous Casting*, Vol 2, Book Crafters, Ins., Chelsea, MI (1984).

[4]   Fasano, A. and Primicerio, M. (Eds.), "Free boundary problems, theory and applications," Vol I–II, Research Notes in Mathematics, 78 and 79, Pitman, 1983.

[5]   Kenmochi,N., *Asymptotic Stability of Solutions to a Two-Phase Stefan Problem with Nonlinear Boundary Condition of Signorini Type*, in Proc. of Conference on Optimal Control of Partial Differential Equations II: Theory and applications (Ed. K.-H. Hoffman, W. Krabs) (1987), Birkhäuser Verlag, Basel.

[6]   E. Laitinen, "On the Simulation and Control of the Continuous Casting Process," Doctoral Thesis, University of Jyväskylä, Department of Mathematics, 1989.

[7]   E. Laitinen, P. Neittaanmäki, *On Numerical Solution of the Problem Connected with the Control of the Secondary Cooling in the Continuous Casting Process*, Control-Theory and Advanced Technology 4 (1988), 285–305.

[8]   E. Laitinen, P. Neittaanmäki, *On Numerical Simulation of the Continuous Casting Process*, Journal of Engineering Mathematics 22 (1988), 335–354.

[9]   Larrecq, M., Birat, J.P., Saquez, C. and Henry, J., *Optimization of casting and cooling conditions on steel continuous casters; Implementation of optimal strategies on slap and bloom casters*, IRSID, ACI 83 RE 1004, Jullet 1983 (1983).

[10]   T. Männikkö, E. Laitinen, P. Neittaanmäki, *Real-time Simulator for the Continuous Casting Process*, in "Proc. of the 6th International Conference on Numerical Methods in Thermal Problems," Swansea, U.K., July 3–7, 1989.

[11]   T. Männikkö, *A Dual Approach to the State Constrained Optimal Control Problems with an Application to the Continuous Casting of Steel*, University of jyväskylä, Department of Mathematics, Reports on Applied Mathematics and Computing 6/1989..

[12]   P. Neittaanmäki, *On the Control of Cooling during Continuous Casting*, in "Numerical Methods in Thermal Problems," (eds. R. W. Lewis, K. Morgan), Proc. of the 4th International Conference, Swansea, U.K., Pineridge Press, July 15–18, 1985, pp. 240–250.

[13]   Nozaki, T. et al, *A secondary cooling pattern for preventing surface cracks of continuous casting slabs*, TISIJ 18 (1978) 6. (1978).

[14]   Okuno, K. et al, *Dynamic spray cooling control system*, Iron and Steel Engineer

(1987).

[15]   Saguez, C., *Contrôle optimal de systemès à frontiere libre*, Thèse, Univ. Technologique Compiegne (1980).

[16]   Wang, Z.G. and Inoue, T., *Analysis of temperature and elastoviskoplastic stresses during continuous casting*, in "Proc. of ICCM-86, Tokyo, VIII-103 – VIII-108," Springer-Verlag, 1986.

Erkki Laitinen
University of Jyväskylä, Department of Mathematics
Seminaarinkatu 15
SF-40500 Jyväskylä
Finland

SOME PROBLEMS OF IDENTIFICATION AND CONTROL ARISING

FROM COOLING ROLLED WIRES

Fredi Tröltzsch

1. INTRODUCTION

In steel mills, rolled materials such as wires, rods or other
profiles must be cooled down from the rolling temperature to a
lower one. This is realized by certain cooling equipment,
where the process of cooling is more or less controllable.
A suitable control can essentially improve the properties of
the produced material.

In this paper we shall discuss some aspects connected with the
mathematical treatment of primary cooling in special water-air
cooling lines having the principal shape indicated below:

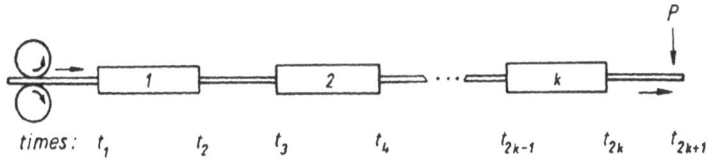

The line is composed of k water-cooling tubes followed by
air-cooling sections. The hot wire, which is supposed to have
circular cross section of radius R passes the line with
constant speed v. Tracking a fixed cross-section during its
passage through the line we obtain the times $t_1, \ldots, t_{2k+1}$
indicated in the figure. We assume symmetry of the temperature
field in the cross-section and denote it by $T = T(t, r)$,
$0 = t_1 \leq t \leq t_{2k+1}$, $0 \leq r \leq R$. As v is usually suffi-

123

Hj. Wacker and W. Zulehner (eds.),
Proceedings of the Fourth European Conference on Mathematics in Industry, 123–132.
© 1991 B.G. Teubner Stuttgart and Kluwer Academic Publishers.

ciently large we can ignore heat conduction in axial direction and obtain the equations

(1.1) $c\varrho \frac{\partial T}{\partial t}(t,r) = \frac{\partial}{\partial r}(\lambda \frac{\partial T}{\partial r}(t,r)) + \frac{1}{r}\lambda \frac{\partial T}{\partial r}(t,r)$

(1.2) $T(0,r) = T_1$

(1.3) $\lambda \frac{\partial T}{\partial r}(t,R) = u_i \alpha(t,T(t,R))[T_w - T(t,R)]$,

$t_{2i-1} < t \leq t_{2i}$ (water cooling),

(1.4) $\lambda \frac{\partial T}{\partial r}(t,R) = \alpha_i'[T_a - T(t,R)]$,

$t_{2i} < t \leq t_{2i+1}$ (air cooling), $i = 1,\ldots,k$, where $T(t,0)$ is to be finite. Here we have c: specific heat, ϱ: specific gravity, λ: heat conductivity, T_w, T_a: mean temperature of water and air, T_1: initial temperature at $t_1 = 0$. The $u_i \in [0,1]$ are controls reflecting the pressure of water entering the tube i (full pressure: $u_i = 1$), α_i' is an artificial heat exchange coefficient derived by linearization from a nonlinear law describing radiation and convection in the air cooling sections (cf. [4]).

The main problem for modelling the cooling process is to find a reliable expression for $\alpha = \alpha(t,T)$ in the water cooling sections. To get an exact formula for α , taking into account all essential parameters of the cooling process, seems to be an impossible task. Therefore we employed different numerical techniques in order to identify α from measurements.

Corresponding results are presented in section 3. Moreover, we investigated some questions of optimal control, about which we report in section 4.

2. EXISTENCE AND UNIQUENESS OF THE SOLUTION TO (1.1)-(1.4)

Before we are going to solve the inverse problem to identify α we must prove existence and uniqueness for the direct problem (1.1) - (1.4). The essence of this problem lies in the corresponding one on a fixed interval $[t_{2i-1}, t_{2i}]$ with given initial condition

$$(2.1) \qquad T(t_{2i-1}, r) = T_i(r) ,$$

where $\qquad T_i(r) = \lim_{t \to t_{2i-1}-0} T(t,r) .$

In general, c , ϱ , and λ depend on the temperature T. This concerns particularly c , which has a "peak" in the range 700 - 800° C due to the carbon change. It is a difficult task to show existence and uniqueness of (1.1) - (1.4) in its full nonlinearity. Therefore we assume for the existence theorem $c = c(r)$, $\varrho = \varrho(r)$, $\lambda = \lambda(r)$ to be smooth and fixed on $[t_{2i-1}, t_{2i}]$. Then there holds the following

THEOREM: Suppose that $T_w < \min_{[0,R]} T_i(r)$, $T_i(r)$ is smooth and $\alpha : [t_{2i-1}, t_{2i}] \times [T_w, \max T_i(r)] \to R_+$ is continuously differentiable with respect to T. Then the problem (1.1), (1.3), (2.1) admits a unique solution on $[t_{2i-1}, t_{2i}]$.

126   F. TRÖLTZSCH

Sketch of the proof: According to the assumptions on c, ϱ,
and λ the system (1.1), (1.3), (2.1) belongs to the class of
semilinear parabolic equations (with nonlinearity only in the
boundary condition). It follows from general considerations
by AMANN [1] that we have a unique (classical) solution T on
a sufficiently small subinterval $[t_{2i-1}, t_{2i-1} + \varepsilon]$, $\varepsilon > 0$.
By the maximum principle the bound $T_w \leq T(t,r) \leq \max_{[0,R]} T_i(r)$
is obtained. Hence the well-known principle "extension or
blow up" leads to global existence on $[t_{2i-1}, t_{2i}]$ (cf. [1]).

Now we suppose $T_w \leq T_a \leq T_1(r)$. Then the existence of a
unique solution T of (1.1) – (1.4), which is a classical
solution on each sub-interval $[t_i, t_{i+1}]$, i = 1,...,2k, and
continuous on $[0, t_{2k+1}] \times [0,R]$, is a conclusion of the theorem.
Here we can either assume c , ϱ , and λ to be independent
of t on $[0, t_{2k+1}]$ or we put $c(r) := c(T(t_i,r))$ on
$[t_i, t_{i+1})$ and proceed analogously for ϱ and λ.

3. IDENTIFICATION OF α

In what follows we assume $u_i = 1$, i = 1,...,k , i.e. we identify
α for full pressure of water.
Let $y_1,...,y_m$ be measured values for T(t,R) at the times
$0 < s_1 < ... < s_m$. We employed two different methods of identification.

3.1. Nonlinear test functions

According to the opinion of engineers we assumed at first one

formula $\alpha = \alpha(T)$ for all tubes, with α monotone decreasing
(the latter is realistic for $T > 200...300^{\circ}$ C). We choose n
functions $\alpha_1,...,\alpha_n$ from $C^1[T_w,T_1]$ fixed such that $\alpha_i > 0$,
$d\alpha_i(t)/dt < 0$ and put

$$\alpha(t) = \sum_{i=1}^{n} x_i \, \alpha_i(T)$$

with unknown coefficients $x_1,...,x_n$. Moreover, we take a
convex closed set $C \subset R^n$ such that α is positive and monotone
decreasing for all $x \in C$.

In this way, we obtain a heavily nonlinear mapping $A : R^n \to R^m$
by $x \mapsto y = (T(s_1,R),...,T(s_m,R)) = Ax$, where T is the
solution of the <u>direct problem</u> (1.1) - (1.4) with $\alpha = \sum_{i=1}^{n} x_i \, \alpha_i$.
In order to identify α we have to solve the <u>inverse problem</u>

(3.1) $Ax = y$

for a given vector y of measurements. We suppose that the
error $\varepsilon \in R^m$ of measurements is bounded by $\|\varepsilon\| \leq \delta$ with
certain $\delta > 0$ and introduce

$$Y_{\delta} = \{x \in C \mid \|Ax-y\|^2 \leq \delta\} .$$

The <u>regularized solution</u> x_{reg} of (3.1) is defined by

(3.2) $\|x_{reg}\|^2 = \min_{x \in Y_{\delta}} \|x\|^2 ,$

according to the well known Tikhonov regularization method
(cf. HOFMANN [3]). To solve (3.2) numerically, we employed a

Lagrange multiplier technique and solved

$$(3.3) \qquad x_\gamma = \arg \min_{x \in C}\{\|Ax-y\|^2 + \gamma\|x\|^2\} \,,$$

where $\gamma > 0$ was determined by the so-called <u>discrepancy principle</u>, i.e. by $\|Ax-y\|^2 = \delta$. Generally, (3.2) and (3.3) are not equivalent, as the Lagrange function corresponding to (3.2) need not have a saddle point. However, such difficulties did not occur in our numerical tests. (3.3) was solved by means of a variant of the Gauss-Newton method (not requiring derivatives). The direct problem (1.1) – (1.4) was treated numerically with a finite difference method (thus instead of the mapping $x \mapsto Ax$ we used a mapping $x \mapsto A_h x$ with a certain discretization parameter h). In our tests functions of the type $\alpha = x_1 + x_2/T$ turned out to be best suited to describe heat exchange coefficients for the cooling line by one formula. This is an interesting fact as this result complies with former experience of engineers.

We investigated also the effect of perturbations by overlaying "exact" values y (generated by solving the direct problem for a given couple (x_1,x_2)) with a Gaussian distributed error ε. Test runs showed that (x_1,x_2) can be reconstructed well up to 5 % measuring error. For further details we refer to KAISER and TRÖLTZSCH [4].

3.2. Assumption of constant heat flux in each cooling tube

The computational effort of the method described above is considerably. Moreover it is our opinion that <u>one</u> expression

$\alpha = \alpha(T)$ for the whole cooling line is too stiff to be adapted to the measurements. The need of a flexible and fast procedure to identify α led to another very natural and simple approach. It is based on the assumption that

(3.4) $\dfrac{\partial T}{\partial r}(t,R) = q_i$

holds approximately with a constant heat flux q_i on $[t_{2i-1}, t_{2i}]$, $i = 1,\ldots,k$. If q_i is determined we have

(3.5) $\alpha(t,T) = \alpha_i(T) = \lambda q_i/(T_w - T)$

for $t \in [t_{2i-1}, t_{2i}]$.

Suppose that α is already identified on $[0, t_{2i-1}]$.

Then we determine the solutions T_1, T_2 of the heat equation (1.1) on $[t_{2i-1}, t_{2i}]$ satisfying

$$T_1(t_{2i-1}, r) = T(t_{2i-1}, r) \;, \qquad \frac{\partial T_1}{\partial r}(t,R) = 0$$

$$T_2(t_{2i-1}, r) = 0 \qquad\qquad , \qquad \frac{\partial T_2}{\partial r}(t,R) = 1$$

(note that T is known on $[0, t_{2i-1}]$). We have $T(t,r) = T_1(t,r) + q_i T_2(t,r)$ on $[t_{2i-1}, t_{2i}]$, and from a measurement y of $T(t,R)$ at $t = t_{2i}$ we obtain

$$q_i = (y - T_1(t_{2i}, R))/T_2(t_{2i}, R) \; .$$

This corresponds to the technique named "exact matching" in BECK, BLACKWELL and CLAIR [2].

A similar method is possible, if only a few measurements at
some arbitrary points located in air-cooling sections are given
(assuming constant heat flux in all tubes between two neighbouring points). This was implemented on a personal computer
and successfully applied by engineers.

EXAMPLE: We regard a cooling line composed of 3 water-tubes
1.56 m in length, separated by air-cooling sections of 0.35 m.
The mean temperature of water was 25° C. For a rod having
initial temperature of 1070° C, diameter of 6.5 mm and a speed
of 50 m/s the following values were obtained (y : measured
temperature after tube i , in °C, α_1, α_2 : α at beginning
and end of tube i , resp., in W/m²K , q : heat flux in K/m
(tab.1)).

tube	y	q	α_1	α_2
1	460	−1.174E06	3.345E04	7.729E04
2	330	−9.003E05	4.542E04	9.169E04
3	250	−7.133E05	4.786E04	9.992E04

Tab. 1   Computed heat flux and heat exchange coefficients

4. SOME REMARKS ON THE CONTROL OF COOLING

The control of cooling of rolled profiles is world-wide an
important problem of research. In our case, after having identified α , the following simple problem is worth to be mentioned:
Which values u_i (having a certain relation to the pressure of
cooling water) must be chosen such that the final temperature

at P is minimal and at all times the differences of temperatures of the rod do not exceed a certain bound Δ ? Thus the
problem is:

Minimize $T(t_{2k},0)$ subject to

$$0 \leq u_i \leq 1 , \quad i = 1,\ldots,k ,$$

$$T(t,0) - T(t,R) \leq \Delta ,$$

$t \in [0,t_{2k+1}]$, where $T(t,r)$ is the solution of $(1.1) - (1.4)$.
For the same data as in the example above we arrived at the
values of table 2:

tube	Δ	u_i	$T(t_{2i},0)$	Δ	u_i	$T(t_{2i},0)$	Δ	u_i	$T(t_{2i},0)$
1	300	0.50	1069	500	0.82	1068	600	0.99	1068
2		0.46	1060		0.70	1038		0.82	1024
3		0.50	1046		0.76	988		0.89	953

Tab. 2   Optimal control with bounded thermal differences

The results indicate that strong bounds Δ would essentially
decrease the feasible speed of cooling.
Problems of this type can be embedded in a more complex class
of optimal control problems for nonlinear parabolic differential
equations. To corresponding theoretical investigations we refer
the reader to TRÖLTZSCH [6], [7].

REFERENCES:

[1]   AMANN, H.: Parabolic evolution equations with nonlinear
      boundary conditions. J. Diff. Equations 72 (1988), 201-269.

[2]   BECK, J.V.; BLACKWELL, B.; CLAIR, C.R.St.: Inverse heat
      conduction. Ill-posed problems. Wiley, New York 1985.

[3]   HOFMANN, B.: Regularization for applied inverse and
      illposed problems. Teubner—Texte zur Mathematik, Vol. 85,
      Teubner, Leipzig 1986.

[4]   HENSEL, A.; TRÖLTZSCH, F.; WOLFERSDORF, L.v.:
      Berechnung der' Abkühlung von Feinstahl und Draht in
      Kühlstrecken. Neue Hütte 25 (1980), 299-301.

[5]   KAISER, T.; TRÖLTZSCH, F.: An inverse problem arising
      in the steel cooling process. Wiss. Z. TU Karl-Marx-Stadt
      29 (1987), 212-218.

[6]   TRÖLTZSCH, F.: Optimality conditions for parabolic
      control problems and applications. Teubner—Texte zur
      Mathematik, Vol. 62, Leipzig 1984.

[7]   TRÖLTZSCH, F.: On convergence of semidiscrete Ritz-
      Galerkin schemes applied to the boundary control of
      parabolic equations with nonlinear boundary conditions.
      To appear in ZAMM.

Fredi Tröltzsch
Technische Universität Karl-Marx-Stadt
Sektion Mathematik
DDR-9010 Karl-Marx-Stadt, PSF 964

Minisymposium: Chemical Engineering

Numerical Simulation of Emulsion Liquid Membrane Permeation

D. Auzinger, A. Ortner, Hj. Wacker
Institut für Mathematik
Johannes-Kepler-Universität Linz
Altenbergerstraße 69, A-4040 Linz, Austria

and

H.J. Bart, R. Marr
Institut für Thermische Verfahrenstechnik und Umwelttechnik
Technische Universität Graz
Inffeldgasse 25, A-8010 Graz, Austria

1. Introduction

Due to an increasing trend towards environment protection recycling technologies get more and more important. Of course these technologies should as well be cheap as efficient. A special problem is the recycling of dissolved heavy metal from dilute aqueous solutions.

Emulsion liquid membrane permeation (ELM) is a very efficient technology for concentrating ionic and nonionic molecules. A big advantage of ELM is its energetic efficieny which is due to the use of chemical potentials for concentrating the interesting species instead of electric or thermal energy. Since it appears that the ELM process is on its way toward becoming an increasingly important unit operation, it is an important engineering task to develop reliable mass transfer models to minimize the experimental work required in developing commercial-scale ELM processes. The models must describe all important process features, but they should not involve too many experimental parameters. Such a model was developed by Bart/Lorbach/Marr et al. [3],[2].

For most of the chemical engineers developing models the resulting mathematical problems should be easy to solve (e.g. using standard software packages), which is an unnecessary restriction, as the co-work of mathematicians and chemical engineers presented in this paper shows.

2. Description of the Process

A liquid membrane is a thin film that selectively permits the passage of a particular component of a mixture. Unlike solid membranes which separate according to size, liquid membranes separate by solubility of the solute or chemical affinity between the solute and a carrier reagent in the membrane.

Emulsion liquid membranes are typically formed by first emulsifying two immiscible phases and then dispersing this emulsion into another continuous phase (referred to as phase III, see figure 1). The emulsion globules achieved behave in a similar way to the dispersed phase in ordinary extraction contactors. The membrane phase (phase II)

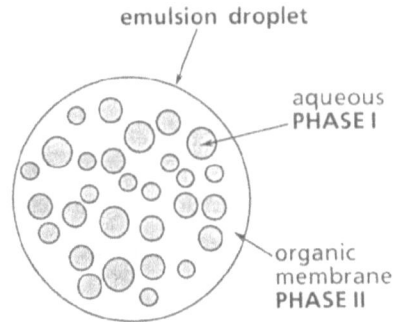

Fig. 1. Emulsion globule

135

Hj. Wacker and W. Zulehner (eds.),
Proceedings of the Fourth European Conference on Mathematics in Industry, 135–145.
© 1991 *B.G. Teubner Stuttgart and Kluwer Academic Publishers.*

may be organic or aqueous, the diameters of the internal emulsion droplets (phase I) typically range between 0.1 and 5.0 μm, those of the emulsion globules between 0.5 mm and 1.5 mm.

Mass transport through the liquid membrane is accomplished by one of two transport mechanisms. The first involves a simple diffusion, the second, which we will consider here, is known as carrier-facilitated transport. This technique is used for membrane-insoluble materials such as metal ions. Figure 2 depicts the transport mechanism for the example of the recovery of copper Cu^{2+} from an aqueous solution. A carrier reactant RH is present in the organic membrane phase, which reacts with a copper ion Cu^{2+} in a ion exchange reaction to form a soluble organic complex (CuR_2) at the external globule interface. This complex then diffuses through the organic membrane phase to the internal interface, where the reverse reaction occurs, caused by a shift of the reaction equilibrium owing to a higher concentration of the counter-ion H^+ in the interior phase. Thus the solute is released into the interior phase droplets. The carrier-counter-ion molecule RH diffuses back to the external interface, thereby completing the process. The driving force of the process is the difference between the activities of the counter-ion in the exterior and interior phases.

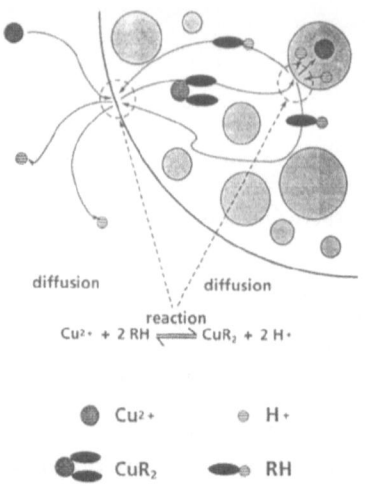

Fig. 2. Carrier-facilitated transport of Cu^{2+} ions through an organic membrane phase.

3. Modelling of Carrier-Facilitated Transport

Owing to the complex physical and chemical structure of carrier-facilitated transport processes the efforts of modelling have been limited mostly to double spherical shell models (for references see [3]). The results predicted by these models do not agree well with experimental results, which is why there was a need for better models. A model taking into account the complexity of diffusion and reaction within emulsion globules was developed by Teramoto and co-workers [7]. Although this model is very descriptive it is tedious to evaluate all the parameters involved experimentally. For this reasons a model with fewer parameters was developed by Lorbach et.al. [2].

We will discuss this model for the separation of copper ions from an aqueous solution, but neither the applicability of the model nor that of the solving algorithm are restricted to this case. The model takes into account the following mass transfer resistances:
- mass transfer resistance for the copper ions in the continuous phase III,
- resistance of the chemical reaction at the phase III - phase II interface,
- diffusional resistance of carrier and carrier-copper-complex within the membrane phase,
- reactional resistance of the stripping reaction at the phase II - phase I interface.

Mass transfer resistances not taken into account are the diffusion of the protons (counter-ion) within phases I and III and the diffusion of the copper-ions within the stripping phase I. This is justified by the much higher diffusivity of the protons and by the small diameter of the phase I droplets. The emulsion is assumed to be a homogeneous mixture of phases I and II.

The droplet concentrations of interest are those of loaded and unloaded carrier within the membrane phase, they are denoted by $[CuR_{2\,II}]$ and $[RH_{II}]$, and the concentration of protons and copper ions within the stripping phase I, denoted by $[H_I^+]$ and by $[Cu_I^{2+}]$. By $[Cu_{III\,int}^{2+}]$ we denote the phase III concentration of the copper ions at the interface III–II. Each of these concentrations should have an additional index "R", as they depend on the droplet diameter, but we omit this index and mention this dependence explicitly whenever necessary. Additional concentrations of interest are the phase III concentrations of copper ions and protons, they are denoted by $[Cu_{III}^{2+}]$ and $[H_{III}^+]$.

Assuming spherosymmetry our model of an emulsion globule of radius R is one dimensional in space. The equations for the membrane phase concentrations $[CuR_{2\,II}]$ and $[RH_{II}]$ are obtained by balancing the change of concentrations in time, radial diffusion and the amount of carrier-copper-complexes consumed respectivly the amount of carrier molecules produced by the stripping reaction (reaction rate $c_{II}n_S$ with c_{II} a constant depending on the proportion of phases I and II and on the phase I droplet radii).

$$\frac{\partial}{\partial t}[CuR_{2\,II}] = \frac{D_C}{r^2}\frac{\partial}{\partial r}\left(r^2\frac{\partial}{\partial r}[CuR_{2\,II}]\right) - c_{II}n_S([CuR_{2\,II}],[RH_{II}],[H_I^+],[Cu_I^{2+}]) \quad (3.1)$$

$$\frac{\partial}{\partial t}[RH_{II}] = \frac{D_R}{r^2}\frac{\partial}{\partial r}\left(r^2\frac{\partial}{\partial r}[RH_{II}]\right) + 2c_{II}n_S([CuR_{2\,II}],[RH_{II}],[H_I^+],[Cu_I^{2+}]) \quad (3.2)$$

These equations must hold for $0 < r < R$ and for each $t > 0$.

At the center of the globule the symmetry conditions

$$\frac{\partial}{\partial r}[CuR_{2\,II}]|_{r=0} = \frac{\partial}{\partial r}[RH_{II}]|_{r=0} = 0 \quad (3.3)$$

must hold, while the radial mass fluxes at the surface must be equal to the amounts of molecules produced or consumed by the extraction reaction, which we describe by the reaction rate n_E. The according conditions, with a_{II} the proportion of phase II volume and emulsion volume, are

$$a_{II}D_C\frac{\partial}{\partial r}[CuR_{2\,II}]|_{r=R} = n_E([CuR_{2\,II}],[RH_{II}],[H_{III}^+],[Cu_{III\,int}^{2+}])(R),$$
$$a_{II}D_R\frac{\partial}{\partial r}[RH_{II}]|_{r=R} = -2n_E([CuR_{2\,II}],[RH_{II}],[H_{III}^+],[Cu_{III\,int}^{2+}])(R). \quad (3.4)$$

In addition, also the mass transport from the bulk of the continuous phase to the emulsion globule surface must be equal to the reaction rate n_E. Using a film model for the copper transport within phase III, the equation

$$k_{III}\left([Cu_{III}^{2+}] - [Cu_{III\,int}^{2+}]\right) = n_E([CuR_{2\,II}],[RH_{II}],[H_{III}^+],[Cu_{III\,int}^{2+}]) \quad (3.5)$$

must hold for each $t > 0$. As already mentioned the transport resistance of the protons due to diffusion can be neglected, thus the proton interface concentration is identified with the proton bulk concentration $[H_{III}^+]$ of phase III.

Due to the stability of the emulsion there is no diffusion of the phase I droplets within the membrane phase and thus (the phase I droplets are very small) no radial diffusion of the copper ions and the protons within the emulsion globule need be assumed. The change in time of the copper ion concentration within the stripping phase (at a fixed position r) must be equal to the copper transport into the stripping phase, the amount of this transport is given by the reaction rate $c_I n_S$. The according equation is

$$\frac{\partial}{\partial t}[Cu_I^{2+}] = c_I n_S([CuR_{2\,II}], [RH_{II}], [H_I^+], [Cu_I^{2+}]). \tag{3.6}$$

Although extraction and stripping reaction are chemicaly identic (a reversible reaction), the mathematical models for the reaction rates n_E and n_S are different as the two reactions take place in two completly different chemical surroundings. The extraction reaction rate is given by

$$n_E([CuR_{2\,II}], [RH_{II}], [H_{III}^+], [Cu_{III\,int}^{2+}]) =$$
$$k_E \left(\frac{[Cu_{III\,int}^{2+}][RH_{II}]}{[H_{III}^+]} - \frac{1}{K_E^{ex}} \frac{[CuR_{2\,II}][H_{III}^+]}{[RH_{II}]} \right) \tag{3.7}$$

with k_E the chemical rate constant and K_E^{ex} the equilibrium constant. For the stripping reaction rate we have

$$n_S([CuR_{2\,II}], [RH_{II}], [H_I^+], [Cu_I^{2+}]) = k_S \left([CuR_{2\,II}][H_I^+] - K_S^{ex} \frac{[Cu_I^{2+}][RH_{II}]^2}{[H_I^+]} \right). \tag{3.8}$$

The phase I proton concentration is obtained by a simple balancing: For each copper ion released into the stripping phase two protons from the stripping phase are required to form the carrier-proton-molecule. Thus, at each time t the equation

$$[H_I^+] + 2[Cu_I^{2+}] = [H_I^+]_0 + 2[Cu_I^{2+}]_0 \tag{3.9}$$

must hold, with $[H_I^+]_0$ and $[Cu_I^{2+}]_0$ being the initial concentrations.

For given initial concentrations $[H_I^+]_0$, $[Cu_I^{2+}]_0$, $[RH_{II}]_0$ and $[CuR_{2\,II}]_0$ and given copper and proton bulk concentrations $[Cu_{III}^{2+}](t)$ and $[H_{III}^+](t)$ the continuous phase equations (3.1)–(3.9) discribe the behaviour of an emulsion droplet of radius R. This model was used to identify some of the problem parameters in comparison to experimental results in a single drop apparatus. It was found that the presented mass transfer model accurately describes the experimental results [2].

4. Modelling of an ELM Column

Figure 3 presents a flow diagram of the ELM process for the separation of copper ions from a dilute aqueous solution. In a first step, the water in oil emulsion is produced. Step two is the permeation step, which takes place within a stirred counterstream column. The waste water (phase III) enters the column at the top ($z = H$) and leaves it at the bottom ($z = 0$) purified. The emulsion enters at the bottom, the droplets ascend, and during their way up the reactive permeation of the copper ions from the waste water into the stripping phase I takes place. At the top of the column the loaded emulsion is separated, in a third step it is split into the membrane phase, which is recycled to step I, and the stripping phase, which is enriched with the copper. The copper concentration of the stripping phase is up to 100 times the initial copper concentration of the waste water ($[Cu_{III}^{2+}]_0$), the waste water copper concentration is reduced to about 1 percent of $[Cu_{III}^{2+}]_0$. The plant is in a steady state, as it is operated continuously.

This paper deals with the processes within the permeation column. The mechanism of flow within a stirred column is very complex, a detailed computation of the flow effects seems rather difficult and not necessary. As permeation columns and extraction columns only differ in the mass transport effects within the ascending droplets (extraction usually works with an organic phase instead of the emulsion, mass transport into the droplets involves diffusion and/or reaction), the models for the hydrodynamics of extraction columns can also be used for permeation columns. The continuous phase is modelled by an ideal plug flow and axial diffusion. As Miyauchi and Vermeulen showed in [4], the nonidealities of the emulsion phase flow due to the nonconstant ascending velocity of the emulsion droplets, the so called dispersion effect, can be modelled by a plug flow with mean velocity v_e and a diffusion like term at least for the case

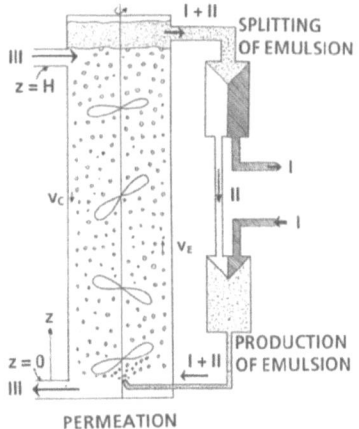

Fig. 3. ELM plant

of rapidly coalescing and breaking droplets. For our case of a very stable emulsion the dispersion model is believed to provide a workable approach for the average concentration at a column level.

Typically, the (one dimensional) dispersion equation for a concentration φ is

$$v_z \frac{\partial}{\partial z}\varphi(r,z) = D_z \frac{\partial^2}{\partial z^2}\varphi(r,z) + \{diffusion\ in\ r\text{-}direction\} + \{source\ terms\},$$

the boundary conditions of a fluid entering at the bottom, which means $v_z \geq 0$, are

$$D_z \frac{\partial}{\partial z}\varphi(r,z)|_{z=0} = v_z\big(\varphi(r,0)-\varphi_0\big) \quad \text{and} \quad D_z \frac{\partial}{\partial z}\varphi(r,z)|_{z=H} = 0.$$

Substituting t in equations (3.1)–(3.9) by z/v_e, combining them with the dispersion equations and balancing the (mean) phase III copper and proton concentration at height z we obtain the following model of a stirred counterstream ELM column for the recovery of copper from waste water:

Emulsion droplet of radius R:
- Mass balance equations $\quad (r \in (0,R), \quad z \in (0,H))$

$$v_e \frac{\partial}{\partial z}[CuR_{2\,II}] = D_e \frac{\partial^2}{\partial z^2}[CuR_{2\,II}] + \frac{D_C}{r^2}\frac{\partial}{\partial r}\big(r^2 \frac{\partial}{\partial r}[CuR_{2\,II}]\big) - c_{II}n_S \tag{4.1}$$

$$v_e \frac{\partial}{\partial z}[RH_{II}] = D_e \frac{\partial^2}{\partial z^2}[RH_{II}] + \frac{D_R}{r^2}\frac{\partial}{\partial r}\big(r^2 \frac{\partial}{\partial r}[RH_{II}]\big) + 2c_{II}n_S \tag{4.2}$$

$$v_e \frac{\partial}{\partial z}[Cu_I^{2+}] = D_e \frac{\partial^2}{\partial z^2}[Cu_I^{2+}] + c_I n_S \tag{4.3}$$

$$[H_I^+] = [H_I^+]_0 + 2([Cu_I^{2+}]_0 - [Cu_I^{2+}]) \tag{4.4}$$

- Boundary conditions in radial direction $\quad (z \in (0,H))$

$$\frac{\partial}{\partial r}[CuR_{2\,II}]|_{r=0} = 0 \qquad\qquad \frac{\partial}{\partial r}[RH_{II}]|_{r=0} = 0 \tag{4.5}$$

$$a_{II}D_C \frac{\partial}{\partial r}[CuR_{2\,II}]|_{r=R} = n_E(R,z) \qquad a_{II}D_R \frac{\partial}{\partial r}[RH_{II}]|_{r=R} = -2n_E(R,z) \tag{4.6}$$

- Boundary layer around the emulsion globule $(z \in (0, H))$

$$k_{III}\left([Cu_{III}^{2+}] - [Cu_{III\,int}^{2+}]\right) = n_E(R, z) \tag{4.7}$$

- Boundary conditions in axial direction $(r \in (0, R))$

$$D_e \frac{\partial}{\partial z}[CuR_{2\,II}]|_{z=0} = v_e\left([CuR_{2\,II}](r, 0) - [CuR_{2\,II}]_0\right)$$

$$D_e \frac{\partial}{\partial z}[RH_{II}]|_{z=0} = v_e\left([RH_{II}](r, 0) - [RH_{II}]_0\right) \tag{4.8}$$

$$D_e \frac{\partial}{\partial z}[Cu_I^{2+}]|_{z=0} = v_e\left([Cu_I^{2+}](r, 0) - [Cu_I^{2+}]_0\right)$$

$$D_e \frac{\partial}{\partial z}[CuR_{2\,II}]|_{z=H} = D_e \frac{\partial}{\partial z}[RH_{II}]|_{z=H} = D_e \frac{\partial}{\partial z}[Cu_I^{2+}]|_{z=H} = 0 \tag{4.9}$$

Continuous phase, waste water (note: $v_w < 0$)
- Mass balance equation $(z \in (0, H))$

$$v_w \frac{\partial}{\partial z}[Cu_{III}^{2+}] = D_w \frac{\partial^2}{\partial z^2}[Cu_{III}^{2+}] - \int_{R_{min}}^{R_{max}} S(R)k_{III}\left([Cu_{III}^{2+}] - [Cu_{III\,int}^{2+}](R)\right)dR \tag{4.10}$$

$$[H_{III}^+] = [H_{III}^+]_0 + 2([Cu_{III}^{2+}]_0 - [Cu_{III}^{2+}]) \tag{4.11}$$

- Boundary conditions in axial direction

$$D_w \frac{\partial}{\partial z}[Cu_{III}^{2+}]|_{z=0} = 0 \tag{4.12}$$

$$D_w \frac{\partial}{\partial z}[Cu_{III}^{2+}]|_{z=H} = v_w\left([Cu_{III}^{2+}](H) - [Cu_{III}^{2+}]_0\right) \tag{4.13}$$

The second right hand side term of equation (4.10) summes up the copper transport into the droplets with different radii R, with $S(R)$ depending on the total surface of droplets with radius R and the proportion of emulsion and continuous phase volumes.

5. Analysis of the Model

The set of equations (4.1)–(4.13) describing the mass transport within a countercurrent ELM column is a system of nonlinear partial differential equations in two space dimensions. They are of elliptic type, but neglecting the dispersion terms (ideal plug flow) the type is parabolic. In that case the third kind axial boundary conditions degenerate to Dirichlet conditions, the Neumann conditions vanish. But the system still is a boundary value problem, as we have the initial conditions for the emulsion at the bottom and for the continuous phase at the top of the column. The parabolic behavior of the counterstreaming phases also dominates the nonideal case, as the dispersion coefficients D_e, D_w are small.

Equations (4.1)–(4.3) (together with the according boundary conditions) and equation (4.7) fix the concentrations of interest of an emulsion droplet of radius R. They are coupled via the nonlinear source terms $\tilde{n}_S$ and via the boundary conditions at the droplet surface. Using (4.4) the proton concentration $[H_I^+]$ can be eliminated from the system, the same is true for $[H_{III}^+]$ using (4.11). As (4.7) is nonlinear (nonlinearity of $\tilde{n}_E$), an elimination of $[Cu_{III}^{2+}{}_{int}]$ is not possible in gereral. The system (4.1)–(4.9) is coupled to the ODE describing the waste water concentration via the integral term of (4.10), the systems describing droplets of different sizes are not coupled directly.

(4.1)–(4.13) is transformed using dimensionless concentrations

$$RH := \frac{[RH_{II}]}{[RH_{II}]_0}, \quad CR := \frac{[CuR_{2II}]}{[RH_{II}]_0}, \quad C_1 := \frac{[Cu_I^{2+}]}{[RH_{II}]_0}, \quad C_{int} := \frac{[Cu_{III}^{2+}{}_{int}]}{[Cu_{III}^{2+}]_0}, \quad C_3 := \frac{[Cu_{III}^{2+}]}{[Cu_{III}^{2+}]_0}$$

and dimensionless space coordinates

$$\rho := \frac{r}{R} \quad \text{and} \quad \zeta := \frac{z}{H}.$$

Equations (4.1)–(4.3) and (4.6)–(4.9) are divided by v_e, equations (4.10), (4.12) and (4.13) by $|v_w|$. Furthermore, the droplet diameter range $[R_{min}, R_{max}]$ is discretized using N_R droplet classes.

The resulting system is of the form $(j = 1, \ldots, N_R,\ 0 < \rho < 1,\ 0 < \zeta < 1)$

$$\frac{\partial}{\partial \zeta} CR^j = D_1 \frac{\partial^2}{\partial \zeta^2} CR^j + \frac{D_2}{\rho^2} \frac{\partial}{\partial \rho}\left(\rho^2 \frac{\partial}{\partial \rho} CR^j\right) - c_1 \tilde{n}_S(CR^j, RH^j, C_1^j) \quad (5.1)$$

$$\frac{\partial}{\partial \zeta} RH^j = D_1 \frac{\partial^2}{\partial \zeta^2} RH^j + \frac{D_3}{\rho^2} \frac{\partial}{\partial \rho}\left(\rho^2 \frac{\partial}{\partial \rho} RH^j\right) + 2c_1 \tilde{n}_S(CR^j, RH^j, C_1^j) \quad (5.2)$$

$$\frac{\partial}{\partial \zeta} C_1^j = D_1 \frac{\partial^2}{\partial \zeta^2} C_1^j + c_2 \tilde{n}_S(CR^j, RH^j, C_1^j) \quad (5.3)$$

$$-\frac{\partial}{\partial \zeta} C_3 = D_4 \frac{\partial^2}{\partial \zeta^2} C_3 - \sum_{j=1}^{N_R} S^j k\left(C_3 - C_{int}^j\right) \quad (5.4)$$

$$k\left(C_3 - C_{int}^j\right) = \tilde{n}_E(CR^j, RH^j, C_3, C_{int}^j)(1, \zeta) \quad (5.5)$$

$$\frac{\partial}{\partial \rho} CR^j|_{\rho=0} = 0 \qquad a_2 D_2 \frac{\partial}{\partial \rho} CR^j|_{\rho=1} = \tilde{n}_E(CR^j, RH^j, C_3, C_{int}^j)(1, \zeta) \quad (5.6)$$

$$\frac{\partial}{\partial \rho} RH^j|_{\rho=0} = 0 \qquad a_2 D_3 \frac{\partial}{\partial \rho} RH^j|_{\rho=1} = -2\tilde{n}_E(CR^j, RH^j, C_3, C_{int}^j)(1, \zeta) \quad (5.7)$$

$$D_1 \frac{\partial}{\partial \zeta} CR^j|_{\zeta=1} = 0 \qquad D_1 \frac{\partial}{\partial \zeta} CR^j|_{\zeta=0} = \left(CR^j(r, 0) - CR_0^j\right) \quad (5.8)$$

$$D_1 \frac{\partial}{\partial \zeta} RH^j|_{\zeta=1} = 0 \qquad D_1 \frac{\partial}{\partial \zeta} RH^j|_{\zeta=0} = \left(RH^j(r, 0) - 1\right) \quad (5.9)$$

$$D_1 \frac{\partial}{\partial \zeta} C_1^j|_{\zeta=1} = 0 \qquad D_1 \frac{\partial}{\partial \zeta} C_1^j|_{\zeta=0} = \left(C_1^j(r, 0) - C_{1,0}^j\right) \quad (5.10)$$

$$D_4 \frac{\partial}{\partial \zeta} C_3|_{\zeta=0} = 0 \qquad D_4 \frac{\partial}{\partial \zeta} C_3|_{\zeta=1} = -\left(C_3(1) - 1\right) \quad (5.11)$$

where $D_1, D_2, D_3, D_4, a_2, c_1, c_2, S^j, k$ are the constants and $\tilde{n}_E, \tilde{n}_S$ the reaction rates resulting from the transformation. It consists of $3N_R$ PDE's, an ODE and N_R algebraic equations.

6. Numerical Solution

As a first step of the numerical treatment the radial direction ρ is discretized using an equidistant grid $\rho_k = (k - 1/2)\Delta_\rho$, $k = 1, \ldots N_\rho$, with $\Delta_\rho = 2/(2N_\rho - 1)$ and approximating the radial diffusion term by

$$\frac{\partial}{\partial \rho}\left(\rho^2 \frac{\partial}{\partial \rho}\varphi\right) \approx \frac{1}{\Delta_\rho^2}\left(\left(\rho + \frac{\Delta_\rho}{2}\right)^2 \varphi(\rho + \Delta_\rho) - 2\left(\rho^2 + \frac{\Delta_\rho^2}{4}\right)\varphi(\rho) + \left(\rho - \frac{\Delta_\rho}{2}\right)^2 \varphi(\rho - \Delta_\rho)\right),$$

which is of order $O(\Delta_\rho^2)$. The resulting system of $3N_R N_\rho + 1$ ODE's and N_R algebraic equations has a very sparse structure, as figure 4 shows.

The axial discretization requires more care, as we have a diffusion-convection problem with counterstreaming phases. Again we use an equidistant grid $\zeta_l = l\Delta_\zeta$, $l = 0, \ldots N_\zeta$, $\Delta_\zeta = 1/N_\zeta$ and a symmetrical second order difference quotient for $\frac{\partial^2}{\partial \zeta^2}\varphi$. To obtain stability, we perform an upwind discretization for the convective term $\frac{\partial}{\partial \zeta}\varphi$, which means approximating the differencial quotient by a difference quotient using only grid points lying backward in flow direction. But as a first order upwind discretization causes high discretization errors for sufficiently coarse grids ("numerical diffusion", [6]), we decided for the second order upwind difference quotient

$$v\frac{\partial}{\partial \zeta}\varphi \approx \frac{|v|}{2\Delta_\zeta}\left(3\varphi(\zeta) - 4\varphi(\zeta - \text{sign}(v)\Delta_\zeta) + \varphi(\zeta - 2 \cdot \text{sign}(v)\Delta_\zeta)\right)$$

with $v = +1$ for the emulsion phase and $v = -1$ for the continuous phase.

As this difference quotient involves the discretization points ζ_{l-2} and ζ_{l-1} besides ζ_l (twostep method), it cannot be applied to the emulsion phase equations at $\zeta_0 = 0$. The same is true for $\zeta_{N_\zeta} = 1$ and the continuous phase. Thus we use the first order upwind discretization for the starting step at ζ_0, which involves only $\zeta_{-1} := -\Delta_\zeta$. ζ_{-1} can be eliminated using the boundary conditions. The resulting discretized system is very sparse (see figure 5), but it also is of a high dimension $(3N_R N_\rho + N_R + 1)(N_\zeta + 1)$ and nonlinear. For $N_R = 5$ droplet classes, $N_\rho = 18$ radial discretization points and $N_\zeta + 1 = 21$ axial discretization points (which seems to be a good choice for our test examples) the system is 5796×5796, the bandwidth is about 550. This dimension of the linear system does not allow a direct solving (CPU time demand about half an hour (COMPAREX 7/78), storage demand about 40 Mbytes, without any pivoting), due to the nonlinearity the system must be solved several times.

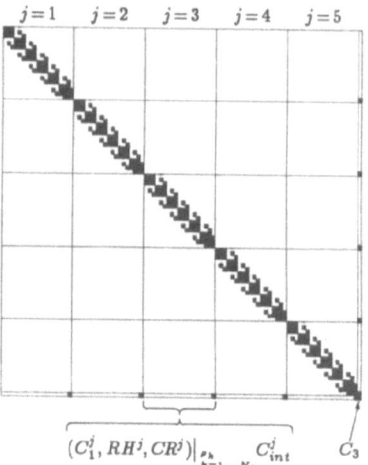

Fig. 4. Structure of the Jacobian of the ODE system right hand side $(N_R = 5, N_\rho = 6)$

The structure of the diagonal blocks D_l, $l = 0, \ldots N_\zeta$, which is the same as the structure of ODE right hand side (figure 4), allows an efficient direct solving of the systems $D_l x_l = b_l$, thus we decided to use a block Gauss-Seidel iteration method. And as the system for the emulsion phase concentrations without axial dispersion ($D_1 = 0$) and without the C_3-component is block triangular, the block Gauss-Seidel method applied to this reduced system is a direct solver provided we perform the method from bottom to top ($l = 0, 1, \ldots, N_\zeta$). The same is true for the discretized C_3-equation without axial diffusion ($D_4 = 0$), Gauss-Seidel performed from top to bottom ($l = N_\zeta, N_\zeta - 1, \ldots, 0$) is a direct solver for the resulting triangular system.

To take this fact resulting from the counter-streaming phases and the upwind discretization of the convection term into account, we change the block Gauss-Seidel direction after each iteration, solving the system for $l = 0, 1, \ldots, N_\zeta$ and for $l = N_\zeta, N_\zeta - 1, \ldots, 0$ alternating. But as it turned out during our numerical tests, it suffices to consider the waste water concentration C_3 only when solving the system from top to bottom (in flow direction of the waste water), with fixed emulsion phase concentrations (in particular fixed interface concentrations). This reduces the computational effort of one bottom-to-top top-to-bottom iteration cycle to about 50 percent, without changing the convergence properties (in particular without an essential increase of the number of iterations required). The nonlinearities of the system (due to the nonlinear

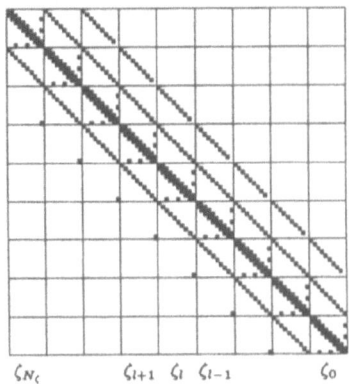

ζ_{N_ζ} ζ_{l+1} ζ_l ζ_{l-1} ζ_0

Fig. 5. Structure of the Jacobian of the fully discretized system

reaction rates) are treated by a Newton linearization (see for instance [1]).

Denoting the waste water copper concentration C_3 at ζ_l by W_l and the emulsion phase concentrations by $E_l := \left[(C_1^j, RH^j, CR^j) \big|_{\substack{\rho_k \\ k=1,\ldots,N_\rho}} \right]_{j=1,\ldots,N_R}$ (for simplicity we neglect C_{int}^j) and the according algebraic equation), the system to solve can be written in the form

$$\frac{1}{2\Delta_\zeta}(3E_l - 4E_{l-1} + E_{l-2}) = \frac{D_1}{\Delta_\zeta^2}(E_{l+1} - 2E_l + E_{l-1}) + \Phi_E(E_l, W_l)$$

$$\frac{1}{2\Delta_\zeta}(3W_l - 4W_{l+1} + W_{l+2}) = \frac{D_4}{\Delta_\zeta^2}(W_{l+1} - 2W_l + W_{l-1}) + \Phi_W(E_l, W_l) \qquad (6.1)$$

for $l = (0, 1,) 2, 3, \ldots, N_\zeta - 2 \ (, N_\zeta - 1, N_\zeta)$.

The left hand sides are the discretized convection terms, the first right hand side term is the discretized diffusion/dispersion term, Φ_E resp. Φ_W stand for the radial effects (diffusion, extraction reaction) and the stripping reaction. Φ_E contains all the nonlinearities of the problem, Φ_W is linear. For $l = 0, 1, N_\zeta - 1, N_\zeta$ the system is slightly different due to the first order starting step for the convection term discretization and the elimination of the unknowns at ζ_{-1} and $\zeta_{N_\zeta+1}$ using the boundary conditions. For the demonstration of the numerical algorithm for solving the discretized system we neglect these boundary effects.

N_ζ \ N_ρ	11	18	25	32
10	30			
20	52	52	53	53
40	108			109

Tab. 1. Numbers of iteration loops

N_ζ \ N_ρ	11	18	25	32
10	9.3			
20	22.9	32.7	45.2	55.2
40	78.9			213.0

Tab. 2. CPU time demand [sec]

Algorithm for solving the discretized system (6.1)

Step 0: given: initial values (E_l^0, W_l^0), $l = 0, 1, \ldots, N_\zeta$, $n := 0$

Step 1: for $l = 0, 1, \ldots, N_\zeta$ compute $(E_l^{n+1}, W_l^{n+\frac{1}{2}})$ from the linear system

$$\frac{1}{2\Delta_\zeta}(3E_l^{n+1} - 4E_{l-1}^{n+1} + E_{l-2}^{n+1}) = \frac{D_1}{\Delta_\zeta^2}(E_{l+1}^n - 2E_l^{n+1} + E_{l-1}^{n+1})$$

$$+ \Phi_E(E_l^n, W_l^n) + \Phi_E(E_l^n, W_l^n) \cdot \begin{pmatrix} E_l^{n+1} - E_l^n \\ W_l^{n+\frac{1}{2}} - W_l^n \end{pmatrix}$$

$$\frac{1}{2\Delta_\zeta}(3W_l^{n+\frac{1}{2}} - 4W_{l+1}^n + W_{l+2}^n) = \frac{D_4}{\Delta_\zeta^2}(W_{l+1}^n - 2W_l^{n+\frac{1}{2}} + W_{l-1}^{n+\frac{1}{2}}) + \Phi_W(E_l^{n+1}, W_l^{n+\frac{1}{2}})$$

Step 2: for $l = N_\zeta, N_\zeta - 1, \ldots, 0$ compute W_l^{n+1} from

$$\frac{1}{2\Delta_\zeta}(3W_l^{n+1} - 4W_{l+1}^{n+1} + W_{l+2}^{n+1}) = \frac{D_4}{\Delta_\zeta^2}(W_{l+1}^{n+1} - 2W_l^{n+1} + W_{l-1}^{n+\frac{1}{2}}) + \Phi_W(E_l^{n+1}, W_l^{n+1})$$

Step 3: $n := n + 1$, if *(end condition)* **Stop**, else **goto 1**

The system matrix at step 1 has the structure shown in figure 4, the linear system is solved directly using Gaussian elimination without pivoting, adapted to the sparsity pattern. For sufficiently small Δ_ζ the system matrix is diagonal dominant, the existence of a triangular decomposition without pivoting is guaranteed.

The total numerical effort of one iteration loop is linear in N_ζ, N_ρ, N_R. The number of iterations required is almost independent from N_ρ and N_R, but it increases with N_ζ. Table 1 gives some numbers of iterations requied for different values of N_ρ and N_ζ, with $N_R = 5$. Table 2 gives the according CPU-time demand on a COMPAREX 7/78 in [sec].

Comparing the results of differently fine discretizations with the results obtained for $N_\rho = 39$ and $N_\zeta = 40$, we get the following image:
- for $N_\zeta = 10$, $N_\rho = 10$ the absolute error of C_3 is already below 10^{-3}, with absolute values between 0.01 and 0.73, while the error of the phase I copper concentration is rather high.
- increasing N_ζ to 20 the error of C_3 is about 10^{-4}, the error of C_1^j is the same as above.
- refining the radial discretization to $N_\rho = 18$, the absolute error of C_1^j is of magnitude 10^{-2}, with absolute values up to 10. The phase II concentrations RH^j and CR^j have absolute values between 0 and 1 resp. 0 and 0.5, the absolute errors are about 10^{-4}.

For further details of discretization and the numerical solution technique see [5].

7. Results

Figures 6–8 give some numerical results. The waste water feed concentration $[Cu_{III}^{2+}]_0$ is 6.25×10^{-3} kmol/m³, its velocity within the column is 3.7×10^{-3} m/s, the column height 3 m. The emulsion (volume proportion of phase I 13.8%) is fed at the bottom, we consider 5 droplet sizes $R_1 = 0.12 \times 10^{-3}$m, $R_2 = 0.18 \times 10^{-3}$m, $R_3 = 0.27 \times 10^{-3}$m, $R_4 = 0.40 \times 10^{-3}$m and $R_5 = 0.60 \times 10^{-3}$m, the ascending velocity is about 17×10^{-3}m/s. The axial diffusion/dispersion coefficients are 10^{-3}m²/s.

The concentrations of the following figures are scaled by $1/[Cu_{III}^{2+}]_0$. Figure 6 shows waste water and the interface concentration for droplet class 3, figure 7 shows the phase III (dashed line) and the mean phase I concentrations for the five droplet classes. The reamining waste water copper concentration is about 1.4% of the feed concentration, the scaled stripping phase concentrations are up to 10 (R_5) and 50 (R_1). Figure 8 gives the radial curve of the scaled phase I concentrations at $\zeta = 0.50$, $\zeta = 0.75$ and $\zeta = 1.00$. There is an almost linear decrease from about 100 (at $\zeta = 1$) to zero within a small, but growing (in ζ) film near the droplet surface.

For the same data, but without axial diffusion/dispersion, the concentration profiles are much steeper, the waste water copper concentration is reduced to 0.3% of the feed concentration.

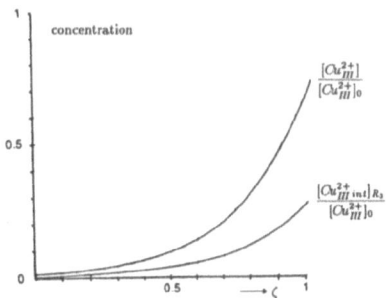

Fig. 6. Phase III copper concentrations

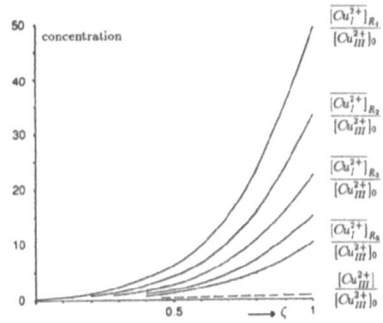

Fig. 7. Phase III and mean phase I copper concentrations for 5 droplet radii.

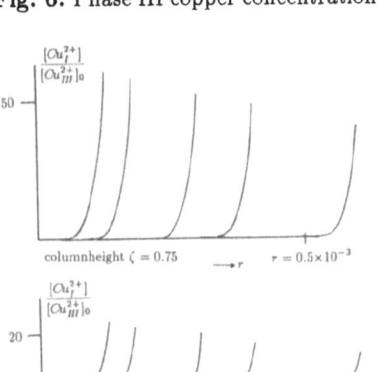

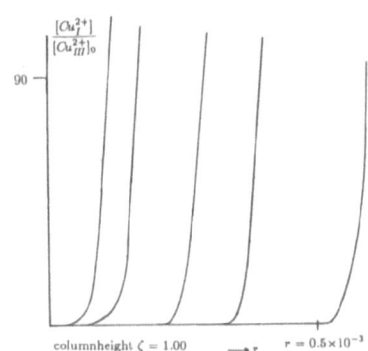

Fig. 8. Phase I copper concentrations at $\zeta = 0.50$, 0.75 and 1.00 for 5 droplet classes.

References

[1] J.E. Dennis, R.B. Schnabel, Numerical Methods for Unconstrained Optimization and Nonlinear Equations, *Prentice-Hall, Inc., Englewood Cliffs, New Jersey* (1983)

[2] D. Lorbach, H.J. Bart and R. Marr, Mass transfer in liquid membrane operation, *Ger. Chem. Eng., 9* (1986) 321–327.

[3] D. Lorbach, R. Marr, Emulsion Liquid Membranes, Part II: Modelling Mass Transfer of Zinc with Bis(2-ethylhexyl)dithiophosphoric Acid, *Chem. Eng. Process., 21* (1987) 83–93.

[4] T. Miyauchi, T. Vermeulen, Longitudinal dispersion in two-phase continuous-flow operations, *Industrial & Engineering Chemistry Fundamentals, 2* (1963) 113–126.

[5] A. Ortner, Flüssigmembran-Permeation in Gegenstromkolonnen, *diploma thesis, Univ. Linz, Math. Deptm.* (1989)

[6] P.J. Roache, Computational Fluid Dynamics, *Hermosa, Albuquerque* (1982)

[7] M. Teramoto, T. Sakai, et. al., Modelling of the permeation of copper through liquid surfactant membranes, *Sep. Sci. Technol., 18* (1983) 735–764.

THE LAY-OUT OF REACTIVE EXTRACTORS

H.J. BART and R. MARR
Techn. University Graz, Inst. Therm. VT,
Inffeldgasse 25, A-8010 Graz, Austria

Abstract. The lay-out procedure for the kinetic dependent reactive systems is similar to the conventional equilibrium based systems. The main difference is in the consideration of the time dependent mass transfer term describing the overall (macro-) kinetics. This term is then incorporated in well known hydrodynamic concepts describing the concentration profile in a technical apparatus. The calculation sequence is demonstrated for liquid/liquid systems with reactive extraction.

Introduction

The lay-out in a homogeneous gas or liquid or even in a heterogeneous gas/liquid system is a standard item in chemical engineering education, even when reactive systems are con- sidered. Also classical physical liquid/liquid extraction systems give no major problems. This is not the case with reactive liquid/liquid systems[1]. The separation of solutes from aqueous solutions is enhanced by the use of reactive liquid ion exchangers. With them it is possible selectively to remove metal ions, organic and inorganic substances like furforal, citric acid, amino acids, Zn, Cu, Cd etc.[2]. They are the key components in liquid-membrane-permeation (LMP) and solvent extraction (SX)[3].

The lay-out sequence is based on three main steps as is the modelling of the

147

Hj. Wacker and W. Zulehner (eds.),
Proceedings of the Fourth European Conference on Mathematics in Industry, 147–158.
© 1991 *B.G. Teubner Stuttgart and Kluwer Academic Publishers.*

*   micro-kinetics
*   macro-kinetics and
*   column performance.

The first step includes all basic chemical knowledge of the
system used. The macro-kinetics considers mass transfer on a
unit interface, e.g. of a single droplet. The column model
combines all effects and takes regard to the hydrodynamic
behaviour of a certain apparatus. This is demonstrated
schematically in Fig.1.

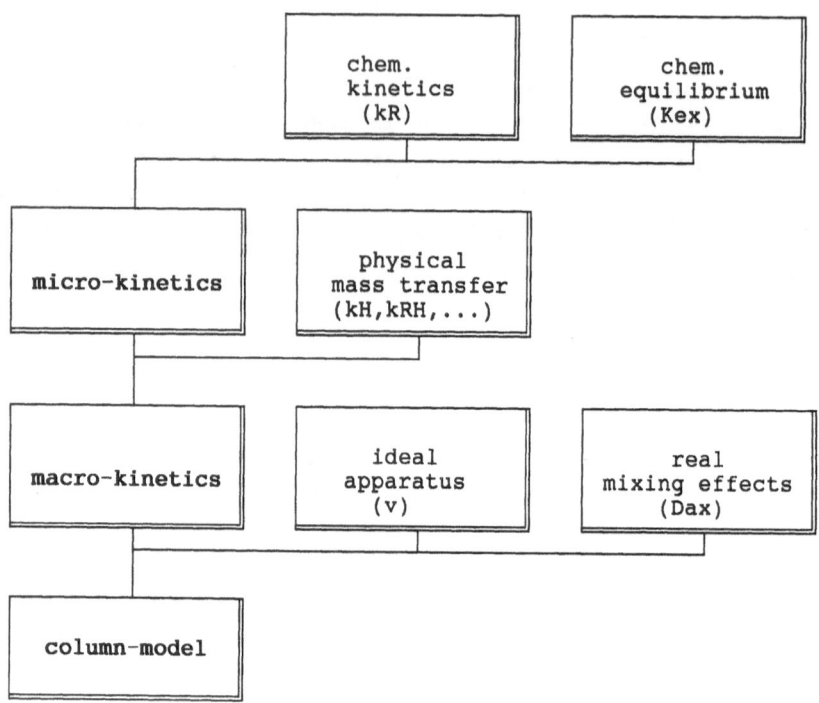

Fig.1. Design sequence in reactive systems

Micro-kinetics

The basis of a formulation of a conversion is based on a stoichiometric reaction as given in Equ.(1) for the extraction of a bivalent metal ion with the ion exchanger RH:

$$Me^{2+} + 2\ \overline{RH} \rightleftharpoons \overline{MeR_2} + 2\ H^+ \qquad\qquad (1)$$

The equilibrium is after the mass action law:

$$K_{ex} = \frac{(\overline{MeR_2}).(H^+)^2}{(Me^{2+}).(\overline{RH})^2} \qquad\qquad (2)$$

The time dependent reaction on the interface (n_R = dMe/dt) is given by the semi-empirical Equ.(3). The concentrations in this equation are interfacial concentrations. The equilibrium coefficient K_{ex} can be estimated in shaking funnel experiments. The kinetic parameters k_F, e, f, g have to be estimated in an apparatus with a constant interface (e.g. Lewis-cell or rising droplet apparatus[1]). The experimental set-up should be carefully chosen in respect to a hydrodynamical or a concentration region where diffusional mass transfer can be neglected[4].

$$\overset{.}{n}_R = k_F \cdot \frac{(Me^*)^e.(\overline{RH^*})^f}{(H^*)^g} - \frac{k_F}{K_{ex}} \cdot \frac{(\overline{MeR_2^*}).(H^*)^{2-g}}{(Me^*)^{1-e}.(\overline{RH^*})^{2-f}} \qquad (3)$$

Macro-kinetics

The macro-kinetic mass transfer in a liquid globule can be described in a simple approach with the two film model of Whitman[5]. A model assumption is to place the chemical reaction directly at the interface. This is plausible since the solute and the ion exchanger are not soluble in the other phase. From this point of view the following set of linear ordinary differential equations will describe the diffusional mass transfer which is represented in Fig.2:

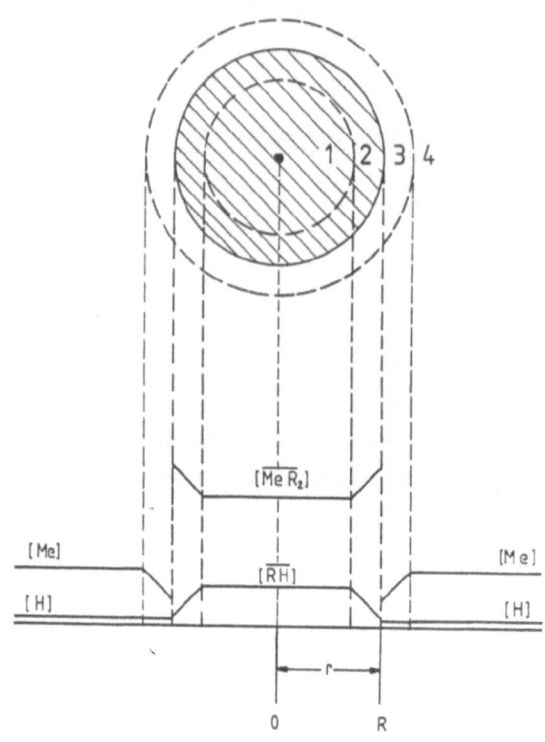

Fig.2 Concentration profile in a droplet with SX: 1 org. bulk, 2 org. film, 3 aqu. film, 4 aqu. bulk.

$$\dot{n}_{H} = k_{H} \cdot (H^{*} - H) \tag{4}$$

$$\dot{n}_{RH} = k_{RH} \cdot (RH - RH^{*}) \tag{5}$$

$$\dot{n}_{MeR2} = k_{MeR2} \cdot (MeR_{2}^{*} - MeR_{2}) \tag{6}$$

$$\dot{n}_{Me} = k_{Me} \cdot (Me - Me^{*}). \tag{7}$$

The total mass flux is a combination of diffusional and chemical resistances. Equations (4) to (7) and (2) must be linked according to stoichiometry as done in Equ.(8) which gives:

$$\dot{n} = \dot{n}_{Me} = \dot{n}_{MeR2} = \tfrac{1}{2} . \dot{n}_{RH} = \tfrac{1}{2} . \dot{n}_{H} = \dot{n}_{R}. \qquad (8)$$

This non-linear system of ordinary differential equations (3 to 7) can be solved numerically with known bulk concentrations. The result is $\dot{n}$ and the unknown interfacial concentrations. The physical diffusion parameters can be experimentally determined in a rising droplet apparatus or evaluated from well-known correlations (Sh-number).

Column model

The estimation of the reaction volumn of a heterogeneous reactor is divided in diameter, D_c, and height, H_c, calculation. The diameter of a column is calculated with the concept of the slip velocity, v_s, which is used in physical extraction[6]. The slip velocity gives the maximum throughput of a specific apparatus and depends on the column internals. It had to be determined experimentally and is known from correlations with well-known and widely used columns[6].

$$D_K = (4.S/ .\dot{v})^{0,5} \qquad (9)$$

with

$$\dot{v} = (0,6 \text{ bis } 0,9). \dot{v}_s. \qquad (10)$$

The height of a column can be evaluated using well known hydrodynamic concepts where the complex chemical mass transfer model on a single droplet is incorporated.

Mass balances over a differential slice of a column including the mass transfer term, ṅ, the ideal convective term, v_c, and the real hydrodynamic term, D_{ax}, (regarding real mixing effects) is formulated according to the dispersion model[7] (compare Fig.1). It is for the continuous phase:

$$D_{ax,c} \cdot \frac{d\ Me}{dz} + v_c \cdot \frac{dMe}{dz} + \dot{n}.a/(1-x_d) = 0 \qquad (11)$$

and for the dispersed phase:

$$D_{ax,d} \cdot \frac{d\ MeR_2}{dz} - v_d \cdot \frac{dMeR_2}{dz} + \dot{n}.a/x_d = 0 \qquad (12)$$

The boundary conditions are described in lit 1 and 6. An integration of equ. (11) and (12) is then done from column height z=0 to z=H_c as described elsewhere[8]. It is possible to introduce any type of kinetics (see Fig.1) for the calculation of the concentration profiles[9].

An example of a model calculation is given in Fig.3. 1,2 g/l copper is operated in a SHE-column[10] with 0,1 m diameter and 2,5 m active height as described elsewhere[11]. Experiments were performed in a wide flow range from 0,0025 to 0,0117 m/s marking the operating limits of the column. Full lines in the picture represent the model calculations. A good representation of column behaviour could be obtained. The agreement between experimental and calculated data is good considering that no parameter was fitted due to the column runs. The parameters for model calculations are given in Tab.1. Parameters describing mass transfer on a single globule were estimated in a Lewis cell and axial dispersion correlations for the SHE-column were from literature[12]. The mean Sauter diameter, d_{23}, was 1,6 mm in both cases. It was estimated with a photoelectric method[13]. Sh_d for rigid spheres was calculated according Brauer[14] and Sh_c according Linton-Sutherland[15]. The

hold up of the column was estimated experimentally with a
hydrostatic method.

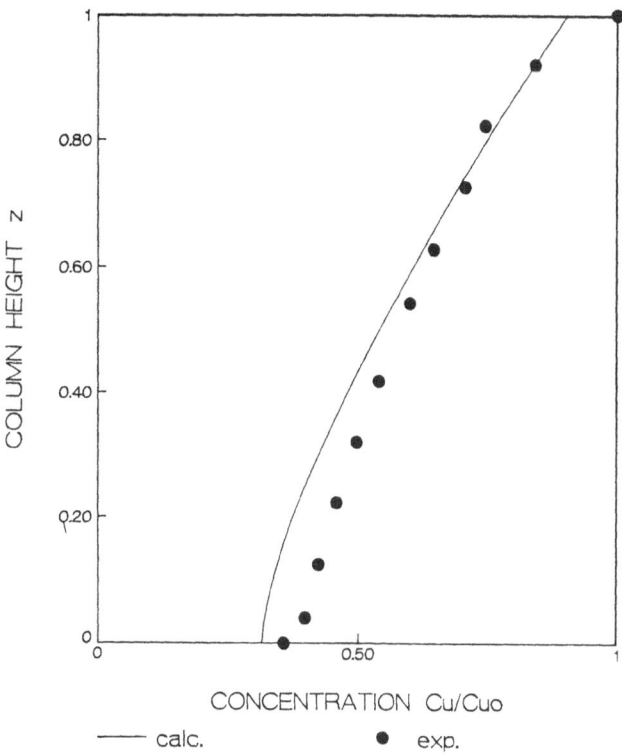

Fig.3.   Concentration profiles with copper extraction

Tab.:1 Mass transfer parameters for SX

$$k_{Cu} = 1.55.10^{-5} \text{ m/s} \quad K_{ex} = 17 \quad K_{st} = 1$$
$$k_{CuR2} = 1.28.10^{-6} \text{ m/s} \quad k_{RH} = 1.94^{-6} \text{ m/s}$$
$$k_F = 2.00.10^{-6} \text{ m/s} \quad e = f = g = 1$$

Model simplifications

A simplified model is convenient for short-cut calculations. Reduction of number of model equations can be done either with a sensitivity analysis or with critical dimensionless numbers. Sensitivity analysis implies variation of the scale of parameters in the full model as described above. Critical cases for a lay out of a technical apparatus can be found and studied. The weight of the parameters and their corresponding equations in regard to column lay-out is easier examined with critical dimensionless numbers[16].

The dominating regime in gas/liquid separations with chemical reactions can be analysed with the Hatta-number[17]. Presumption is that the solute is physically soluble in both phases. In contrary, the chemical reaction in reactive liquid/liquid systems is always on the interface, since the solute is insoluble in the counterphase. It can be extracted only after a transformation in a soluble compound via the chemical reaction. Since the reaction is at a defined phase boundary the process is comparable with catalytic gas/solid surface reactions. The critical II. Damköhler-number describes such system behaviour[18]. Three different DaII-numbers can be derived from Fig.2 neglecting a diffusional resistance of the protons.

Equ.13 gives the relation of diffusion resistance to the chemical reaction resistance of the aqueous solute, Equ.14 of the free and Equ.15 of the loaden ion exchanger. DaII numbers <<1 give evidence of kinetic reaction regime and DaII >>1 of a dominating diffusion resistance. Tab.2 gives DaII-numbers for different concentrations of the solute

$$Da_{II,Me} = \frac{k_F \cdot (Me)^{e-1} \cdot (\overline{RH})^f}{k_{Me} \cdot (H)^g} \qquad (13)$$

$$Da_{II,RH} = \frac{k_F \cdot (Me)^e \cdot (\overline{RH})^{f-1}}{k_{RH} \cdot (H)^g} \qquad (14)$$

$$Da_{II,MeR_2} = \frac{k_F \cdot (Me)^{e-1} \cdot (\overline{RH})^{f-2}}{k_{MeR_2} \cdot K_{ex} \cdot (H)^{g-2}} \qquad (15)$$

.

Tab.2   DaII-numbers (Initial Concentrations in mol/L)

$(Me)_0$	0.002	0.018	0.157
$(RH)_0$	0.090	0.090	0.400
$(H)_0$	0.100	0.010	0.001
$(MeR_2)_0$	0.000	0.000	0.000
$Da_{II,Me}$	0.12	1.20	53.3
$Da_{II,RH}$	0.02	1.90	162.3
Da_{II,MeR_2}	0.10	0.01	0.0
regime	reaction	mixed	diffusion

The calculation procedure of the apparatus is then very simple. As an example, reaction regime (Equ.3) dominates and axial dispersion ($D_{ax}=0$) can be neglected (depends on Bo-number[19]) so Equ.11 reduces to:

$$\frac{v_c \cdot (1-X_d)}{a} \cdot \frac{dMe}{dz} = k_F \cdot \frac{(Me)^e \cdot (RH)^{-f}}{(H)^g} \qquad (16)$$

This term can be easily evaluated. The procedure is similiar done with dominating diffusional resistance (one of Equ 4 to 7 dominates according to Da_{II}).

Conclusions

The estimation of column height is possible by the use of the presented model. The calculation can be done straight forward having experimentally evaluated the chemical mass transfer parameters. All physical and hydrodynamic parameters can be obtained from correlations in literature according to the system conditions and apparatus used.

The models can be used for studies predicting the critical limits of column behaviour. Especially, the optimal conditions of pilot plants can be calculated. In the presented manner model simplifications can be done in order to get a simple analytical formula which can be used for a short-cut calculation procedure.

Acknowledgements

The authors are grateful for the grant of Austrian Science Foundation and the Jubiläumsfonds of the Austrian National-bank and the Christian Doppler Foundation which made this work possible.

Notation

a   specific interfacial area (m^2/m^3)
e,f,g   reaction order
D_{ax} dispersion coefficient (m^2/s)
D_c column diameter (m)
H_c column height (m)
k_F forward reaction constant
K_{EX} equilibrium constant
k   mass transfer constant (m/s)
n   specific molar flux $(kmol/m^2.s)$
v   specific flow (m/s)
X_d dispersed phase hold-up
z   dimensionless height

Subscripts: c   continuous

                d   dispers

                *   at interface

                -    org. phase

                0   initial

Literature cited

[1]   Bart, H.J., <u>Metallsalzextraktion - Stoffaustausch und Reaktionstechnik</u>, dbv-Verlag, 1988, Graz

[2]   Bart, H.J., ÖChemZ 3 (1989) 74-79

[3]   Marr, R., Bart, H.J., Draxler, J., Sep. Sci. & Technol. (1989) in press

[4]   Bart, H.J., Bauer, A., Marr, R., Chem. Eng. Technol. 10 (1978) 291-296

[5]   Whitman, W.G., Chem. and Met. Eng., <u>29</u>, 147 (1923).

[6]   <u>Handbook of Solvent Extraction,</u> Ed. T.C. Lo et al., Wiley-Interscience, New York, 1983

[7]   Danckwerts, P.V., Chem. Eng. Sci.2 (1953)1

[8]   Bauer, A., Bart, H.J., Marr, R., in Proc. Chem. Eng. Fundamentals 87 EFCE Event 349, April 1987, Giardini Naxos Italy, 97-103

[9]   Bart, H.J., Bauer, A., Marr, R., EFCE Publ. Ser. No. 347, Inst. Chem. Eng. Rugby, England 1987, 291-307

[10]   Marr, R., Husung, G., E-Pat. 0048 239

[11]   Gaubinger, W., Husung, G., Marr, R., Ger. Chem. Eng. 6 (1983) 74-79

[12]   Gaubinger, W., Husung, G., Marr, R., Chem.-Ing.-Tech. 54 (1982) No2, 850-851

[13]   Pilhofer, T., Miller, H.D., Chem.-Ing.-Tech. 44 (1972) 295-300

[14]   Brauer, H., <u>Stoffaustausch einschließlich chemischer Reaktion</u>, Sauerländer, Aarau 1971

[15]   Linton, M., Sutherland, K.L., Chem. Eng. Sci., 12 (1960) 214-229

[16]   Bart, H.J., Marr, R., Bauer, A., Chem.-Ing.-Tech. (1989) in press

[17]   Levenspiel, O., <u>The Chem. Reactor Omnibook</u>, OSU Book Stores Inc., 1984, Oregon

[18]   Baerns, M., Hofmann,H., Renken, A., <u>Chem. Reaktionstechnik</u>, G. Thieme Verlag, Stuttgart, 1987

[19]   <u>Recent Advances in Liquid-Liquid Extraction</u>, Ed.: C. Hanson, Pergamon Press, Oxford, 1971

The simulation of a co–current bubble reactor

by

Heikki Haario and Ilkka Turunen

1.Introduction. We consider the modelling of gas–liquid reactions in a cocurrent packed bubble reactor: both gas and liquid are fed in from the bottom of the column and removed from the top. The liquid froms a continuous phase in the column and the gas flows in the form of bubbles through the liquid phase. The desired product accumulates in the liquid as a result of a chemical reaction taking place both in the bulk liquid and at the gas/liquid interface. The column is operated continuously and is therefore in a steady state.

The concentrations of the reacting components vary in the vertical direction but are supposed to be constant with respect to the horizontal space variable. The mass transfer from gas to liquid and the reaction at the interface are modelled by the 'film' theory . The theory assumes a thin layer around each bubble, in which the diffusion and reaction create nonhomogeneous horizontal concentration profiles. The situation is modelled as being one–dimensional in space at each height. For a presentation of the film theory see [3].

Denote by $c = (c_1, ..., c_n)'$ the vector of the concentrations of the reacting components inside the film. The equations for our model can be thought as obtained by replacing the time coordinate in a conventional reaction/diffusion equation by the vertical space coordinate. The reaction/diffusion in the film is thus generically given by the equations

$$(1.1) \qquad D_i \frac{\partial^2 c_i}{\partial z^2} - R_i(c) = v \frac{\partial c_i}{\partial h}, \quad i = 1, ..., n$$

where z denotes the horizontal spatial coordinate of the film and h is the height coordinate of the column. The vector D gives the diffusion coefficients, R the

159

Hj. Wacker and W. Zulehner (eds.),
Proceedings of the Fourth European Conference on Mathematics in Industry, 159–168.
© 1991 B.G. Teubner Stuttgart and Kluwer Academic Publishers.

reaction rates, and v is the velocity term $v = \partial h/\partial t$. The system is coupled with a set of ODE's describing the reactions taking place in the bulk liquid. If C denotes the bulk concentrations the additional ODE's typically have the form

$$(1.2) \qquad -D_i a \frac{\partial c_i}{\partial z} - R_i(C) = v \frac{\partial C_i}{\partial h}, \quad i = 1, ..., n$$

with the interfacial area parameter a. Non–volatile components are characterized by the boundary conditions $\partial c_i/\partial z = 0$ at the gas boundary of the film. The initial conditions for $h = 0$ are given so as to describe the concentrations at the bottom of the column where the gas is brought in the liquid.

From the practical point of view, the above model serves as a tool for the design and optimization of the reactor. Some of the parameters are known, some should be estimated. Typically the diffusion coefficients can be assumed as known. The values of the bulk concentrations at certain heights in the column can be measured. From this information one should be able to estimate such unknown values as a or the mass transfer coefficients. A reliable way to compute the solution of the system together with the boundary conditions is thus required.

2.The model More specifically, we consider the following situation. Two reactants A, B are initially, at the bottom of the column, in different phases. Reactant A is a component in the gas phase and reactant B is a substance dissolved in the liquid. According to the film theory, a thin liquid layer with thickness ℓ_1 is assumed to exists on the liquid side of the gas/liquid interface. Denote by u and v the concentrations of A and B in this film, where the diffusion and part of the chemical reaction take place. In a more exact treatment, a similar film should be assumed on the gas side to explain the diffusion of reactant A from bulk gas to the interface. However, this can be neglected because the diffusion in the gas phase is fast enough to cause a constant concentration of A in the gas bubbles. The dispersion effects in the vertical direction of the column can also be neglected, due to the packing of the column. The concentrations of A and B in the bulk liquid are denoted by U and V.

For $h > 0$, A and B react and diffuse in the film according to the model

(2.1)
$$D\frac{\partial^2 u}{\partial z^2} - kuv = v_l\frac{\partial u}{\partial h}$$
$$E\frac{\partial^2 v}{\partial z^2} - kuv = v_l\frac{\partial v}{\partial h}$$

The component B is non–volatile. In addition, the boundary conditions for $h > 0$ describeing the flux of A into the bulk liquid can be written in the form

$$\frac{\partial v}{\partial z} = 0, \qquad \text{at} \quad z = 0$$

(2.2) $\quad -Da(1-\epsilon_p)\dfrac{A_{cr}}{\dot{L}}\dfrac{\partial u}{\partial z} - k(1-\epsilon_p)(1-\epsilon_G)\dfrac{A_{cr}}{\dot{L}}UV = \dfrac{\partial U}{\partial h} \qquad \text{and}$

$\quad -Ea(1-\epsilon_p)\dfrac{A_{cr}}{\dot{L}}\dfrac{\partial v}{\partial z} - k(1-\epsilon_p)(1-\epsilon_G)\dfrac{A_{cr}}{\dot{L}}UV = \dfrac{\partial V}{\partial h} \qquad \text{at} \quad z = \ell_1.$

Here k is the reaction rate constant, D and E denote the diffusion coefficients and a gives the interfacial area, i.e. the total area of the bubbles at a given height divided by the volume of gas and liquid at that height. By ϵ_p we denote the volume fraction of the packing in the column. The gas hold–up, the volume fraction of gas at each height in the column, is given by the function ϵ_G. The function ϵ_G depends on the specific way in which the packing of the column is done and thus can not be modelled exactly. The empirical modelling of ϵ_G in our test column is described below. The function v_l gives the velocity of the liquid in the column. The constant A_{cr} represents the area of the column cross–section and the constant $\dot{L}$ gives the liquid flowrate.

The value u^* of the concentration of A at the gas boundary depends in a non–linear way, also as described below, on the partial pressure of A in the gas phase.

The change of the pressure p in the vertical direction is given by the equation

(2.3)
$$\frac{\partial p}{\partial h} = -\rho g(1 - \epsilon_G)$$

where ρ is the density of the liquid and g is the gravitational constant. This equation (i.e., the ϵ_G–dependency) was experimentally verified in the process conditions in question. Due to the low velocity there is no dynamic pressure loss. The function ϵ_G is modelled as an empirical polynomial

(2.4)
$$\epsilon_G = a_3 v_s^3 + a_2 v_s^2 + a_1 v_s + a_0$$

where v_s is the superficial velocity of the gas given by $v_s = \dot{V_g}/A_{cr}$ and $a_3, ... a_0$ are constants determined by separate experiments . The ideal gas law $pV = nRT$ yields for the gas stream $\dot{V_g}$

$$(2.5) \qquad \dot{V_g} = (\dot{n}_A + \dot{n}_N)\frac{RT}{p}$$

where the function $\dot{n}_A$ is the upward molar stream of A and the constant $\dot{n}_N$ gives the molar stream of the non–reacting inert gas. The decrease of $\dot{n}_A$ in the vertical direction is determined by the horizontal flux of A into the liquid, the flux given by the gradient of A at the gas boundary $z = 0$:

$$(2.6) \qquad D\frac{\partial u}{\partial z}aA_{cr}(1 - \epsilon_p) = \frac{\partial \dot{n}_A}{\partial h}.$$

The gas solubility could be approximated by Henry's law, $u^* = p_A/H$ where H is the Henry's constant and $p_A = \dot{n}_A/(\dot{n}_A + \dot{n}_N)$ is the partial pressure of A. However, experiments revealed that a better correlation was given by

$$(2.7) \qquad u^* = a' + b'p_A$$

with numerical constants a', b'. The value of u^* is needed in computing the flux term in the equation (6).

The interfacial parameter a also depends on the pressure and molar stream. The correlation can be modelled in several ways, one possibility is to set $a = \alpha\sqrt{v_s}$, where the superficial velocity $v_s = \dot{V_g}/A_{cr}$ is given by (2.5).

It remains to fix the initial conditions at $h = 0$. The value of $\dot{L}$ is a fixed constant, the value of the gas stream $\dot{V_g}$ at the bottom is known. From the gas law $p_A\dot{V_g} = \dot{n}_A RT$ we get

$$\dot{n}_A = \frac{y_A\dot{V_g}}{RT}p$$

where the mole fraction y_A of A and the pressure at $h = 0$ are known.

It can be assumed that the concentration of A in the liquid at $h = 0$ is zero so the initial conditions at $h = 0$ for $0 < z \le \ell_1$ in the film can be given as

$$(2.8) \qquad \begin{aligned} u &\equiv 0, & v &\equiv v^b \\ U &= 0, & V &= v^b \end{aligned}$$

where v^b is the known initial bulk concentration of B. For $z = 0$ we take $u = u^*$ by (7) and set $v = v^b$.

Thus we arrive at a diffusion/reaction model leading to a system of parabolic PDE's. This system is coupled with a set of ODE's describing the boundary conditions, the reaction taking place in the bulk liquid. Finally, taking the gas stream and the changing pressure into account introduces two additional components to the system.

Next we normalize the equations in a dimensionless form with respect to z and h. Let H denote the height of the column. New variables x, y are introduced by setting

$$h = yH, \quad 0 \le y \le 1$$
$$z = x\ell_1, \quad 0 \le x \le 1$$

The collected system of equations then takes the form

(9)
$$u_y = D_1 u_{xx} - k_1 uv$$
$$v_y = D_2 v_{xx} - k_1 uv$$
$$U_y = -D_3 u_x|_{x=1} - k_1 UV$$
$$V_y = -D_4 v_x|_{x=1} - k_1 UV$$
$$(\dot{n}_A)_y = D_5 u_x|_{x=0}$$
$$p_y = -D_6$$

with

$$D_1 = \frac{DH}{\ell_1^2 v_l}$$
$$D_2 = \frac{EH}{\ell_1^2 v_l}$$
$$D_3 = \frac{k_l^A a H}{(1 - \epsilon_G) v_l}$$
$$D_4 = \frac{k_l^B a H}{(1 - \epsilon_g) v_l}$$
$$D_5 = k_l^A a H A_{cr}(1 - \epsilon_p)$$
$$D_6 = H \rho g (1 - \epsilon_G)$$

where $v_l = \dot{L}/A_{cr}(1 - \epsilon_p)(1 - \epsilon_G)$, $k_1 = kH/v_l$, $k_l^A = D/\ell_1$, $k_l^B = E/\ell_1$, and ϵ_G and a are functions of $v_s = (\dot{n}_A + \dot{n}_N)RT/pA_{cr}$: ϵ_G given as the polynomial (2.4) and $a = \alpha\sqrt{v_s}$.

The boundary conditions are given by

$$v_x|_{x=0} = 0$$

$$u|_{x=0} = u^* = a' + b'\frac{\dot{n}_A}{\dot{n}_A + \dot{n}_N}$$

$$u|_{x=1} = U$$

$$v|_{x=1} = V$$

and the initial conditions at $y = 0$ by

$$p = p_0$$

$$\dot{n}_A = \frac{\dot{V}_G y_A p_0}{RT}$$

$$u = 0$$

$$U = 0$$

$$v = v^b$$

$$V = v^b$$

where all the right hand sides represent known values.

3. Theoretical properties of the system. The system of equations given above is an extension to that considered in [1]. By using the approach of semigroup theory it could be shown that the system in [1] had a unique solution analytic in both variables. The present model differs from that of [1] as it has non–constant boundary values u^* and variable diffusion coefficients (due to the term v_l). The theoretical properties of the system can be studied along the lines given in [2], Chapter 3. Details of the study will be presented elsewhere.

4. Description of the algorithm We want to employ an algorithm which would be as simple as possible, enabling the user to compute the solution without a PDE-solver and minimizing the need of writing code. Our choice is to use the 'method of lines', to make a discretization with respect to the x variable and to solve the resulting system of ODE's with respect to y with an ordinary ODE solver.

The most straightforward way of computing a numerical solution for the equation of the type $c_{xx} = f$ is to use the Taylor expansion for c at a points x_{j-1}, x_j and x_{j+1} and write

$$\frac{c_{j-1} - 2c_j + c_{j+1}}{\delta^2} = c_j^{(2)} + \frac{1}{12}\delta^2 c_j^{(4)} + \dots$$

Substituting on the right–hand side the formula $c_j'' = f_j$ and omitting $O(\delta^2)$ we apply this formula to the system (2.9).

We discretize the interval $[0,1]$ by n points $x_j = (j-1)\delta$, $j = 1,...,n$, such that $x_n = 1$. Thus for a component c we have the equations

$$(2) \qquad D\frac{c_{j-1} - 2c_j + c_{j+1}}{\delta^2} = r_j + \frac{\partial c_j}{\partial y}$$

with $j = 2,...,n-1$, where r is the reaction term. We note that this discretization may lead to inaccuracies in computing the flux terms c_x at $x = 0$ or $x = 1$. An alternative is to use a discretization more dense at the end points of the interval $[0,1]$. In any case, we get a set of ordinary differential equations for the values of c at the discretization points. The matrix 'components of $c \times$ at discretization points x' is numbered columnwise, i.e. so that the c–components at a given discretization point will be consequtive components of the ODE.

We still have to take into account the boundary conditions and the ODE's for U, V, $\dot{n}_A$ and p. The boundary condition $\partial v/\partial x = 0$ is most simply taken care of by the trick of 'false boundaries', i.e. by requiring that v be symmetric across the boundary.

The ODE's for U, V $\dot{n}_A$ and p are given in (2.9). The flux terms are approximated by the difference computed at the points x_1, x_2 or x_{n-1}, x_n. The flux term at $x = 0$ is computed as $u_x \simeq (u(x_2) - u^*)/\delta$.

Thus we arrive at a system of ODE's of the form $\dot{c} = F(c)$ for the values of the components c at the discretization points. If np denotes the number of the discretization points, we have a system with $2 * np + 1$ equations. With a typical value $np = 20$ the solution of the system takes a few CPU–second with a VAX8600. The system can be solved by standard solvers; in the examples below we have used DIVPAG from IMSL.

5. The model versus experiments. The model was tested by comparing the results with those obtained from an industrial full–scale column. The conditions in

the column correspond exactly the assumptions made in the theoretical part of the
modelling. The coefficients in the equation (2.4) were experimentally found for this
column. The parameters k_i^A and α were estimated on the basis of the experimental
results. Samples were taken from three different heights of the column. The values
of k_i^A and α were determined by a least squares fitting.

Some comparisons between the results given by the model and the data measured
when running the process in different conditions are shown in Figures 1,2 and 3.
Although the fit can be considered as rather good, oval–shaped contour plots for k_i^A
and α show that some uncertainty remains in the values of those parameters.

References

[1] H. Haario, T. Seidman: *Reaction and diffusion at a gas/liquid interface.* Pro-
        ceedings of the minisymposium on numerical methods for semiconduc-
        tors and magnets, ed. by P. Neittaanmäki. Report 42, University of
        Jyväskylä, 1988

[2] D. Henry:*Geometric theory of semilinear parabolic equations.* Lecture Notes in
        Mathematics, 840, Springer–Verlag, New York,1981.

3   Westerterp,K.R., van Swaaij, W.P.M., Beenackers,A.A.C.M: Chemical reactor
        design and operation. Wiley, 1984.

Heikki Haario   Ilkka Turunen
Kemira Espoo Research Center   Kemira Oy, Oulu Research Laboratory
Luoteisr!nne 2   P.O. Box 171
02270 Espoo, Finland   90101 Oulu, Finland

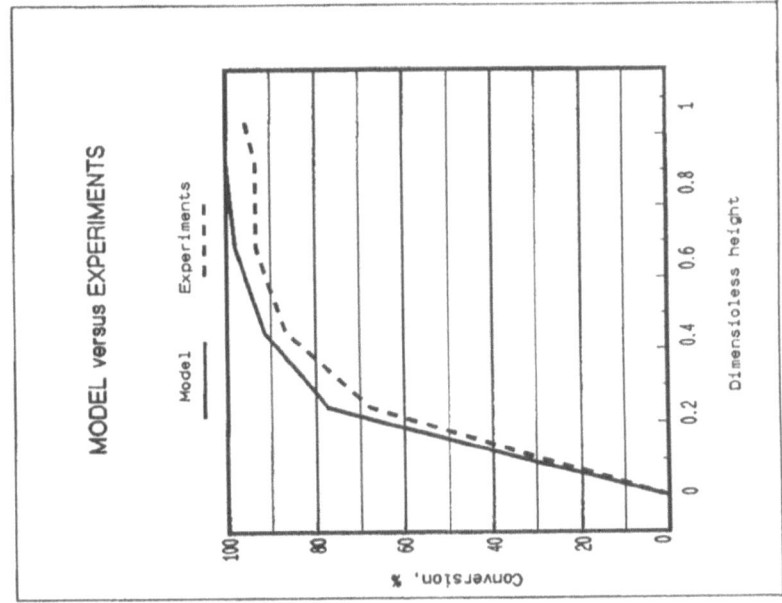

fig. 2

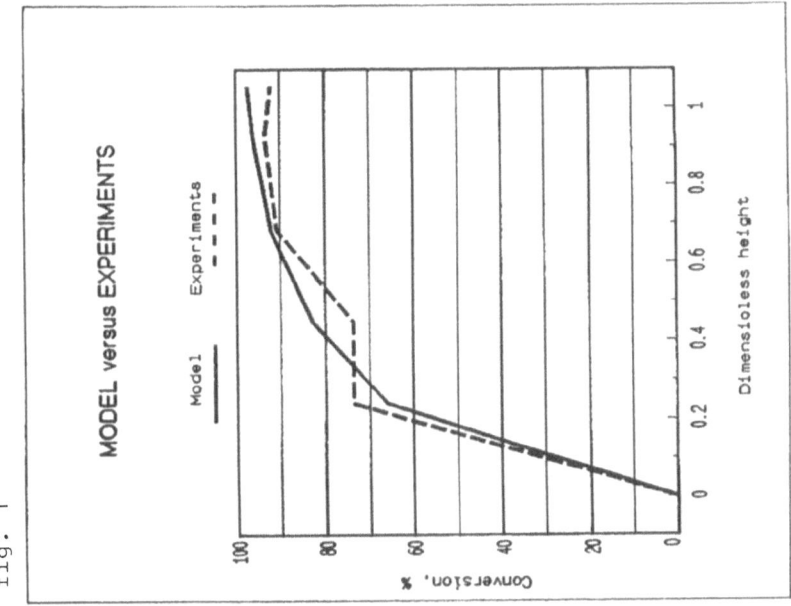

fig. 1

fig. 3

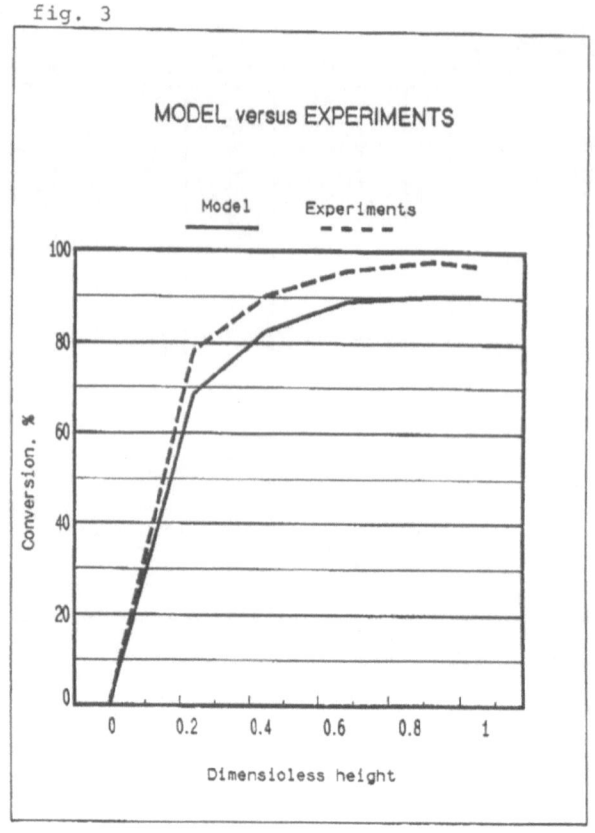

A NUMERICAL MODEL FOR CALCULATION OF A
COUNTER CURRENT HEAT EXCHANGER

F.Kokert*, M.Leitner**, L.Peer**, Hj.Wacker**

1. Problem description

We consider a countercurrent heat exchanger (see fig. 1)

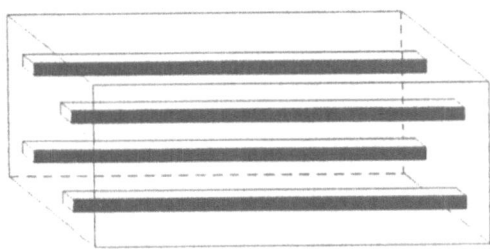

Fig. 1

Inside the heat exchanger there is a number of parallel tubes through which mass streams are flowing. In the general case we have tubes with countercurrent streams. Between the tubes there is no mass transfer but only heat transfer. We do not consider any reactions within a stream though we allow for liquid-volatile phase transitions. The temperature and the consistency of each mass stream when entering the heat exchanger is known. We want to determine the temperature of the streams leaving the heat exchanger.

Our approach here is only based on balancing heat along the length of the heat exchanger. No fluid dynamics is included. For a possible alternative, the methods of cells, see [1]. For another approach see [2].

2. The Mathematical Model

2.1. Parameters

Feedstream Parameters

N	...	number of feed streams	
M_i	...	number of components of stream i	
(f_{ij})	...	molar composition of feed stream i	
	...	$(1 \leq j \leq M_i)$	$[mol/s]$
T_i^f	...	temperature of feed stream i	$[K]$
P_i^f	...	pressure of feed stream i	$[bar]$

* VOEST-ALPINE STAHL, SF/FAT3, 4031 LINZ, AUSTRIA
** University of Linz, Mathematical Department, 4040 LINZ, AUSTRIA

Hj. Wacker and W. Zulehner (eds.),
Proceedings of the Fourth European Conference on Mathematics in Industry, 169–178.
© 1991 B.G. Teubner Stuttgart and Kluwer Academic Publishers.

Parameters of the heat exchanger

L	...	length of the heat exchanger	$[m]$
U_i	...	circumference of tube i	$[m]$
ΔP_i	...	pressure drop of stream i	$[bar]$
K_i	...	heat exchange coefficient of stream i	$[kW/m^2/K]$
d_i	...	direction of stream i	
	...	$+1$ ... from left to right	
	...	-1 ... otherwise	
K_w	...	heat exchange coefficient of the wall	$[kW/m^2/K]$
U_W	...	circumference of the wall	$[m]$
T_A	...	ambient temperature	$[K]$

Model parameter

l	...	actual distance from the origin	$[m]$
$H_i(l)$	...	enthalpy of stream i at distance l	$[kJ/s]$
$T_i(l)$	...	temperature -"-	$[K]$
$P_i(l)$	...	pressure -"-	$[bar]$
$T_w(l)$	...	temperature of the wall at position l	$[K]$
$H(T, P, (f_j))$	...	enthalpy as function of temperature,	
	...	pressure and molar composition	

Assumptions:

I) The pressure of each stream decreases linearly with l resp. (L-l)

II) Because of the high conductivity of the heat exchanger material (e.g. metal) we may assume the wall temperature to be constant over the whole crosscut at each l.

III) We neglect heat transfer by radiation and heat transfer by conduction in the direction where the mass stream moves.

2.2. The Model

For the variance of the enthalpy of stream i in dependence on l we get

$$(1) \qquad \frac{dH_i(l)}{dl} = K_i \cdot U_i \cdot d_i \cdot \left(T_w(l) - T_i(l)\right) \qquad (1 \leq i \leq N)$$

where we have $\quad H_i(l) = H\left(T_i(l), P_i(l), f_{i1} \cdots f_{iM_i}\right)$

To close the system (1) we add the overall energy balance

$$(2) \qquad \sum_{i=1}^{N} \frac{H_i(l)}{dl} + K_w \cdot U_w \cdot \left(T_w(l) - T_A\right) = 0$$

From (1), (2) the following explicite system is derived

$$(3) \qquad \frac{dT_i(l)}{dl} = \frac{K_i \cdot U_i \cdot d_i \left(T_w(l) - T_i(l)\right) - \frac{\partial H\left(T_i(l), P_i(l), (f_{ij})\right)}{\partial P} \cdot \frac{dP_i(l)}{dl}}{\frac{\partial H\left(T_i(l), P_i(l), f_{ij}\right)}{\partial T}}$$

$$(1 \leq i \leq N)$$

with $T_w(l) = \dfrac{\sum_{i=1}^{N} K_i \cdot U_i \cdot T_i(l) + K_w \cdot U_w \cdot T_A}{\sum_{i=1}^{N} K_i \cdot U_i + K_w \cdot U_w}$

Therefore it remains to solve the two point boundary value problem:

$$(*) \begin{cases} \frac{dT(l)}{dl} = f(l, T(l)) & T(l) = \big(T_1(l), \cdots T_N(l)\big) \\ T_i^f = \begin{cases} T_i(o) & d_i > 0 \\ T_i(L) & d_i < 0 \end{cases} & 1 \le i \le N \end{cases}$$

The main difficulty arises from the possibility that there might occur phase transitions in a stream. In these cases the enthalpy function of the temperature (see fig. 2) is no more continuously differentiable (mixtures) or has even jumps (pure substances). Hence the right hand side of $(*)$ is only piecewise continous. The Multiple Shooting Technique therefore may be applied in a suitable way.

fig. 2   Enthalpy

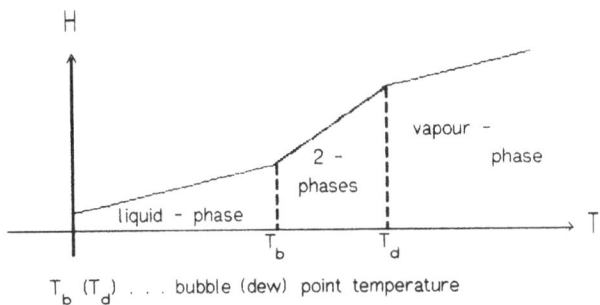

T_b (T_d) ... bubble (dew) point temperature

3. Numerical Solution

3.1. The main strategy
The resulting two point boundary value problem is solved by the classical Multiple Shooting Technique.
We discretize: $0 = l_0 < l_1 < \ldots < l_M = L$ and demanding continuity for the solutions we have to solve the nonlinear equational system

(4)
$$F_j(t_j, t_{j+1}) := T(l_{j+1}, l_j, t_j) - t_{j+1} = 0 \qquad (1 \le j \le M)$$
$$F_M(t_1, t_M) := \begin{cases} t_{i1} - T_i^f = 0 & d_i > 0 \\ t_{iM} - T_i^f = 0 & d_i < o \end{cases} \qquad (1 \le i \le N)$$

t_j stream temperatures at l_j
$T(l_{j+1}, l_j, t_j)$: solution of the initial value problem (5) at $l = l_{j+1}$.

(5)
$$\begin{cases} \frac{du}{dl}(l) = f(l, u(l)) \\ u(l_j) = t_j \end{cases} \qquad l \in [l_j, l_{j+1}]$$

We solve (4) by Newton's Method with damping:

$$(6) \qquad \begin{cases} \frac{\partial F}{\partial t}(t^{(k)})\Delta t^{(k)} &= -F(t^{(k)}) \\ t^{(k+1)} &= t^{(k)} + \delta_k \Delta t^k \end{cases}$$

i.e. we have to perform the following steps:

i) evaluation of F
ii) evaluation of the Jacobian
iii) solution of a linear system
iv) damping

ad i) To integrate the IVP (5), we use trapezoidal rule with quadratic extrapolation. For later purposes we go into detail: at each (inner) discretization point we have to solve the nonlinear system of equations:

$$g(T^+) := T^+ - T - \frac{h}{2}(f + f^+) = 0$$

where
$$l^+ := l + h \qquad \begin{array}{ll} T := T(l, l_j, t_j) & T^+ := T(l^+, l_j, t_j) \\ f := f(l, T) & f^+ := f(l^+, T^+) \end{array}$$

Solving iteratively for T^+ by Newton's Method again we get

$$(7) \qquad T^{(k+1)+} = T^{(k)+} - \left[\frac{\partial g(T^{(k)+})}{\partial T}\right]^{-1} g(T^{(k)+})$$

where

$$\frac{\partial g}{\partial T}(T^{(k)+}) = I_N - \frac{h}{2} * \frac{\partial f^+}{\partial T}$$

To start the process we use the Euler predictor: $T^{(0)+} := T + hf$
Numerical tests show that 2 - 3 iterations suffice without any need for a damping procedure

3.2. Bubble Points and Dew Points - Phase Transition

In (3) the enthalpy has to be differentiated with respect both to temperature and pressure. Here we have to observe that the enthalpy is an only piecewise differentiable function of the temperature.

At the transition points pure liquid-vapour/liquid (bubble point) and pure vapour-vapour/liquid (dew point) we only have directional derivatives w.r.t. temperature. This implies that depending on the actual phase of a stream we might have up to 3 different right hand sides in (3).

To decide which phase we are in we should know both the bubble point and the dew point of our actual position. For given pressure and composition both points are determined by solving a nonlinear system of equations. Because a step size control is used the computation has to be performed in dependence on the actual position.

To reduce the considerable numerical amount both the bubble point and dew point are only determined (in advance) at the fixed outer discretization points l_j. At the integration points we use linear interpolation.

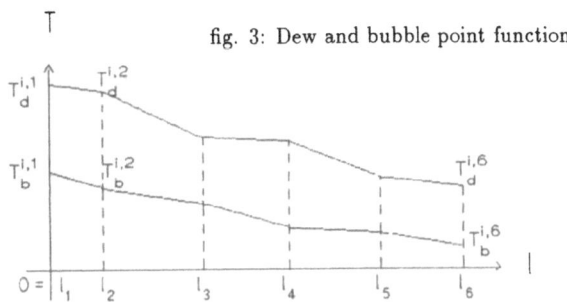

fig. 3: Dew and bubble point functions

To handle phase transitions we proceed as follows. Assume that at point l stream i is in the liquid phase and at the next step l^+ it is in vapour-liquid phase. (see fig. 4)

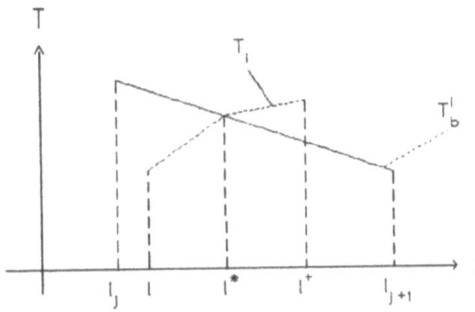

fig. 4 Phase transition

By continuty there is a (unknown) point l^*, where a (first) phase transition takes place. Equation (7) describes how the temperature T^+ at l^+ depends on l^* – for ease of demonstration we confine ourselves to only one point of transition here.

$$(7) \qquad T^+ = T + \int_l^{l^*} f^1(x, T(x, l_j, t_j))dx + \int_{l^*}^{l^+} f^2(x, T(x, l_j, t_j))dx$$

l^* ...   transition point (here: bubble point)
f^1 ...   liquid phase function of f
f^2 ...   liquid vapour phase

To determine l^* for a stream i which changes phase we have to solve the following nonlinear system:

$$(7a) \qquad b(x) := T_i(x, l_j, t_j) - \left(T_{i,b}^j + \frac{x - l_j}{l_{j+1} - l_j} \cdot (T_{i,b}^{j+1} - T_{i,b}^j)\right) = 0$$

Solution is conventiently done by the one-dimensional Newton's method. We have

(7b)
$$\frac{\partial b}{\partial x} = f_i(x, T(x)) - \frac{1}{l_{j+1} - l_j}\left(T_{i,b}^{j+1} - T_{i,b}^j\right)$$

where we used our differential equation and linear interpolation. Numerical experiences show that Newton's iteration may be finished after two iterations at most.

3.3. Determination of the Jacobian

We have the wellknown Multiple Shooting structure for the Jacobian: a block diagonal matrix with the unit matrix $(-I_N)$ as upper (block) diagonal matrices. Due to the special boundary conditions the resulting blocks A and B are (N, N) diagonal blocks with 1 or 0 resp. 0 or 1 in the diagonal. (fig. 5)

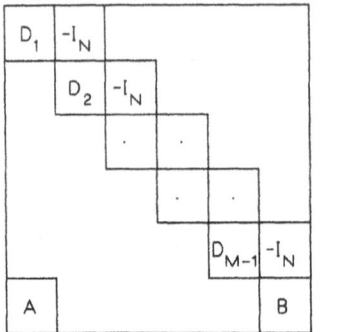
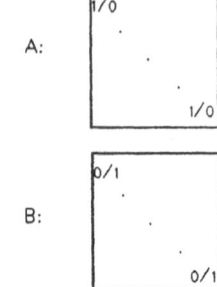

Fig. 5

It remains to evaluate the diagonal blocks $\quad D_j := \frac{\partial F_j}{\partial t_j} = \frac{\partial T}{\partial t_j}$

where T solves (5). To determine $\frac{\partial T}{\partial t_j}$ we differentiate both sides of (5) and we get the following linear IVP:

(8)
$$\frac{\partial \tilde{T}(l)}{dl} = \frac{\partial f(l, T(l))}{\partial T} \cdot \tilde{T}(l) \quad \text{with} \quad \tilde{T}(l_j) := I_N$$

where we use the abbreviation $\tilde{T} := \frac{\partial T}{\partial t_j}$

In the case of a phase transition we derive $\tilde{T}(l)$ from the following relations.

$$T(l, t_j)|_{l=l^*} = t_j + \int_{l_j}^{l^*(t_j)} f^1(x, T(x, l_j, t_j))dx \tag{9}$$

All terms in (9) are differentiable with respect to t_j. We get

$$\tilde{T}(l)|_{l=l^*} = I_N + \int_{l_j}^{l} \frac{\partial f^1}{\partial T}(\cdots)dx\Big|_{l=l^*} + \underbrace{f^1\left(l^*(t_j), T(l^*,l_j,t_j)\right)\frac{dl^*(t_j)}{dtj}}_{(*)} \qquad (10)$$

Differentiating (10) with respect to l we get the IVP (8). Hence the solution of (8) at l^* must be corrected by means of $(*)$ to get

$$\tilde{T}(l^*); \quad i.e.: \tilde{T}(l^*) \rightarrow \tilde{T}(l^*) + f^1(\ldots) * \frac{dl^*(t_j)}{dt_j} \qquad (11)$$

$$\uparrow$$
$$\text{from (8)}$$

Now for $l > l^*$ we have

$$\tilde{T}(l) = \tilde{T}(l^*) + \int_{l^*(t_j)}^{l} \frac{\partial f^2(x, T(x,l_j,t_j))}{\partial T} * \tilde{T}(x)dx - f^2(l^*, T(l^*,l_j,t_j)) * \frac{dl^*(t_j)}{dt_j} \qquad (12)$$

From (12) we get the IVP (13) (differentiating both sides of (12) with respect to l):

$$\left.\begin{array}{l} \dfrac{d\tilde{T}(l)}{dl} = \dfrac{\partial f^2(l, T(l,l_j,t_j))}{dT} * \tilde{T}(l) \\[2mm] \tilde{T}(l^*) = \tilde{T}(l^*) - f^2(\cdots)\dfrac{dl^*(t_j)}{dt_j} \end{array}\right\} \qquad (13)$$

$$\uparrow$$
$$\text{From (11)}$$

There remains the problem of calculating $\frac{dl^*(t_j)}{dt_j}$

To see the dependence on t_j we rewrite (7a)

$$b\left(l^*(t_j)\right) := T_i(l^*(t_j),l_j,t_j) - \left(T_{i,b}^j + \frac{l^*(t_j)-l_j}{l_{j+1}-l_j} * (T_{i,b}^{j+1} - T_{i,b}^j)\right) = 0 \qquad (14)$$

Again by differentiating both sides of (14) with respect to t_j we get:

$$f_i^1\left(l^*(t_j),l_j,t_j\right) \cdot \frac{dl^*(t_j)}{dt_j} + \tilde{T}\left(l^*(t_j)\right) - \frac{1}{l_{j+1}-l_j}(T_{i,b}^{j+1} - T_{i,b}^j) * \frac{dl^*(t_j)}{dt_j} = 0 \qquad (15)$$

$$\uparrow$$
$$\text{From (8)}$$

By means of the linear equation (15) $\frac{dl^*(t_j)}{dt_j}$ may easily by calculated

To solve (8) we again use the trapezoidal rule. With an analagous notation as in (7) we get:

$$(9) \qquad (I - \frac{h}{2}\frac{\partial f^+}{\partial T})\tilde{T}^+ = [I + \frac{h}{2}\frac{\partial f}{\partial t}]\tilde{T}$$

In (9) we have the same system matrix as in (7) which is already calculated and factorized. Hence the numerical effort to determine $\tilde{T}(l)$ is reduced to the solution of an already decomposed linear system

3.4. Solution of the linearized System: $J_F \Delta t = F$

Using classical Multiple Shooting Technique we get by recursion a system with the following matrix:

$$D := A + BD_{M-1} \cdots D_1$$

In our field of application the condition numbers of D might be quite large, e.g. up to 10^{30} and only useless results are produced. Therefore instead of any recursion we used direct factorization of the large system exploiting the sparse structure of J_F. Though the numerical amount increases by 500% compared to the classical recursive technique all examples so far work well.

3.5. Gradients of the Product Streams

Heat exchangers are often found as units of larger chemical plants. The simulation of a flowsheet oriented sequential modular algorithm requires both product stream calculation of the involved units and the determination of their gradients with respect to some parameters connected to the unit and the feedstreams. Computation is done on the same lines as for the determination of $\tilde{T}$ in (8).

Using $\overset{\approx}{T}(l) := \frac{\partial T}{\partial a}$ where a is the parameter vector (e.g. length l, k-values, composition of the input streams etc.) we get

$$(16) \qquad \frac{d\overset{\approx}{T}}{dl}(l) = \frac{\partial f(l, T(l, l_j, t_j, a), a)}{\partial T}.\overset{\approx}{T}(l) + \frac{\partial f(l, T(), a)}{\partial a}$$

$$\overset{\approx}{T}(l_j) \quad = 0$$

Again we make use of the already factorized matrix. To determine the gradients of the product streams we differentiate (4). Using the abbreviation $\overset{\approx}{t}_j := \frac{\partial t_j}{\partial a}$ we have to solve the linear system

$$J_F.\overset{\approx}{t} = \begin{bmatrix} -\overset{\approx}{T}(l_2) \\ \vdots \\ -\overset{\approx}{T}(l_M) \\ \frac{\partial T'}{\partial P_a} \end{bmatrix}$$

Again the system matrix is already factorized (compare (4))
The gradients of the temperatures of the product streams are given by:

$$\frac{\partial T_i}{\partial a} = \begin{cases} \overset{\approx}{t}_{iM} & d_i > 0 \\ \overset{\approx}{t}_{i1} & d_i < 0 \end{cases}$$

For further details on the numerical solution procedure see [3], [4].

4. Numerical Results

For the following numerical examples a heat exchanger with phase transition is chosen.

Input data (compare fig. 6)

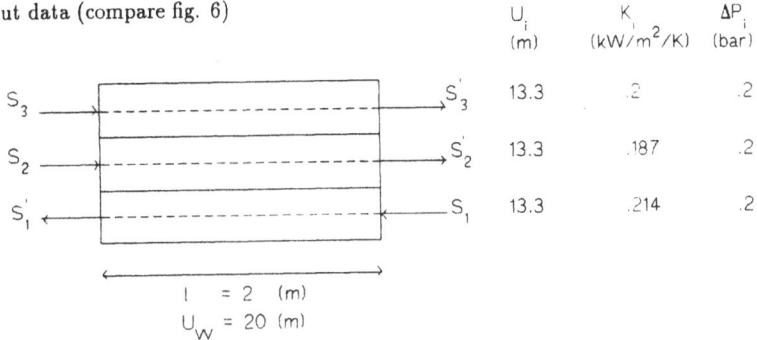

	U_i (m)	K_i (kW/m²/K)	ΔP_i (bar)
S_3'	13.3	.2	.2
S_2'	13.3	.187	.2
S_1'	13.3	.214	.2

$l = 2$ (m)
$U_w = 20$ (m)

Fig. 6

Feedstream data:

	S_1	S_2	S_3
Temp. ($^\circ K$)	190	95	105
Pressure (bar)	6	6	6
Argon (mol/s)	10	10	10
Nitrogen (mol/s)	70	70	70
Oxygen (mol/s)	20	20	20
phase condition	pure vapour	pure liqid	pure vapour

Results:
Product stream data: (composition unchanged)

	S_1'	S_2'	S_3'
Temperature	135.88	99.34	127.52
phase condition	vapour	vapour liquid *)	vapour

*) phase transition in tube 2 at $l^* = 0.474m$

Number of multiple shooting iterations: 10
(discretezation points: 7)
CPU-time demand (without gradients): 4'17"
Calculation of gradients: 53" ($\sim$ 20 %) (COMPAREX 7/78)

References:

[1] K. Stephens: Modellierung von Rohrbündelwärmetauschern mit Hilfe der Zellenmethode, Inst. Bericht Univ. Stuttgart, Inst. f. Techn. Thermaldynamik u. Therm. Verfahrenstechnik. Stuttgart, Juni 1989, (1 - 26)

[2] A. Pignotti: Matrix Formalism for Complex Heat Exchangers, Journal of Heat Transfer, VOL 106, May 1984 (352 - 360)

[3] M. Leitner: Numerical Simulation of the Steady State of Counter Current Heat Exchangers, Dipl.Thesis, Univ. Linz, Math.Dept. (1988), pp 1 - 42 (in German)

[4] L. Peer: Simulation and Optimization of Chemical Plants, Doct. Thesis, Univ. Linz, Math. Dept. (1989) (in German)

Contributed Lectures

MODELLING OXYGEN SUPPLY IN SPHEROIDS OF TUMOR CELLS

H. Acker, J. Dehnhardt, T. Küpper, C. Rohner

1. Introduction

We discuss a mathematical model to describe the physiology of
tumor cells. The mathematical results are compared with experimental measurements carried out at the Max-Planck-Institut für
Systemphysiologie. It is the main objective to find out if the
oxygen consumption is independent of the oxygen concentration
or if it is reduced in the case of a low oxygen supply. This is
important to know as a change of the physiology goes along with
a different response to radiation.

The study is carried out for spheroids (ball-shaped conglom-
erations) of tumor cells as they permit precise experimental
control of essential parameters governing the physiology.
Furthermore their simple geometric structure is very well
adapted to mathematical modelling.

2. Experimental set-up

The spheroids are ball-shaped configurations with a diameter
in the range of 400 μm to 800 μm. During the experiment they are
fixed on a glass plate immersed in a well-stirred liquid medium.
Using O_2-sensitive and pH-sensitive electrodes, both O_2-pressure
(PO_2) and pH-Values have been measured along different channels
within the spheroid. Experimental studies show (Acker and
Carlson [1], Grunewald [5]):

     (i)   the O_2-consumption by the microelectrode during
             measurements can be neglected

     (ii)   temporal changes during measurements can be neg-
             lected

     (iii)   profiles measured on different channels leading
             approximately to the centre coincide; it should
             be noted that is very difficult in practice to hit
             the center perfectly. In real measurements a va-
             nishing gradient of the concentration (as it
             should happen when passing through the center of
             a radialsymmetric situation) is hardly met.

Hj. Wacker and W. Zulehner (eds.),
Proceedings of the Fourth European Conference on Mathematics in Industry, 181–186.
© 1991 B.G. Teubner Stuttgart and Kluwer Academic Publishers.

The profiles depend on the amount of oxygen outside the spheroid.
A strong reduction leads to areas with vanishing O_2-pressure
(dead cores), eventually causing necrotic zones.

Although the spheroids can be assumed to be ball-shaped, the
concentrations need not be radially symmetric as there is less
diffusion near the bottom due to the experimental set-up. Rather
than dealing with a much more complicated
unsymmetric problem requiring numerical   fig.1
computations, we try to overcome these
difficulties by considering an artificially
enlarged spheroid for the modelling,
leading to a radially symmetric problem.
This is justified by improved approximations of PO_2-profiles
(Wiesecke [8]).

3. The mathematical model

We use a mathematical model which relates the measured data
(PO_2 and pH profiles) to the O_2-consumption rate and the H^+-
production. This relation will show that there is in fact a
decrease of the oxygen consumption for small concentrations.

The O_2-concentration u is obtained from the PO_2-values via
Henry's law as $u = \alpha PO_2$, where α denotes the solubility coefficient of oxygen. The $[H^+]$-concentration v is given by $v = 10^{-pH}$.
Following Britton [2] the physiology is described by a coupled
reaction-diffusion system

(1)
$$u_t - D_u \Delta u + F_1(u,v) = 0$$
$$v_t - D_v \Delta v - F_2(u,v) = 0$$
$$(x \ \varepsilon \ \Omega)$$

where $\Omega \subset \mathbb{R}^3$ denotes the spheroid. Here D_u and D_v are the
diffusion coefficients, the functions F_1 and F_2 describe consumption of O_2 and production of H^+ resp. For simplicity they
are taken to be independent of the spatial variable explicitly.
In the steady state case ($u_t=v_t=0$) which will be considered here,
division by D_u and D_v leads to the system:

$$- \Delta u + \frac{1}{D_u} \cdot F_1(u,v) = 0$$

(2)

$$- \Delta v - \frac{1}{D_v} \cdot F_2(u,v) = 0$$

It is assumed that the functions F_1 and F_2 satisfy the following rules:

(i)   The oxygen consumption is nondecreasing as a function of the oxygen concentration, i.e. $\partial F_1/\partial u \geq 0$, and it is constant in the area of saturation.

(ii)   The $[H^+]$-concentration may reduce the O_2-consumption, i.e. $\partial F_1/\partial v \leq 0$, but not completely.

(iii)   Reduction of O_2-concentration implies an increase of $[H^+]$-production, i.e. $\partial F_2/\partial u \geq 0$.

(iv) $[H^+]$-production is independent of the $[H^+]$-concentration, i.e. $\partial F_2/\partial v \equiv 0$.

The assumptions are matched by a simple ansatz

$$F_1(u,v)D_u = a(u)f(v), \quad F_2(u,v)/D_v = +b(u)g(v)$$

with $a' \geq 0$ and $b' \leq 0$. The functions a,b,f,g are shown in fig. 2, they are piecewise linear and can be completely characterized by the parameters $u_o, u_1, v_o, Q_o, Q_1, C_1, C_2$ and P_o. The values of u and v in the medium surrounding the spheroid are denoted by M_u and M_v respectively.

fig. 2

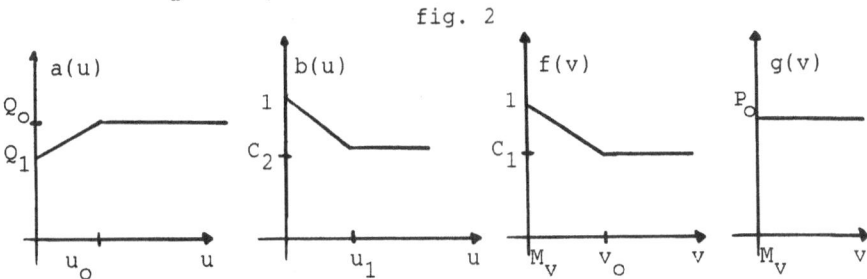

Mathematical results concerning existence and regularity of the solutions of system (2) are given by Diaz and Hernandez [4] if a and b are monotonically nondecreasing (maximal monotone graphs). Using similar techniques these have been extended by Wiesecke [8] to include the situation under consideration, in particular to a non-increasing function b(u).

The restriction to radial symmetry leads to the system

$$-u''(r) - \frac{2}{r} u'(r) + a(u(r)) \cdot f(v(r)) = 0,$$

(3) $0 < r < R$

$$-v''(r) - \frac{2}{r} v'(r) - b(u(r)) \cdot g(v(r)) = 0,$$

with the boundary conditions

$$u'(0) = 0, \quad u(R) = M_u, \quad v'(0) = 0, \quad v(R) = M_v.$$

Using the technique of lower and upper solutions one obtains the following range-domain implications which can be used to check hypotheses concerning the functions F_1, F_2.

If $Q_1 C_1 \leq a(u) \cdot f(v) \leq Q_o$ and $P_o \geq b(u) \cdot g(v) \geq C_2 P_o$ and $u(0) > 0$ then the nonlinear terms of (3) can be estimated by constants and lower resp. upper solutions are given explicitly:

(4)
$$\underline{u}(r) := \frac{1}{6} Q_o \cdot (r^2 - R^2) + M_u \leq u(r) \leq \frac{1}{6} Q_1 C_2 \cdot (r^2 - R^2) + M_u =: \bar{u}(r).$$

$$\underline{v}(r) := \frac{1}{6} C_2 P_o \cdot (R^2 - r^2) + M_v \leq v(r) \leq \frac{1}{6} P_o \cdot (R^2 - r^2) + M_v =: \bar{v}(r).$$

Hence, if the measured data do not match these requirements, this is a clear indication that F_1 and F_2 do not obey the assumptions.

Via the same method lower solutions can also be constructed in the case that dead cores occur.

4. The scalar case

We can make use of the fact that for constant $f(v)$, $v \geq v_o$, the system can be decoupled. For u one obtains the equation

(5) $$- u'' - \frac{2}{r} u' + C \cdot a(u) = 0$$

with the boundary conditions $u'(0) = 0$, $u(R) = M_u$. The solution of this scalar equation can then be substituted into the equation for v.

For the radially symmetric problem it is easy to see that dead cores $N(u) := \{r: u(r) \equiv 0\}$ can only occur if the growth of $a(u)$ is superlinear in 0 (see [3]); that is $\Psi(o) < \infty$ where

$$\Psi(\tau) := \frac{3}{2} \int_o^\tau j(s)^{-1/3} ds \text{ with } j(s) := \int_o^s a(u) du.$$

Besides a strong decrease of the consumption for low concentrations a balance between the size of the spheroid and the overall consumption is required for the occurence of dead cores; i.e. $N(u) = \emptyset$ for $R < \Psi(M_u)$ and $N(u) \neq \emptyset$ for $R \geq \sqrt{3} \, \Psi(M_u)$.

If a is a monotonically increasing and concave function
then the sets $\Omega_c = \{ x \in \Omega : u(x) \leq c\}$ are convex for all $c \geq 0$
(Kawohl [6]). Further using lower and upper solutions construc-
ted for corresponding balls $B_R \leq \Omega \leq B_{\bar{R}}$ it can be shown that
small deviations from a ball-like configuration have only little
influence on the solution (Wiesecke [8]). These estimates
support the approach to consider a radialsymmetric problem which
will be most helpful for the purpose of identification.

5. Identification in the scalar case

For any concentration in the medium the concentration u(r)
within the tumor is available through measurements. The parameters that characterize the functions a,b,f,g in fig. 2 can be
adapted through a least squares fit. Numerical computations
based on different series of measurements have been carried out
by C. Rohner [7]. The parameter Q_o captures the oxygen consumption within the area of saturation. If the concentration M_u is
sufficiently large, then the oxygen supply within all of the
spheroid can be assumed to be above saturation.
For that reason Q_o can be determined by a separate identification procedure. The parameters Q_1 and u_o describe the possible
change of the physiological state. Their determination requires
low concentrations M_u. It turns out that they are not uniquely
determined by the least squares fit, but that the solution u is
in good agreement with the measured data for suitable combinations of Q_1 and u_o as described in fig. 3.

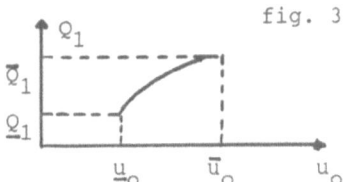

fig. 3

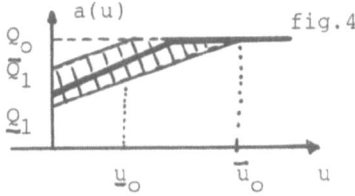

fig. 4

From the physiological point of view this implies that the
present model does not allow to discriminate between an early
and slow decay or a late and rapid one. A reasonable function
a may be taken within the shadowed region of fig. 4. But at
least one important fact can clearly be established by this
model: the oxygen consumption decreases for small concentrations.

6. References

[1] Acker, A.; Carlson, J.: Relations between pH, oxygen
    partial pressure and growth in cultured cell spheroids.
    Int. J. Cancer 31 (1988), 715-720

[2] Britton,N.F.: Reaction-diffusion equations and their
    applications to biology. London: Academic Press 1986

[3] Diaz, J.I.: Nonlinear Partial Differential Equations
    and Free Boundaries, Vol.I London: Pitman 1985 =
    Research Notes in Math., Bd. 106

[4] Diaz, J.I.; Hernandez, J.: On the existence of a free
    boundary for a class of reaction-diffusion systems,
    SIAM J. Math. Anal. 15 (1984), 670-685

[5] Grunewald, W.: Diffusionsfehler und Eigenverbrauch der
    Pt-Elektrode bei PO_2-Messungen im steady state.
    Pflügers Arch. 320 (1970), 24-44

[6] Kawohl, B.: When are solutions to nonlinear boundary
    value problems convex? Comm. Partial Diff. Eq. 10
    (1985), 1213-1225

[7] Rohner, C.: Parameteridentifikation bei nichtlinearen
    Reaktions-Diffusions-Problemen. Diplomarbeit, 1989

[8] Wiesecke, J.: Eine freie Randwertaufgabe zur Modellierung
    der Sauerstoffversorgung in Geweben.
    Diplomarbeit Dortmund, 1987

Prof.Dr.Tassilo Küpper   Prof.Dr. H. Acker
Jürgen Dehnhardt
Claudia Rohner

Inst.f.Angewandte Mathematik   Max-Planck-Institut
Universität Hannover   für Systemphysiologie
Welfengarten 1   Rheinlanddamm 201
D-3000 Hannover 1   D-4600 Dortmund 1

ON A NUMERICAL METHOD FOR VARIABLE HEAT FLOW THROUGH EXTERNAL WALLS

K. Audenaert and R. Van Keer

1. Introduction

1.1. An important topic in air conditioning in buildings concerns the heat flow through an external wall under the daily variation of the outside air temperature, combined with an indoor air temperature kept at a fixed value. As a wall structure, used in practice, consider a single cavity construction, shown below.

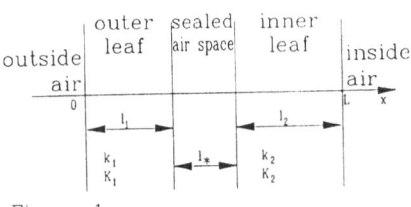

Figure 1

For this structure Pratt considers a model problem, with practical relevance, see [2] section 6.2, concerning the transient heat flow through the wall, caused by a single step-function change in the outdoor air temperature, initially the system being in a steady state

The wall is assumed to conduct heat in one dimension only, orthogonal to the parallel wall surfaces. Moreover in this (popular) model the enclosed airspace is characterised by a thermal conductance h and zero capacity, giving rise to the usual transition B.C.'s of "non-perfect thermal contact" between the two leaves. Finally in [2] the conductivities and capacities of the two leaves are constant functions.

1.2. In this paper we consider a more realistic wall structure with non homogeneous outer and inner leaves, allowing for space dependent conductivities and capacities. Likewise the case of two (or more) sealed airspaces is included, the capacity of the air no longer being idealised to be zero. Neither the eigenfunctions method of [1], chapter 9, nor the $\mathcal{L}$–transformation method, followed in [2], can be used.

We apply a finite element–finite difference method to calculate the inside surface temperature $\theta_{is}(t)$, to obtain the time constant τ of the wall, measuring its response to a sudden drop in the outdoor air temperature. On numerical and physical grounds a simple formula is found to be effective for analyzing the influence of different wall structures on the value of τ.

For the present approach it is crucial to leave from an other formulation of the heat conduction problem, in which, contrary to [2], there is a single unknown in the whole wall. This, in turn, may be rewritten in a weak variational form in a standard way. The physically important situation of a sinus excitation of the outdoor air temperature is incorporated in the general problem considered. The corresponding time lag and attenuation may be obtained.

Hj. Wacker and W. Zulehner (eds.),
Proceedings of the Fourth European Conference on Mathematics in Industry, 187–191.

2. An alternative formulation

The air space (or an other isolation sheet) in Fig.1 is treated as a third layer, being in perfect contact with the outer and inner leaf of the wall, cfr [1], p. 390. We consider the general problem of determining a smooth function $u(x,t)$, $0 \le x \le l_1 + l_2 + l^* = L$, $t \ge 0$, satisfying

$$(2.1) \quad \frac{\partial}{\partial x}(k \cdot \frac{\partial u}{\partial x}) = w.\frac{\partial u}{\partial t}, \quad \text{a.e. in } (0,L), t > 0$$

$$(2.2) \quad k \cdot \frac{\partial u}{\partial x} = h_0 \cdot [u(x,t) - \alpha(t)], \quad x = 0, t > 0$$

$$(2.3) \quad k \cdot \frac{\partial u}{\partial x} = h_i \cdot [\beta(t) - u(x,t)], \quad x = L, t > 0$$

$$(2.4) \quad u(x,0) = u_0(x) \quad \text{a.e. in} (0,L)$$

$$(2.5) \quad \text{Continuity of } u \text{ and the flux } k.\frac{\partial u}{\partial x} \text{ at } x = l_1 \text{ and } x = l_1 + l^*, t > 0$$

(P)

Here $k(x)$ and $w(x)$ are sufficiently regular functions of x, measuring the conductivity and capacity respectively. h_0 and h_i are the (constant) transmission coefficients at the outer and inner surface of the wall, the temperature of the outdoor and indoor air temperature being $\alpha(t)$ and $\beta(t)$ respectively. After a simple substitution the model problem in [2], p.164, may be seen to correspond to the case $u_0 = 0, \beta = 0, \alpha = -\theta_0$ (constant), where the outdoor air temperature suddenly drops from θ_0 to zero at $t = 0$.

3. Variational form. Fully discrete approximation

3.1. Denoting by $V = H_1(0,L) \hookrightarrow C^0([0,L])$ the first order Sobolev space on $(0,L)$, (P) is readily found to be formally equivalent, see [4] section 5.2, to the problem of determining u such that

$$u(x,t) \in H_1(0,L), \frac{\partial u}{\partial t} \in L_2(0,L), \text{ a.e. } t > 0; u(x,0) \in L_2(0,L)$$

$$(w.\frac{\partial u}{\partial t}, v) + B(u,v) = f_t(v) \equiv h_0.\alpha(t).v(0) + h_i.\beta(t).v(L), \quad \forall v \in V$$

$$u(x,0) = u_0(x) \quad, \text{ a.e in } (0,L)$$

(P_v)

where $(.,.)$ is the $L_2(0,L)$-inner product and

$$B(v_1, v_2) = (k \cdot \frac{dv_1}{dx}, \frac{dv_2}{dx}) + h_0 \cdot v_1(0) \cdot v_2(0) + h_i \cdot v_1(L) \cdot v_2(L), \quad \forall v_1, v_2 \in V$$

(P_v) makes sense when $w(x), k(x) \in L_\infty(0,L)$. Moreover for the physical problem w and k may assumed to be bounded below by a strictly positive constant, while $\alpha, \beta \in H_1(0,T)$, where $T > 0$ is arbitrary.

(P_v) may be shown to fit into the rigorous framework of well posed abstract parabolic problems, the solution being sufficiently regular under suitable conditions on the data, met in practice. See e.g. [5], chapter IV.

3.2. Introduce a Lagrange finite element subspace V_h of V and an approximation $u_{0h} \in V_h$ of u_0. Next let Δt be a time step and $t_n = n.\Delta t, n \in \mathbf{N}$, be time points. Finally $\mu \in [0,1]$ is a parameter. From (P_v) a μ–family of finite difference–finite element approximations $u_h^n \in V_h$ of $u(x, t_n)$ may be obtained in a standard way, see e.g. [4], section 5.3.

4. Numerical results

4.1. To compare the present approach to the one of [2], we consider a single cavity brick wall of 280 mm with identical inner and outer leaf, the thermal characteristics being indicated in Fig. 2, and the enclosed air space being taken into account with $w_* = 0, k_* = h.l_*$. See [1] p.128 for the underlying equivalence. We use a linear Lagrange finite element space with 10 equal elements in each leaf and one element in the air space. Moreover a backward finite difference scheme (i.e. $\mu = 1$) is used with $\Delta t = 60$ sec.

Fig. 2 shows the response of the inside surface temperature (i.e. at $x = L$) relative to the causal drop through θ_0 in the outdoor air temperature. The time constant τ, i.e. the time needed for $\dfrac{\theta_{is}}{\theta_0}$ to reach 63 percent of its limit value, may easily be read off.

Figure 2 $h_i = 8.11 W/m^2 \mathrm{K}$ $l_1 = l_2 = 112mm$ =Pratt's method

$h_0 = 18.9 W/m^2 \mathrm{K}$ $k = 0,865 W/m\mathrm{K}$ = method above

$h = 5.68 W/m^2 \mathrm{K}$ $K = 0,542 \times 10^{-6} m^2/s = k/w$

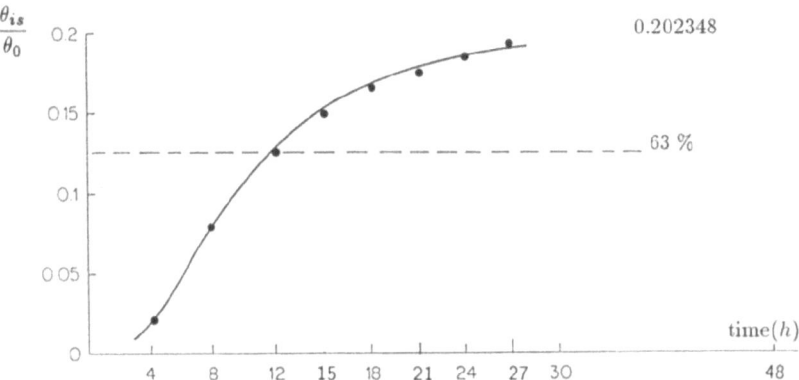

4.2. As an extension of Pratt's model consider a "lumped" model of a wall with N homogeneous layers with resistivity $r_m = 1/k_m$ and capacity $w_m, 1 \leqslant m \leqslant N$

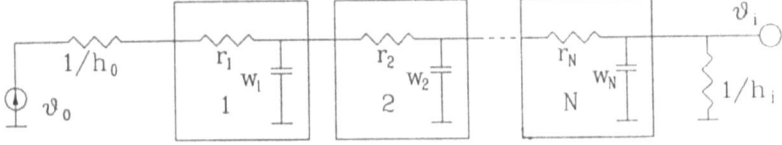

From the numerical results τ is found to be almost linearly depending on the capacities,

(4.1)
$$\tau \simeq \sum_{m=1}^{N} w_m . r_m^*$$

From identification arguments in the thermal network analysis the coefficients r_m^* turn out to be

$$r_m^* = \left([r_m^{(l)}]^{-1} + [r_m^{(r)}]^{-1} \right)^{-1}$$

where $r_m^{(l)}$ is the total resistivity up to the mth lump and $r_m^{(r)}$ is the total resistivity to the right of it,

$$r_m^{(l)} = h_0^{-1} + \sum_{j=1}^{m} r_j , \quad r_m^{(r)} = \sum_{j=m+1}^{N} r_j + h_i^{-1}$$

The analogon of (4.1) for a wall with $r(x)$ and $w(x)$ varying arbitrarily with x is obvious.

4.3. Extensive tests reveal that the estimated value τ_e of τ, obtained by these formulae, are in good agreement with the numerical approximation τ_n, found by the finite difference–finite element method. The relative deviation is of the order 0.01, not reaching 0.05. Results are given below for a few walls with h_0 and h_i in Fig.2 (case I below).

	$\theta_{is}(+\infty)/\theta_0$	τ_n (s)	τ_e (s)	$(\tau_n - \tau_e)/\tau_n$
I	0.202348	41280.0	39575.7	0.041286
II	0.404262	12540.0	12335.7	0.016295
III	0.436437	12380.0	12263.0	0.009456

Wall II is a one layer wall with total resistivity $(7.72)^{-1}$ and total capacity 178 000. Wall III is a three layer wall of which the outer, the middle and the inner leave have respective total resistivities 23^{-1}, 50^{-1} and 23^{-1}, while the corresponding total capacities are 70 000, 50 000 and 70 000.

5. Concluding remarks

5.1. The approach above is appropriate to deal with the model problem of [2] and, equally simple, with considerably more realistic problems. In addition note that the functions α and β in (2.2)–(2.3) only affect the first and last entry of the load matrix in the linear finite element systems (corresponding to the two relevant nodes in the mesh).

The choice $\alpha(t) = \alpha_0 \cdot \sin(\omega t), \beta(t) = \theta_i, u_0 = 0$ in (2.2)-(2.4) correspond to a second model problem of [2]. We readily obtain the physically important time-lag (delay) and attenuation of the simple harmonic variation in the outside air temperature in its passage through the wall to the inside surface. See Fig. 3 for the time-lag and Fig. 4 for the attenuation versus the period, for the cavity brick wall of §4.1.

5.2. The estimate τ_e, given by (4.1) or its continuous counterpart, is computationally simple, but yet is sufficiently accurate for practical purposes. Moreover the transparancy of the formula permits to "think" with it, i.e. to readily obtain qualitative information

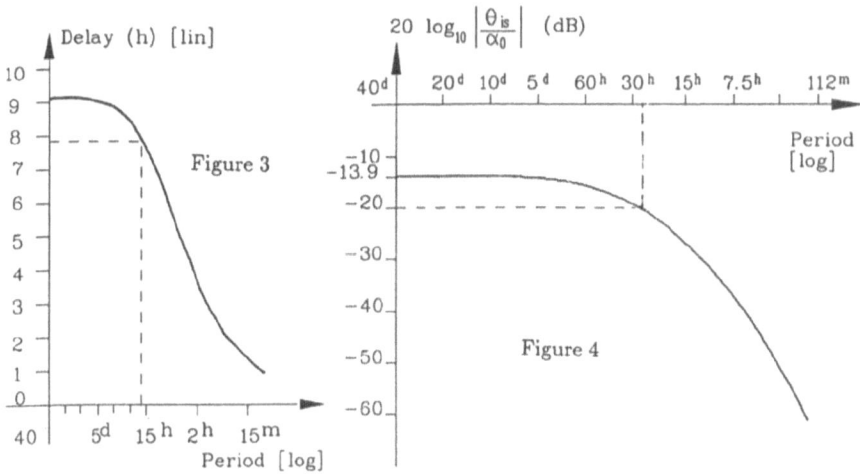

from it. For instance, assume that two materials are available with equal resistivities but different capacities. A thin layer of the material with the highest capacity will give maximal τ_e if it is sandwiched between two layers of the other material. Indeed $r^*(x)$ may be found to reach its maximum value at the middle of the wall.

5.3. An additional feature of the present approach is that the stability properties (relative to μ, h and Δt) of the fully discrete method of §4, as well as its error estimates (with respect to $h, \Delta t, \mu$ and the degree of the finite element mesh) may be well established, see e.g.[3] for some results in this respect.

Acknowledgments

We thank the referee for his critical comments and valuable suggestions.

References

[1] Mikhailov,M.D. ;Özişik, M.N.:Unified Analysis and Solutions of Heat and Mass Diffusion. N.Y. : J. Wiley 1984

[2] Pratt, A.W. : Heat Transmission in Buildings. Chichester : J. Wiley 1981

[3] Van Keer, R. : On a semi-discrete and a fully discrete Galerkin method for a class of parabolic problems. In Sendov, B. (ed): Numerical Mathematics and Applications. Bulgarian Academy Sciences (1989) 535-539

[4] Wait, R. ; Mitchell, A.R. : Finite Element Analysis and Applications. Chichester : J. Wiley 1985

[5] Wloka, J. : Partial differential equations. Cambridge : Cambridge U.P. 1985

K. Audenaert and R. Van Keer, Seminar Mathematical Analysis, Faculty of Applied Sciences, R.U.G., St. Pietersnieuwstraat 39, B-9000 Ghent, Belgium

Optimization of a Solar Domestic Hot-Water System

D. Auzinger, M. Bergmayr, W. Kapfer, B. Schürz, Hj. Wacker
Institut für Mathematik, Johannes Kepler-Universität Linz
Altenbergerstraße 69, A-4040 Linz, Austria

1. Problem Description

In Austria solar heating is mainly used to provide for hot water. In order to achieve an economical attractivity of solar hot-water systems, the total costs of providing one family with hot water over a time of 20 to 30 years should be minimized. On the one hand, a solar system which can provide for the hot-water all over the year (even in winter) is very expensive, on the other hand, a small and thus cheaper system requires an auxiliary heating facility, which causes high operating expanses.

Fig. 1 shows a scheme of a solar hot-water system with an auxiliary heating facility. The black surface of the collector absorbes the direct radiation from the sun, but also the diffuse radiation from all over the sky, gets hot and thus heats the fluid within the collector. As soon as the collector fluid reaches a certain temperature, it is pumped through an heat exchanger in order to heat the water within the boiler. The hot water required is taken from the top of the boiler (where its temperature is highest). As soon as the temperature of the water at the upper part of the boiler decreases under a certain value, the auxiliary heater is started to provide for the hot water demand.

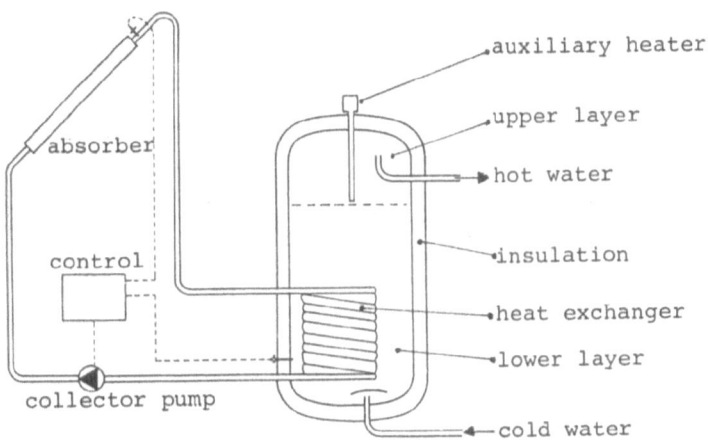

Fig. 1: Scheme of a Solar Domestic Hot-Water System with Auxiliary Heater

Mathematically we get a box constrained, nonlinear optimization problem. There are several parameters to be determined in order to minimize the total costs which are the sum of the prime costs and the operating costs of solar system and auxiliary heater required to provide for a given daily hot-water demand. The most important ones are collector area (A_c), orientation of the collector (angles β and γ) and dimensions of the hot-water tank $(r, h_u$ and $h_l)$.

193

2. The Mathematical Model

We describe the behavior of the system by a system of ordinary differential equations (ODE's). At least three time-dependent temperatures and an ODE for each of them must be taken into account: the temperature of the absorber plate, denoted by $T_A(t)$, and the temperatures of two different layers of hot water within the boiler, denoted by $T_u(t)$ (upper boiler part) and $T_l(t)$ (lower part).

$$\frac{d}{dt}T_A(t) = \frac{1}{m_A c_A}\left[A_c \cdot (\tau_a(t,\beta,\gamma)H_{tot}(t,\beta,\gamma) - V(t,T_A)) - Q_{sol}(t,T_A,T_l,A_c)\right] \quad (E1)$$

$$\frac{d}{dt}T_u(t) = \frac{1}{2\pi r^2 h_u c_w}\left[Q_{aux}(t,T_u) - Q_{u,loss}(t,T_u,r,h_u) - 2\pi r^2 \lambda_w(T_u - T_l)\right] \quad (E2)$$

$$\frac{d}{dt}T_l(t) = \frac{1}{2\pi r^2 h_l c_w}\left[Q_{sol}(t,T_A,T_l,A_c) - Q_{l,loss}(t,T_l,r,h_l) + 2\pi r^2 \lambda_w(T_u - T_l)\right] \quad (E3)$$

ODE $(E1)$ balances the energy of the absorber plate: $A_c \tau_a H_{tot}$ is the energy input gained by radiation, depending on collector area, angle of incidence of the solar radiation, glass transparency, clearness of the sky ..., V denotes the specific energy losses of the absorber (depending on outdoor temperature and wind speed) and Q_{sol} the heat going into the boiler when the collector pump is active. Hourly weather data (total radiation, outdoor temperature and wind speed) for a (typical) year are required.

ODE's $(E2)$ and $(E3)$ for the boiler temperatures balance the heat Q_{sol} coming from the collector, the heat input Q_{aux} of the auxiliary heater, heat losses Q_{loss} through the boiler wall and the heat exchange between the water layers.

The terms Q_{sol} and Q_{aux} depend on the status of the system (collector pump respectivly auxiliary heating active or not). Each change of the status, e.g. switching off the collector pump, causes a discontinuity of the ODE right hand side. The switching points are determined by the zeros of switching functions. For instance, the switching function ϕ_1 for activating the collector pump is given by

$$\phi_1(T_A,T_l) := \begin{cases} T_A - T_l - \Delta T_0 & \text{for collector pump active} \\ T_A - T_l - \Delta T_1 & \text{for collector pump switched off.} \end{cases}$$

For $\phi_1 \geq 0$ the collector pump is active ($I_{sol} = 1$), for $\phi_1 < 0$ it is switched off ($I_{sol} = 0$), and thus $Q_{sol} = 0$. Another switching function $\phi_2(T_u)$ controlles the auxiliary heating, and there is even a third switching function $\phi_3(T_u,T_l)$ which determines wether the water within the boiler is assumed to have only one temperature ($T_u(t) = T_l(t)$) or there are two layers with different temperatures (this is because for physical reasons the lower layer must not be warmer than the upper layer). The use of status dependent switching functions ϕ_1 and ϕ_2 prevents the system from switching infinitly often.

Hot water consumption takes place each hour. It is taken from the upper layer, cold water is added to the lower layer.

The objective $C(A_c,\beta,\gamma,r,h_u,h_l)$ is the sum of the prime costs $C_o(A_c,r,h_u,h_l)$ and the costs of the running of the system (including auxiliary heating) over a period of 30 years (30 times the costs for one year). The operating costs over one year are calculated by integrating

$$\frac{d}{dt}C_r(t) = I_{sol} \cdot C_{pump} + Q_{aux} \cdot C_{aux} \quad (E4)$$

with C_{pump} being the costs for running the collector pump and C_{aux} being the energy costs of the auxiliary heater. To determine the objective we have to solve the ODE system $(E1) - (E4)$.

3. The Numerical Model
To solve our ODE system

$$x'(t) = \begin{cases} f_1(t,x) & \text{for } \phi(t,x) \geq 0 \\ f_2(t,x) & \text{for } \phi(t,x) < 0 \end{cases} \tag{1}$$

with discontinuities of the right hand side we use the Trapezoidal rule

$$x_{j+1} = x_j + h/2 \cdot [f(t_j, x_j) + f(t_{j+1}, x_{j+1})]$$

and solve for $x_{j+1} \approx x(t_{j+1})$ by Newtons method. As we use hourly weather data our basic integration stepsize is $h = 1[\text{h}]$. This gives $365 \times 24 = 8760$ integration steps for one year. In order to get a consistent approximation of (1) we have to calculate the zeroes t^* of the switching functions (discontinuities of the ODE right hand side) and use these switching points t^* as additional discretization points. Finding the zeroes of ϕ itself is a nonlinear problem, which is solved by Newtons method. Thus, each switching point calculation requires some integration steps, and as there are up to 5000 switching points per year (and even more) we have to perform up to 30.000 integration steps per year.

For the (first) derivatives of the objective C with respect to the optimization parameters $p \in \{A_c, \beta, \gamma, r, h_u, h_l\}$ we have: $\partial/\partial p\, C = \partial/\partial p\, C_0 + \partial/\partial p\, C_r(T)$ with $T = 1[\text{year}]$. The determination of $\partial/\partial p\, C_r(T, p) = (x_p(T, p))_{4^{th}\,component}$ with $x_p(t, p) := \partial/\partial p\, x(t, p)$ is done by simultaniously solving (1) and the sensitivity ODE's

$$\frac{\partial}{\partial t}(x_p(t,p)) = \begin{cases} \frac{\partial}{\partial p} f_1(t, x(t,p), p) & \text{for } \phi(t, x(t,p)) \geq 0 \\ \frac{\partial}{\partial p} f_2(t, x(t,p), p) & \text{for } \phi(t, x(t,p)) < 0, \end{cases} \tag{2}$$

which are obtained by derivation of (1) with respect to p.

NOTE: also the switching points t^* are functions of the parameters p, which causes additional correction terms for x_p at each switching point t^*. Assuming that $\phi(t, x(t,p)) \geq 0$ for $t \in [\underline{t}, t^*]$ and $\phi(t, x(t,p)) < 0$ for $t \in \,]t^*, \overline{t}]$ with $\underline{t}, \overline{t}$ fixed and $t^* \in \,]\underline{t}, \overline{t}[$, we obtain by integrating (1) over $[\underline{t}, \overline{t}]$:

$$x(\overline{t}, p) = x(\underline{t}, p) + \int_{\underline{t}}^{t^*(p)} f_1(t, x(t,p), p)dt + \int_{t^*(p)}^{\overline{t}} f_2(t, x(t,p), p)dt, \tag{3}$$

with $t^*(p)$ defined by

$$\phi(t^*(p), x(t^*(p), p)) = 0. \tag{4}$$

Differentiation of (3) with respect to p yields

$$x_p(\overline{t}, p) = x_p(\underline{t}, p) + \int_{\underline{t}}^{t^*(p)} \frac{\partial}{\partial p} f_1(t, x(t,p), p)dt + f_1(t^*, x(t^*, p), p) \cdot \frac{\partial t^*(p)}{\partial p}$$
$$+ \int_{t^*(p)}^{\overline{t}} \frac{\partial}{\partial p} f_2(t, x(t,p), p)dt - f_2(t^*, x(t^*, p), p) \cdot \frac{\partial t^*(p)}{\partial p}, \tag{5}$$

the correction term for x_p at switching points thus being

$$[f_1(t^*, x(t^*, p), p) - f_2(t^*, x(t^*, p), p)] \cdot \frac{\partial t^*(p)}{\partial p}. \tag{6}$$

$\frac{\partial t^*(p)}{\partial p}$ is defined by (7), obtained by differentiating (4) with respect to p:

$$0 = \frac{\partial}{\partial p}\phi\big(t^*(p), x(t^*(p), p)\big) = \frac{\partial \phi}{\partial t}\cdot\frac{\partial t^*}{\partial p} + \frac{\partial \phi}{\partial x}\cdot\Big(\frac{\partial x}{\partial t}\cdot\frac{\partial t^*}{\partial p} + \frac{\partial x}{\partial p}\Big). \tag{7}$$

With respect to the high numerical amount of function and gradient evaluation we use a Sequential Quadratic Programming (SQP) technique with a Quasi Newton approximation of the Hessian for solving

$$min\{C(p) \mid a_i \le p_i \le b_i, \ i = 1,\dots 6\}. \tag{8}$$

4. Numerical Results

4.1. Sensitivity Analysis: Influence of storage insulation

Thickness of insulation [cm]	15	10	5
Costs [ATS]	93308	95317	104244

That means that a relatively cheap increase of insulation reduces the costs up to 10 %.

4.2. Optimization

Basic Assumption: 1m² collector area: 3.000 ATS
                   1m³ storage volume: 15.000 ATS

Case 1: auxiliary heating costs: 1.25 ATS/kWh
         daily hot water demand of the customer: 300 l at 50°C

Case 2: auxiliary heating costs: 1.70 ATS/kWh
         daily hot water demand of the customer: 300 l at 50°C

Case 3: auxiliary heating costs: 1.70 ATS/kWh
         daily hot water demand of the customer: 500 l at 50°C

		A_C	β	γ	v	h_u	h_2	costs [ATS]
(1)	Starting value	12	45	0	0.3	0.8	0.8	134.264
	Optimal solution	8.46	48.5	3.6	0.25	0.3	1.27	121.487
(2)	Starting value	12	45	0	0.3	0.8	0.8	167 197
	Optimal solution	10.4	47.5	2.7	0.25	0.3	1.41	153 341
(3)	Starting value	12	45	0	0.5	0.8	0.8	225.901
	Optimal solution	15.4	48.5	3.5	0.25	0.3	1.9	213.464

For details and further results see [1], [2].

5. References

[1] W. Kapfer, *Optimale Dimensionierung einer Niedertemperatur-Solaranlage für die Warmwasserbereitung*, Diploma-thesis, Univ. Linz, Math. Deptm. (1989)
[2] B. Schürz, *Optimale Dimensionierung einer Solaranlage für die Warmwasserbereitung mit Nachheizung*, Diploma-thesis, Univ. Linz, Math. Deptm. (1989)

The fluttering of fibres in airspinning processes

M. Bäcker, H. Neunzert, S. Younis

1. Introduction

The motivation for this research is derived from a practical
problem which manufacturers of plastic fleece encounter.
Artificial fibres are drawn in an airspinning process in which
liquid plastic is extruded into a vertical air duct.
A high speed air stream is used to draw the fibres which form
the fleece at the end of the duct on a swinging conveyor belt.
The desired tangential forces acting on the fibres imply
increasing the velocity of the air stream. In it's turn, this
increase leads to an amplification of the transversal motion
of the fibres which anyway exists due to turbulency influence.
The fibres begin to flutter rapidly resulting in clumping,
breakoff and hence in quality losses and frequent maintainance
of dirty production units.
The aim of this research is the basic study of the flutter
effects. The whole scale problem - a large number of hot and
thus still deformable plastic fibres in a high speed subsonic
turbulent flow in three dimensions - is in fact too
complicated to begin with.
Therefore we consider a single fibre in a parallel two
dimensional flow. Clearly, we deal now with a different
topology. In this case the fibre represents a one dimensional
boundary in a two dimensional flow domain.
Nevertheless, we deal with a moving boundary problem where the
motion of the boundary (the fibre..) affects the flow, the
fluid in it's turn produces the forces driving the boundary
itself.
An equation governing the motion of the fibre is therefore
needed. The flow is assumed to be governed by the Euler or
Navier-Stokes equations. The interaction between the fluid and
the fibre must also be studied to achieve a correct coupling
of both motions.

Hj. Wacker and W. Zulehner (eds.),
Proceedings of the Fourth European Conference on Mathematics in Industry, 197–205.
© 1991 *B.G. Teubner Stuttgart and Kluwer Academic Publishers.*

2. Governing equation of the fibre

Let the motion of a point on the fibre be given through the vector function

$$\vec{r}(s,t) = \begin{pmatrix} x(s,t) \\ y(s,t) \end{pmatrix}$$

where t is the time and s is a curve parameter.

Further let T(s,t) be the magnitude of the tensile force in the fibre at the position $\vec{r}(s,t)$, let $\vec{F}(s,t)$ be the external force acting on the fibre and $\rho(s,t)$ be the mass density at the same point.

Balancing the forces acting on a finite segment of the fibre yields the following equation of motion

$$(1) \qquad \rho \cdot \frac{\partial^2 \vec{r}}{\partial t^2} = \vec{F} + \frac{1}{\left\| \frac{\partial \vec{r}}{\partial s} \right\|} \frac{\partial}{\partial s} \left(T \frac{\frac{\partial \vec{r}}{\partial s}}{\left\| \frac{\partial \vec{r}}{\partial s} \right\|} \right)$$

The absense of bending moments is assumed i.e. the fibre is considered to be fully elastic.

For a homogenious mass distribution we can substitute

$$\rho(s,t) = \frac{\rho_0}{\left\| \frac{\partial \vec{r}}{\partial s} \right\|} .$$

Without loss of generality we set $\rho_0 = 1$ and get

$$(1a) \qquad \frac{\partial^2 \vec{r}}{\partial t^2} = \left\| \frac{\partial \vec{r}}{\partial s} \right\| \vec{F} + \frac{\partial}{\partial s} \left(T \frac{\frac{\partial \vec{r}}{\partial s}}{\left\| \frac{\partial \vec{r}}{\partial s} \right\|} \right) .$$

(1a) represents two equations for the three unknown functions x, y and T. Thus one more equation for T is needed. Two possibilities will be considered:

(i) The tensile force T follows a constitutive law (Hooke's law)

(2) $T(s,t) = T(\epsilon) = c \cdot \epsilon$

ϵ is the relative elongation of the fibre defined as

$$\epsilon := \left\| \frac{\partial \vec{r}}{\partial s} \right\| - 1$$

c can be seen as an elasticity constant of the fibre. Note
that for $\epsilon = 0$ the parameter s is just the arc length of the
fibre. Further, for $\epsilon = 0$ we get $T = 0$ i.e. the fibre is not
prestressed.

Substituting (2) in (1a) we get

(1b) $$\frac{\partial^2 \vec{r}}{\partial t^2} = (\epsilon+1)\vec{F} + \frac{\partial}{\partial s}\left(\frac{T(\epsilon)}{\epsilon+1} \frac{\partial \vec{r}}{\partial s} \right).$$

a quasilinear system of second order differential equations.

(ii) We suppose the fibre to be inextensible. The additional
equation we need in this case is:

(3) $\left\| \frac{\partial \vec{r}}{\partial s} \right\| = 1$

The modification of (1b) for this case yields

(1c) $$\frac{\partial^2 \vec{r}}{\partial t^2} = \vec{F} + \frac{\partial}{\partial s}\left(T \frac{\partial \vec{r}}{\partial s} \right)$$

Equations (1c) and (3) represent a differential algebraic
system for the functions x, y and T with (3) being an indirect
'algebraic' equation for the tensile force.

A short discussion of the systems (1b), (2) and (1c), (3)

Case (i)

As a system of second order differential equations, this case
is simpler to deal with.

The system is transformed into a four dimensional system of
first order differential equations.

The eigenvalues of this system are

$$\pm \sqrt{\frac{\overline{cc}}{1+c}} \quad \text{and} \quad \pm \sqrt{c},$$

all real and different for $c \geq 0$. Under this condition the system becomes strictly hyperbolic. In the case of $c = 0$ the system degenerates parabolically.

As boundary conditions - valid in both cases - we get for the upper and fixed point

$$\frac{\partial \vec{r}}{\partial t} (t,0) = 0,$$

For the free end we expect the absense of tensile forces i.e.

$$c = 0, \qquad \left\| \frac{\partial \vec{r}}{\partial s} \right\| = 1.$$

Additionally we have to define four initial conditions for

$$\frac{\partial \vec{r}}{\partial s} \quad \text{and} \quad \frac{\partial \vec{r}}{\partial t}.$$

A brief discussion of the characteristics of the system shows that these conditions define a well posed problem for (1b)

Case (ii)

This case is indeed hardly accessable for a theoretical discussion. In fact we could express T as a function of x and y.

If we take $p(s,t)$ to be the pressure difference on both sides of the fibre and thus define the resulting external force

$$\vec{F} = p \left(\frac{\partial \vec{r}}{\partial s} \right)^{\perp} \qquad \text{we get:}$$

$$T = \frac{1}{x_{ss}y_s - y_{ss}x_s} \left(P - (y_{tt}x_s - x_{tt}y_s) \right).$$

The tensile force is thus a function of the external force, of the inertia forces and of the radius of curvature of the fibre. The system (1c) is in no way a quasilinear one.

3.) Numerical solution of the fibre equation

The cases (i) and (ii) differ now significantly. In case (i)
it is quite natural to apply a method of characteristics using
the characteristics corresponding to the eigenvalues $\pm\sqrt{c}$.
A careful discussion reveals only difficulties at the free end
of the fibre where $c = 0$. To complete the method we have to
take into consideration the degenerate equation at this end:

$$\frac{\partial^2 \vec{r}}{\partial t^2} = \vec{F} + c\,\frac{\partial c}{\partial s}\cdot\frac{\partial \vec{r}}{\partial s}.$$

In case (ii) we can develop a finite difference method using
equations (1c) and (3). The basic idea is very simple:
equation (3) becomes

$$(3')\qquad \frac{\partial}{\partial t}\,\left\|\frac{\partial \vec{r}}{\partial s}\right\|^2 = x_s x_{st} + y_s y_{st} = 0.$$

(3') and (1c) were transformed into difference equations so
that it is possible to advance the solution from the time
level i to i + 1. Centred difference quotients were formulated
for terms with second derivatives, backward ones for $\frac{\partial}{\partial s}$ and
forward differences for the time derivatives $\frac{\partial}{\partial t}$.
The discretization of equations (1c) and (3') follows at the
i-th. time level. The system of difference equations for
$\vec{r}_{i+1}$, and T_i can now be exactly solved.
The scheme is implicit in the s direction and explicit in
time. Numerous examples were computed with both methods, some
considering gravity effects and thus solving a slightly
modified equation of motion. These simulations are documented
on a video film

4.) Computation of the flow for a given fibre motion

We assume an isentropic compressible and inviscid flow in two
dimensions goverened by the Euler equations along with an
equation of state. The lack of friction is a significant
restriction in this case. Tangential forces resulting from
skin friction are missing now. Only normal forces stemming
from pressure differences on both sides can now be exerted on
the fibre by the flow. In fact, for our problem - the study of

flutter effects, this is a sufficient approximation.

Our main interest is in the transversal motion of the fibre i.e. the motion of the boundary within the flow domain.

Suppose now that we fix the fibre during a time increment Δt, compute the flow at time $t + \Delta t$ and then allow the fibre to move. By proceeding this way we usually get 'holes' in the computational domain for the next time step. These 'holes' must somehow be filled.

A way to avoid this difficulty is to let the fibre move during the time step Δt: while computing the flow the fibre moves locally uniformly according to the pressure differences which were present at the beginning of the time step.

Consequently, for the computation of the flow we have to apply a method which works also for locally uniformly moving boundaries.

The method we apply, a so called kinetic method, principally has the desired performance. An important advantage of the method lies in it's time structure.

A major difference between kinetic methods and finite difference methods is the fact that kinetic methods are inherently continuous in time, which is of major importance for the moving boundary we deal with.

A short description of the method

The Euler equations are hyperbolic partial differential equations governing the temporal and spatial evolution of the state variables density ρ and momentum m of a fluid in the isentropic case.

The more general equation governing a gas motion is the so called Boltzmann equation which is an integro-differential equation for the distribution function $f(x, \xi, t)$ of a gas. f is thus a function of position, of velocity and of time. The variables $\rho(x,t)$ and $m(x,t)$ correspond to the zeroth and first moments of f respectively.

$$\rho(x,t) = \int_{\mathbb{R}^2} f(x, \xi, t) d\xi, \quad m(x,t) = \int_{\mathbb{R}^2} \xi f(x, \xi, t) d\xi.$$

These definitions are the basis for all kinetic methods (see

[1], [2], [3]) principally following the scheme:

```
┌─────────────────────────────────────────────────┐
│  given: flow variables at time   nΔt             │
├─────────────────────────────────────────────────┤
│           ρ(x,nΔt),  m(x,nΔt)                     │
└─────────────────────────────────────────────────┘
```

choose a class of   define the distribution
distribution func-   function so that the zeroth
tions such as Max-   and first moments correspond
well functions   to ρ and m respectively

$$f_n(x, \mathfrak{f}, n\Delta t)$$

temporal evolution   free flow eventually with
of f_n reflection at the boundary

$$f_n(x - \mathfrak{f}\Delta t, \mathfrak{f}, n\Delta t)$$
(at the boundary somewhat more complicated)

computation of the
moments

$$\rho(x, (n+1)\Delta t) = \int_{\mathbb{R}^2} f_n(x - \mathfrak{f}\Delta t, \mathfrak{f}, n\Delta t) \, d\mathfrak{f}$$

$$m(x, (n+1)\Delta t) = \int_{\mathbb{R}^2} \mathfrak{f} f_n(x - \mathfrak{f}\Delta t, \mathfrak{f}, n\Delta t) \, d\mathfrak{f}$$

The kinetic methods developed according to this concept differ
in fact only in the class of distribution functions they use
for the redifinition.
Here f(x, $\mathfrak{f}$, t) is always given as

$$f(x, \mathfrak{f}, t) = G(\rho(x,t), u(x,t), \mathfrak{f})$$

(u = $\frac{m}{\rho}$ being the flow velocity).
The function G must fullfill some consistency conditions, but
freedom in the choice of this function still remains.
The first functions to come in mind are the maxwellians which
disadvantagely have no compact support in $\mathfrak{f}$. This fact
increases the computational effort significantly.

The class of functions we use was suggested by S. Kaniel [2] and has a compact support in $\mathfrak{f}$.

The corresponding function G has a simple form:
For a monoatomic gas one gets

$$G(\rho,u,\mathfrak{f}) = \begin{cases} g = const \text{ for } |u-\mathfrak{f}| \leqslant \sqrt{n} \cdot c \\ 0 \text{ else;} \end{cases}$$

where n is the space dimension and $c = \sqrt{\dfrac{\partial p}{\partial \rho}}$ is the local speed of sound.

Numerical aspects of the Kaniel method

The difficulties occuring in this method are associated with the computation of the moments of the distribution function after the drift i.e. in evaluating the integrals

$$\int_{\mathbb{R}^2} \begin{bmatrix} 1 \\ \mathfrak{f} \end{bmatrix} f_n(x-\mathfrak{f}\Delta t, \mathfrak{f}, n \cdot \Delta t) d\mathfrak{f}$$

(somewhat different expressions at the boundary).

To achieve this, spatial discretization is neccesary. We use a rectangular equidistant grid with the cells c_{ij}. The moments are averaged over the cell c_{ij} i.e. we define

$$\rho_{ij}((n+1)\Delta t) = \frac{1}{\text{Area}(c_{ij})} \int_{c_{ij}} \int_{\mathbb{R}^2} f_n(x-\mathfrak{f} \cdot \Delta t, \mathfrak{f}, n \cdot \Delta t) d\mathfrak{f} dx,$$

m_{ij} is defined analogously.

For cells within the flow domain (no boundary cells) the values ρ_{ij} and m_{ij} can be computed after the redefinition through a balance of the so called moment fluxes (which can principally be computed analitically) between c_{ij} and it's eight neighbours. The Courant-Friedrich-Lewy condition implies that a cell interacts only with it's direct neighbors.
For CFL $\geqslant$ 1 the interaction of a cell c_{ij} with indirect neighbor cells increases the computational effort significantly.
For boundary cells specular reflection is applied as a boundary condition. For cells neigboring the moving fibre the

distribution function is approximated through discrete measures i.e. using simulation particles which are reflected at the moving boundary according to it's position at the expected moment of collision. Exactly at this point, the advantages of the kinetic method are evident.

The method is being implemented.

Bibliography

[1] Cercignani, C.: The Boltzmann equation and it's applications, Applied mathematical sciences vol. 67, Springer Verlag 1988.

[2] S. Kaniel: Approximation of the hydrodynamic equations by a transport process.
    In: Approximation Methods for Navier-Stokes Problem. Lecture Notes in Mathematics No. 771, 272-286, Springer-Verlag (1980).

[3] B. Perthame: Relations Between Boltzmann Equations and Boltzmann schemes for Gas Dynamics: Preprint, Paris 1989

[4] C. Cristescu: V. Dynamic Plasticity. North-Holland Series in Applied Mathematics and Mechanics, Amsterdam (1967).

[5] M. Bäcker, K. Dreßler: A Kinetic Method for Strictly nonlinear Conservation Laws, Preprint, Kaiserslautern, 1989.

A Numerical Model for Electromagnetic Casters *

O. Besson, J. Bourgeois, P.-A. Chevalier, J. Rappaz and R. Touzani

ABSTRACT

We briefly describe a numerical model we have developed to simulate electromagnetic casting processes. The model is two-dimensional. The set of partial differential equations is given and the numerical methods are outlined. A numerical experiment describing the case of an Aluminium ingot is given in order to show the feasibility of the model.

1.   INTRODUCTION

In several metal casting technologies, Electromagnetic Casting (EMC) appears now as an interesting alternative to the traditional continuous casting processes. Indeed, since the actual mould is *electromagnetic*, the high temperature environment is more easily handled and furthermore, the resulting metal ingots are smoother. Fig. 1 gives the principle of an EMC setup.

From a mathematical viewpoint, the major difficulty in modelling such a problem resides in the fact that various coupled phenomena are involved and even more such phenomena have different scales which causes numerical difficulties. We shall however describe, in this communication, a numerical model where some of these effects are neglected. Namely, we restrict ourselves to electromagnetic, hydrodynamic and free surface interactions, this last one including surface tension effects. Let us notice that a more detailed description of the model with further numerical results can be found in [1] and that other approaches to this problem have been developed by other authors (see for instance [5,6]).

2.   THE EQUATIONS

2.1.   The Electromagnetic Problem

Assume we are in presence of a set of n infinite cylindrical conductors $\Lambda_1, \Lambda_2, ..., \Lambda_n$ of which the intersections with the plane Ox_1x_2 are denoted by $\Omega_1, \Omega_2, ..., \Omega_n$ and the generating lines are parallel to the x_3-axis. The domains Ω_j, j = 1,...,n are assumed to be bounded, connected and disjoined and their boundaries are denoted respectively by $\Gamma_1, \Gamma_2, ..., \Gamma_n$. We set

$$\Omega = \bigcup_{j=1}^{n} \Omega_j, \qquad \Gamma = \bigcup_{j=1}^{n} \Gamma_j.$$

---

* This research was partially supported by the swiss "Commission pour l'Encouragement de la Recherche Scientifique" and the Alusuisse Company.

Hj. Wacker and W. Zulehner (eds.),
Proceedings of the Fourth European Conference on Mathematics in Industry, 207–212.
© 1991 *B.G. Teubner Stuttgart and Kluwer Academic Publishers.*

The inductor is assumed to be travelled by an alternating current parallel to the x_3-axis with a "moderate" frequency $f = \omega/2\pi$. The system of the n conductors, the permeability of which is supposed to be μ_0, and the vaccum is then governed by the classical Maxwell's equations (Cf. [4] for instance) in the whole space $\mathbb{R}^3$, the main feature of the model consisting in neglecting displacement currents and assuming x_3-invariance and time periodicity of the solution.

Let $\mathbf{u}$ denote the flow velocity vector in the fluid region and let σ denote the electric conductivity. Using the fact that the magnetic induction $\mathbf{b} = (b_1, b_2)$ is divergence-free, we conclude that there exists a scalar potential

$$\phi : \mathbb{R} \to \mathbb{C} \text{ such that } b_1 = \frac{\partial \phi}{\partial x_2}, \; b_2 = -\frac{\partial \phi}{\partial x_1} .$$

After some developments, we get the system of equations (See [1]) :

(1) $- \Delta\phi + \mu_0\sigma \, (\mathbf{u}\cdot\nabla\phi + i\omega \, \phi - C_k) = 0$ in Ω_k, $k = 1,2,...,n$,

(2) $\Delta\phi = 0$ in $\mathbb{R}^2 \setminus \bar{\Omega}$,

(3) $[\phi] = [\frac{\partial \phi}{\partial n}] = 0$ on Γ,

(4) $\phi_2(\mathbf{x}) = d \log |\mathbf{x}| + 0 \, (|\mathbf{x}|^{-1})$ $|\mathbf{x}| \to +\infty$,

where $[\cdot]$ stands for the jump of a function on the boundary Γ and $\dfrac{\partial}{\partial n}$ denotes the normal derivation and where C_k is a constant on Ω_k.

2.2.   The Hydrodynamic Problem

In the liquid region, Ω_L say, the resulting Lorentz force field is responsible for the fluid flow. Hence, we shall consider in Ω_L the incompressible Navier-Stokes equations with a body force term :

(5) $- 2 \, \text{div} \, \mathbf{D}(\mathbf{u}) + \rho\mathbf{u}\cdot\nabla\mathbf{u} + \nabla p = \mathbf{f} + \rho\mathbf{g}$ in Ω_L,

(6) $\text{div} \, \mathbf{u} = 0$ in Ω_L,

(7) $\mathbf{u} = 0$ on Γ_0,

(8) $\mathbf{u}\cdot\mathbf{n} = 0$ on Γ_M,

(9) $\sigma(\mathbf{u},p)\cdot\mathbf{t} = 0,$ on Γ_M.

In the above equations, the boundary Γ_L of Ω_L is partitioned into Γ_M and Γ_0, where Γ_M represents the meniscus and Γ_0 is the interface with the solid (Cf. Fig. 2), the vector $\mathbf{t}$ denoting the unit tangent vector to Γ_M. The tensor σ is the Cauchy stress tensor

$$\sigma(\mathbf{u},p) = 2\,\mathbf{D}(\mathbf{u})\cdot\mathbf{n} - p\mathbf{n}, \quad \mathbf{D}(\mathbf{u}) = \frac{\eta}{2}\,(\nabla\mathbf{u} + \nabla\mathbf{u}^T).$$

The function p is the pressure; the positive constants η and p denote respectively the dynamic viscosity and the density of the fluid. The vectors $\mathbf{f}$ and $\mathbf{g}$ denote respectively the Lorentz and the gravity forces. Conditions (8), (9) stand for a slip boundary condition on the free boundary ($\mathbf{n}$ is the outward unit normal to Γ_M).

A simple calculation shows that (Cf. [1]) :

$$\mathbf{f} = \frac{\sigma\omega}{2}\,(\phi_I\,\nabla\phi_R - \Phi_R\,\nabla\Phi_I) - \frac{\sigma}{2}\,((\mathbf{u}\cdot\nabla\phi_R)\,\nabla\phi_R + (\mathbf{u}\cdot\nabla\phi_I)\,\nabla\phi_I,$$

where ϕ_R, ϕ_I denote respectively the real and the imaginary part of the potential ϕ.

2.3.   The Free Boundary Problem

As in [2], the free boundary is assumed to satisfy the so-called *Laplace-Young* equation which states that the jump of the normal traction on the boundary Γ_M is proportional to the curvature of Γ_M. Namely :

(10) $\tau\kappa + \sigma(\mathbf{u},p)\cdot\mathbf{n} + C = 0$ on Γ_M,

where τ, κ and C denote respectively the surface tension, the curvature of Γ_M and an (unknown) constant.

3.   THE NUMERICAL METHODS

The electromagnetic problem (1)-(4) is solved by a coupled boundary element / finite element procedure. That is, the exterior potential is represented by an integral equation on the boundary Γ; then a mixed formulation similar to that of [3] is used to discretize the resulting equations.

The hydrodynamic problem is solved via a penalty finite element technique. This one consists in perturbing the incompressibility constraint (6) and the slip boundary condition (8) and then using a standard finite element method to discretize these equations.

The free boundary problem uses a less standard technique. In fact, using the Newton's method to linearize equation (10) would imply to solve auxiliary problems similar to (1)-(4) and (5)-(9) the dependence of the normal tractions on the meniscus position being nonlocal. Instead, we use a modified Newton iterative method where the nonlocal part of the normal tractions is taken at the previous iteration. This leads to a much cheaper algorithm.

The fully coupled MHD free boundary problem is actually solved by performing the following steps :

1. Give a domain Ω^0 (and therefore a meniscus shape Γ_M^0);

2. Set n := 0;

3. Generate a finite element mesh of the whole domain;

4. Compute the potential ϕ^n and the body forces;

5. Solve the Navier-Stokes equations and compute the normal tractions on the meniscus;

6. Update the meniscus shape Γ_M^{n+1} (and therefore Ω^{n+1}) with a standard finite difference technique;

7. If the meniscus shape is "sufficiently close" to the previous one, then Stop;

8. Otherwise, set n := n + 1 and go to 3.

5.   A NUMERICAL EXAMPLE

We have designed, for the purpose of this test, a *schematic* Aluminium EMC device. We have sketched on Fig. 2 the various parts of the domain Ω. Namely :

Subdomain	Medium	Conductivity $\sigma[\Omega^{-1}m^{-1}]$
Ω_L	Liquid Aluminium	4.083×10^6
Ω_S	Solid Aluminium	9.653×10^6
Ω_2	Inductor (Copper)	5.89×10^7
Ω_3	Screen (Stainless steel)	1.37×10^6
Ω_4	Cooling device (Aluminium)	9.653×10^6

The physical date are :

$J_0 = 6000$ A,   f = 2500 Hz, $\mu_0 = 4\pi \times 10^{-7}$ H/m,

$\eta = 2,35$ Kg(ms)$^{-1}$,   p = 2350 Kg/m^3,   r = 1 N/m,

where J_0 is the prescribed effective current in the inductor.

Fig. 3 shows the successive meniscii obtained after each iteration. It turns out that the convergence is reached after 13 iterations, the initial guess being relatively far from the converged solution. The velocity vector in the fluid region is depicted in Fig. 4.

REFERENCES

[1] Besson, O.; Bourgeois, J.; Chevalier, P.-A.; Touzani, R.; Rappaz, J. : *Numerical Modelling of Electromagnetic Casting Processes.* To appear in J. Comput. Phys. (1990).

[2] Cuvelier, C. : *Some numerical methods for the computation of capillary free boundaries governed by the Navier-Stokes equations.* Technical Report 87-69, TU Delft (1987).

[3] Johnson, C.; Nedelec, J.-C. : *On the coupling of boundary integral and finite element methods.* Math. of Comp., 35, N° 152 (1980), 1063-1079.

[4] Landau, L.; Lifchitz, E. : *Electrodynamics of Continuous Media*. Pergamon Press, London (1960).

[5] Sakane, J.; Li, B.K.; Evans, J.W. : *Mathematical modelling of meniscus profile and melt flow in electromagnetic casters*. Metallurgical Transactions B, 19B (1988), 397-408.

[6] Weber, J.C.; Buxmann, K.; Von Kaenel, R.; Plata, M. : *New applications for the electromagnetic casting process*. Edited by L.G. Boxall (Warrendale, PA : The Metallurgical Society, Light Metals) (1988), 508.

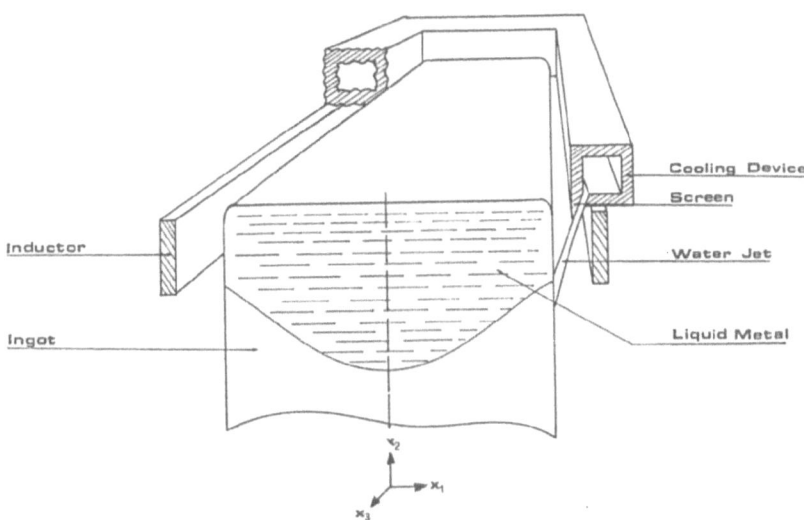

Fig. 1 : Principle of an EMC setup.

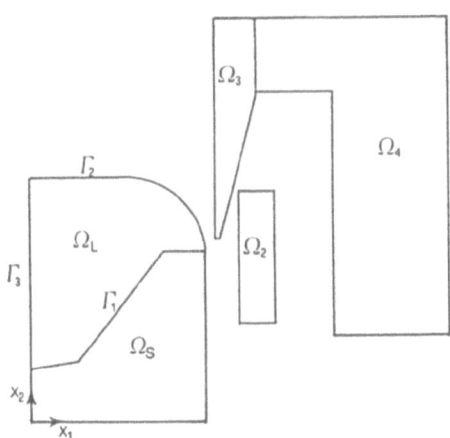

Fig. 2 : Detail of the conductors.

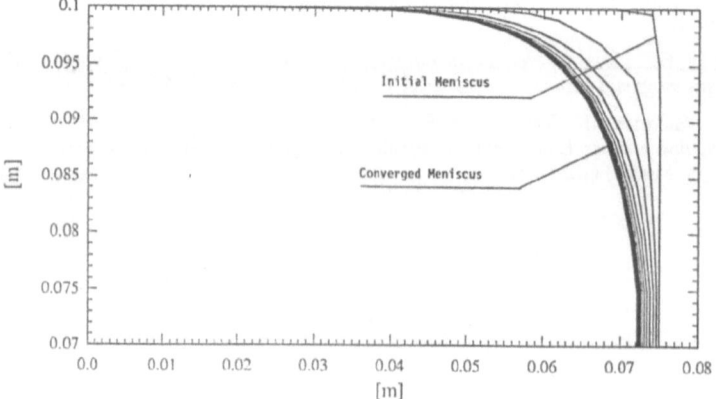

Fig. 3 : Iterated meniscus shapes.

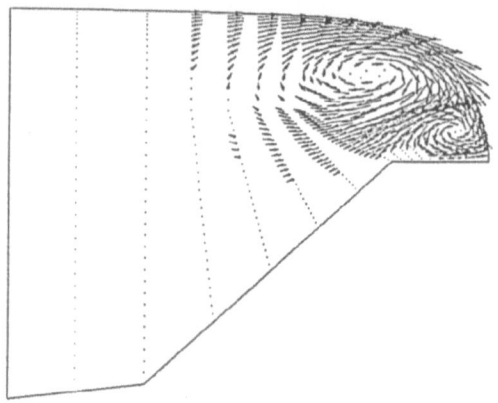

Fig. 4 : Melt flow.

Département de Mathématiques, Ecole Polytechnique Fédérale de Lausanne, CH 1015 Lausanne (Switzerland)

On Consistent Initialization of Differential-Algebraic Equations with Discontinuities

L. Brüll and U. Pallaske

I. Introduction:

In general, most of the time-dependent mathematical models in chemical engineering are given in the form of a linear implicit differential algebraic equation (DAE) system:

$$A(z) \cdot \dot{z} = f_1(z, u)$$
$$0 = f_2(z, u) \tag{1.1}$$

$$z(0) = z_0$$

$$f_1 : R^n \times R^l \to R^k$$
$$f_2 : R^n \times R^l \to R^{n-k}: \quad f_1, f_2 \; smooth$$
$$rg(A(z)) = k$$

$$A(z) = \begin{pmatrix} a_1(z) \\ a_2(z) \\ \vdots \\ a_k(z) \end{pmatrix} , a_i^T(z) : R^n \to R^n.$$

Here, z denotes the vector of state variables, whereas u contains the control variables which are used to control the chemical plant described by (1.1). These control variables can be changed (within suitable limits) arbitrarily from outside the plant. In this paper we study the influence of control variables on systems of the form (1.1), especially when discontinuities (e.g. bang-bang controls) occur.

In the following we will assume that the original system (1.1), or more generally the full implicit system

$$f(\dot{z}, z, t, u) = 0 \tag{1.2}$$

$$z(0) = z_0$$

has global (or differential) index 1 (see [3]). Since a lot of problems (not only in chemical engineering) have at least index 2, this assumption implies that a suitable index-reduction algorithm (see [1]) was used first to reduce

213

Hj. Wacker and W. Zulehner (eds.),
Proceedings of the Fourth European Conference on Mathematics in Industry, 213–217.
© 1991 B.G. Teubner Stuttgart and Kluwer Academic Publishers.

the index of the system. Connections with the perturbation index (see [4]) are discussed at the end of this paper.

Usually, a mathematical model is defined in terms of the underlying differential and algebraic equations, possible boundary conditions and information concerning the initial values of the problem. For smooth equations with smooth control variables a unique solution will exist. Taking the case of discontinuous control variables we assume in the following that $u\mid_{[0,t_0)}, u\mid_{(t_0,\infty)}$ are differentiable functions with continuous derivatives and $u^+ := u(t_0 + 0) \neq u^- := u(t_0 - 0)$. Let us denote $z^- := z(t_0 - 0)$. To describe the value of the state variable z^+ after the jump in the control variables from u^- to u^+ we consider (for fixed ε_0 and $\varepsilon \leq \varepsilon_0$) the set of regularizations on the interval $[t_0, t_0 + \varepsilon]$:

$$\mathcal{F}_\varepsilon := \{u_\varepsilon \epsilon C^1(R_0^+, R^l) \mid u_\varepsilon(t) = u(t) \quad for \quad t < t_0$$
$$u_\varepsilon(t) = u(t - \varepsilon) \quad for \quad t > t_0 + \varepsilon\}.$$

Further, we denote by $z_\varepsilon(t, u_\varepsilon)$ the unique solution of the initial value problem

$$f(\dot{z}, z, t, u_\varepsilon) = 0$$

$$z(0) = z_0$$

Definition:

z^+ is called the genuine initial value (corresponding to z^- and the jump of u at t_0) iff:

there exists $M_0 > 0$ such that for all $M \geq M_0$ the limit

$$z^+ = \lim_{\varepsilon \to 0} z_\varepsilon(t_0 + \varepsilon, u_\varepsilon)$$

exists for all regularizations $u_\varepsilon \epsilon \mathcal{F}_\varepsilon$,$\max_{t\epsilon[t_0,t_0+\varepsilon]} \mid u_\varepsilon(t) \mid \leq M$ of u.

Example:
For ordinary differential equations

$$\dot{z} = f(t, z, u) \tag{1.3}$$

$$f : R \times R^n \times R^l \to R^n \quad continuous, Lipschitz - continuous \ in \ z$$

we have

$$z^+ = z^-. \tag{1.4}$$

This can be seen by integrating equation (1.3) to get

$$z_\varepsilon(t_0 + \varepsilon) = z^- + \int_{t_0}^{t_0+\varepsilon} f(s, z_\varepsilon(s), u_\varepsilon(s))ds. \tag{1.5}$$

From the assumptions it follows that the right-hand side $f(t, z_\varepsilon, u_\varepsilon)$ is uniformly bounded in $[t_0, t_0 + \varepsilon]$, and hence, taking the limit in (1.5), yields (1.4).

II. The genuine initial value and DAE systems:

In this section we will describe sufficient conditions for a DAE system to guarantee the existence of the genuine initial value z^+. Further, we will give an example of index 1 where z^+ does not exist. This shows again that a remarkable difference exists between ODE systems and DAE systems (even in the index-1 case).
The main result in this section is the

Theorem:

Consider the linear implicit DAE system (1.1) and assume that f_1, f_2 are bounded, differentiable functions with continuous derivatives, Lipschitz-continuous with respect to z. Assume further that for each row $a_i(z), i = 1, \cdots, k$ there exists a non-zero continuous differentiable function $g_i(z) : R^n \to R$, such that $g_i(z) \cdot a_i(z)$ has a potential $p_i(z)$ (i.e. $\frac{d}{dt}(p_i(z)) = g_i(z) \cdot a_i(z) \cdot \frac{d}{dt} z$) $(i = 1, \cdots, k)$.
Then:
z^+ exists and can be calculated from

$$p_i(z^+) = p_i(z^-), \quad i = 1, \cdots, k$$

$$f_2(z^+, u^+) = 0.$$

Proof:
Consider:

$$v := (v_1, \cdots, v_k)^T,$$
$$F_1 := (g_1 \cdot f_{1,1}, \cdots, g_k \cdot f_{1,k})^T,$$

where

$$f_1 = (f_{1,1}, \cdots, f_{1,k})^T.$$

Then system (1.1) is equivalent to the explicit index-1 DAE system:

$$\dot{v} = F_1(z, u) \tag{2.1}$$
$$0 = v_i - p_i(z), \quad i = 1, \cdots, k \tag{2.2}$$
$$0 = f_2(z, u). \tag{2.3}$$

Integrating equation (2.1) with an arbitrary regularization u_ε yields:

$$v_\varepsilon(t_0 + \varepsilon) = v^- + \int_{t_0}^{t_0+\varepsilon} F_1(z_\varepsilon(s), u_\varepsilon(s)) ds \tag{2.4}$$
$$0 = v_{i,\varepsilon}(t_0 + \varepsilon) - p_i(z_\varepsilon(t_0 + \varepsilon)), \quad i = 1, \cdots, k \tag{2.5}$$
$$0 = f_2(z_\varepsilon(t_0 + \varepsilon), u_\varepsilon(t_0 + \varepsilon)) \tag{2.6}$$

Again, because of the assumptions, the solution $(v_\varepsilon, z_\varepsilon)$ is bounded on $[t_0, t_0 + \varepsilon]$. Hence from (2.4) we conclude that v^+ exists and

$$v^+ = v^- \tag{2.7}$$

holds.

Substituting (2.7) into (2.5) and using the index-1 assumption (i.e. (2.5),(2.6) are uniquely solvable for $z_\varepsilon(t_0 + \varepsilon)$), we get the existence of z^+ and the equations

$$v_i^+ = p_i(z^+), \quad i = 1, \cdots, k$$
$$0 = f_2(z^+, u^+).$$

This proves the theorem.

Remarks:

Obviously, the proof can be carried out under much weaker conditions on the right hand sides f_1, f_2.

In order to show that z^+ need not exist for linear implicit DAE systems which do not have integrating factors $g_i(z)$ for some of the rows $a_i(z)$, we discuss the following

Example:

Consider the three-dimensional problem

$$\dot{x} + z \cdot \dot{y} = 0 \tag{2.8a}$$
$$y = u(t) \tag{2.8b}$$
$$z = v(t) \tag{2.8c}$$

where (x, y, z) denotes the state variable vector and u, v are the control variables. Obviously, (2.8) has global index 1.

Assume $u(t) = v(t) = H(t)$, where

$$H(t) = \begin{cases} 0 & t \le 0 \\ 1 & t > 0 \end{cases}$$

denotes the Heavyside-function, and consider the solution $x(t) = y(t) = z(t) = 0$ for $t \le 0$, i.e. $x^- = y^- = z^- = 0$.

Using an arbitrary sequence $u_\varepsilon(t)$ of regularizations of u, and putting $v_\varepsilon(t) = u_\varepsilon(t)$, equation (2.8a) now reads:

$$x_\varepsilon(t_0 + \varepsilon) = x^- - \frac{1}{2} \cdot \int_{t_0}^{t_0+\varepsilon} \frac{d}{ds}(u_\varepsilon^2)ds,$$

and hence

$$\lim_{\varepsilon \to 0} x_\varepsilon(t_0 + \varepsilon) = x^- - \frac{1}{2}.$$

If we repeat the calculation, but put $v_\varepsilon(t) = u_\varepsilon^2(t)$, we get

$$\lim_{\varepsilon \to 0} x_\varepsilon(t_0 + \varepsilon) = x^- - \frac{1}{3}.$$

Hence the genuine initial value does not exist.

Remarks:
1. If an index-1 system with $n = 2$ is given, the genuine initial value always exists, since every smooth function $a(x, y)$ of two variables has an integrating factor.
2. Example (2.8) has global index $d_i = 1$, but perturbation index $p_i = 2$. In [2] it is shown that $d_i \leq p_i \leq d_i + 1$ always holds and that $d_i = p_i$ is true in the case of a linear implicit system (1.1) in integral form. Obviously, this result can be generalized to systems (1.1) with integrating factors for each row.

III. Conclusions:

The results of chapter II show that the treatment of linear implicit DAE systems (1.1) with discontinuous control variables may be pointless, if a transformation into an explicit DAE system is not possible. On the other hand if integrating factors exist, these have to be known for the calculation of the genuine initial value. Hence, in this case, it will often be preferable to transform the implicit system (1.1) into explicit form and to solve the resulting explicit DAE system.

References:

[1]   Bachmann, R.; Brüll, L.; Mrziglod, Th.; Pallaske, U.:
       A contribution to the numerical treatment of differential algebraic
       equations arising in chemical engineering, Dechema-Monographs,
       Vol. 116 (1989).
[2]   Gear, C. W.:Differential algebraic equations, indices, and integral
       algebraic equations, Report No. UIUCDCS-R-89-1505,
       Uni. Illinois (1989).
[3]   Gear, C.W.; Petzold, L.: ODE methods for the soluiton of
       differential/algebraic systems, SIAM J. Numer. Anal. 21 (1984).
[4]   Hairer, E.; Lubich, Ch.; Roche, M.: The numerical solution of
       diierential-algebraic systems by Runge-Kutta methods, Report,
       Uni. Geneve (1988).

Author's adress: L. Brüll and U. Pallaske, Bayer AG, Abt.: AV-IM-AM,
5090 Leverkusen, West Germany.

ON THE GENERATION OF OPTICAL SOLITONS

J. Burzlaff, Dublin.

Summary: Data transmission by optical solitons is discussed briefly. Then, the injection of solitons into an optical fibre is studied by solving a linear eigenvalue problem associated with the nonlinear Schrödinger equation for three special families of input pulses.

Nonlinear pulse propagation in a monomode dielectric fibre can, to leading order in the ratio of width of the frequency spectrum to carrier frequency, be described by the nonlinear Schrödinger equation [6].

$$(1) \qquad i\frac{\partial u}{\partial t} + \frac{\partial^2 u}{\partial x^2} + \rho|u|^2 u = 0.$$

Here u is the envelope function of the pulse, ρ is a constant measuring the strength of the nonlinearity, and x and t are proportional to $\bar{t} - \bar{x}/v_g$ and $\bar{x}$, respectively, where $\bar{x}$ and $\bar{t}$ are the space-time coordinates, and v_g is the group velocity. The reason for the nonlinear term is the Kerr effect, which takes the alignment of the dipoles due to the propagating pulse into account, and the Raman effect, i.e. the excitation of electrons and the subsequent emission of radiation. In both cases, the pulse influences its own propagation.

Already in 1973, it was noticed [3] that eq. (1) admits a solution of the form

$$(2) \qquad u(x,t) = \mu \, \text{sech}\,[\mu(x + \chi t)] \exp\{i[-\chi t + \frac{1}{2}(\mu^2 - \chi^2)t]\}$$

(u and χ are arbitrary constants) and that it would be therefore possible for a pulse (a so-called optical soliton) to propagate along a fibre without any change in shape. In this form of propagation, the nonlinearity exactly cancels the dispersive broadening. The experimental verification followed in 1980 [7], opening new exciting possibilities for data communication. The most recent achievement [8] is repeaterless soliton transmission over more than 4000 km in an all optical communication system where the loss is periodically compensated by Raman gain.

Furthermore, soliton pulses with soliton number 2 can be used in a feed back loop of variable length to produce very narrow sech shaped laser pulses of controlled width

Hj. Wacker and W. Zulehner (eds.),
Proceedings of the Fourth European Conference on Mathematics in Industry, 219–223.
© 1991 B.G. Teubner Stuttgart and Kluwer Academic Publishers.

[9]. Also for this reason the question of what type of laser can inject a soliton into a fibre is important. We address this question or the more general question: what is the soliton number N of a given input pulse? A partial, rigorous answer to this question can be given for three families of input pulses $u(x,0)$, namely (i) square shaped pulses [1,2], (ii) sech x shaped pulses [11], and (iii) $\exp(-|x|)$ shaped pulses [1,2].

In all three cases, the same formula holds:

(3)
$$N = \left\langle \frac{1}{2} + \frac{A}{\pi} \right\rangle,$$

where

(4)
$$A = \int_{-\infty}^{\infty} |u(x,0)|\,dx,$$

and $\langle \cdots \rangle$ denotes the integer smaller than the argument.

The analysis starts with the observation [13] that the soliton number N is equal to the number of L^2-integrable solitons of the eigenvalue problem.

(5)
$$\begin{pmatrix} i\frac{d}{dx} & u(x,0) \\ -u^*(x,0) & -i\frac{d}{dx} \end{pmatrix} \begin{pmatrix} v_1 \\ v_2 \end{pmatrix} = \lambda \begin{pmatrix} v_1 \\ v_2 \end{pmatrix}.$$

(u^* is the complex conjugate of u.) We find this number by determining $a(\lambda)$ for real λ from the asymptotic form of the solution $\vec{v}$,

(6)
$$\begin{pmatrix} e^{-i\lambda x} \\ 0 \end{pmatrix} \underset{x \to -\infty}{\longleftarrow} \begin{pmatrix} v_1 \\ v_2 \end{pmatrix} \underset{x \to +\infty}{\longrightarrow} a(\lambda) \begin{pmatrix} e^{-i\lambda x} \\ 0 \end{pmatrix} + b(\lambda) \begin{pmatrix} 0 \\ e^{i\lambda x} \end{pmatrix},$$

by analytically continuing $a(\lambda)$ into the complex upper half-plane and by determining the zeros of a.

In the first case [1,2], the input pulse is $u(x,0) = -iq(x)$, where

(7)
$$q(x) = \begin{cases} 0 & (|x| > \frac{\alpha}{2}) \\ \beta & (|x| \le \frac{\alpha}{2}), \quad \beta > 0 \end{cases}$$

The equations (5) are easily integrated on three different intervals and the solutions are matched at $\pm \alpha/2$. This yields

(8) $\qquad a(\lambda) = \dfrac{e^{i\lambda \alpha}}{\sqrt{\beta^2 + \lambda^2}} (i\lambda \sin \sqrt{\beta^2 + \lambda^2}\, \alpha - \sqrt{\beta^2 + \lambda^2} \cos \sqrt{\beta^2 + \lambda^2}\, \alpha).$

Finding the eigenvalues amounts to finding the zeros of a. Attempting this for $Re\ \lambda \neq 0$ leads to a negative result. The zeros are therefore purely imaginary. We set $\eta := Im\ \lambda$ and $\rho := \sqrt{\beta^2 - \eta^2}\, \alpha$ and are left with solving the equations

(9) $\qquad\qquad\qquad \rho^2 + \alpha^2 \eta^2 = \alpha^2 \beta^2, \quad \eta = -\dfrac{\rho}{\alpha} \cot \rho,$

for $\rho, \eta > 0$. Studying the intersections of the ellipses with the branches of the cotangent function given by eqs. (9), we obtain formula (3) in the first case.

In the second case [11], $u(x,0) = \gamma \operatorname{sech} x$ holds. This leads to the hypergeometric equation for v_1 and v_2. Using the known asymptotic behaviour of the solutions,

(10) $\qquad\qquad\qquad a(\lambda) = \dfrac{\Gamma^2(-i\lambda + \frac{1}{2})}{\Gamma(-i\lambda + \frac{1}{2} + \gamma)\Gamma(-i\lambda + \frac{1}{2} - \gamma)}$

can be derived. The zeros of a for $Im\ \lambda > 0$ are the poles of $\Gamma(-i\lambda + 1/2 - \gamma)$. Therefore, $\lambda = i(\gamma + 1/2 - r), r = 1, 2, \cdots, N$, are the eigenvalues, where N is determined by the condition $\gamma + 1/2 - r > 0$. Again, formula (3) follows.

In the third case [1,2], $u(x,0) = -i\beta e^{-\alpha|x|}$ holds which leads to Bessel's equations for v_1 and v_2. The known asymptotic behaviour of Bessel's functions and matching at $x = 0$ is used to derive

(11) $\qquad\qquad\qquad a(\lambda) \sim \dfrac{J_{\nu+1}^2(\frac{\beta}{\alpha}) - J_\nu^2(\frac{\beta}{\alpha})}{J_\nu(\frac{\beta}{\alpha})Y_{\nu+1}(\frac{\beta}{\alpha}) - Y_\nu(\frac{\beta}{\alpha})J_{\nu+1}(\frac{\beta}{\alpha})},$

with $\nu = -1/2 - i\lambda/\alpha$. The condition the eigenvalues have to satisfy is therefore

(12) $\qquad\qquad\qquad\qquad J_\nu \left(\dfrac{\beta}{\alpha}\right) = \pm J_{\nu+1} \left(\dfrac{\beta}{\alpha}\right).$

That condition (12) cannot be satisfied for non-real ν with $Re\ \nu > -1/2$ can be proved as follows [10]: Assume that $J_\nu \pm J_{\nu+1}$ has a real zero s for non-real ν. Then using the Mittag-Leffler expansion [12,p.497]

$$(13) \qquad 1 \pm \sum_{n=1}^{\infty} \frac{2s}{j_{\nu n}^2 - s^2} = 0,$$

and therefore,

$$(14) \qquad \sum_{n=1}^{\infty} \frac{Re j_{\nu n} \, Im j_{\nu n}}{|j_{\nu n}^2 - s^2|^2} = 0$$

follows, where $j_{\nu n}$ are the zeros of $s^{-\nu} J_\nu(s)$. This equation cannot hold because $Re\, j_{\nu n}/Im j_{\nu n} \gtrless 0$ for $Re\lambda \lessgtr 0$ for all n with $Im\, j_{\nu n} \neq 0$ [4], and because there are $j_{\nu n}$ with $Re\, j_{\nu n} \neq 0$ and $Im\, j_{\nu n} \neq 0$.

We are left with studying the points of intersection of J_ν and $\pm J_{\nu+1}$, which we denote as $s_n(\nu)$, for real order $\nu = \eta/\alpha - 1/2 > -1/2$. It is easy to prove that labelling the points of intersection by $s_n(\nu)$ makes sense because, if ν changes the number of points of intersection stays the same, and s_n changes continuously with ν. Furthermore, $s_n \to \infty$ for $n \to \infty$ and for $\nu \to \infty$, and s_n increases monotonically with ν [10, consequence of Lemmas 2.3 and 2.5 in ref. 5]. This implies that $s_n(-1/2) = (2n - 1)\pi/2$ determines the soliton number, which, in terms of A, turns out to be again given by eq. (3).

Some work has also been done on the super-Gaussian pulse [1]

$$(15) \qquad u(x,0) = A_0 \exp[-(1/2)(1 - i\alpha)(x/\sigma)^{2m}],$$

which is a realistic pulse from a semiconductor laser. No rigorous result has been derived and due to the complicated form of the pulse it is not expected that formula (3) can be proved or disproved in this case rigorously. Therefore, instead of going into more detail we end with the main message of our work: The three rigorously solvable cases lead to formula (3) which shows that the injection of soliton pulses is surprisingly insensitive to the shape of the pulse.

References

[1] Breen, K.: Special solutions of the optical soliton eigenvalue problem. Dublin City University M.Sc. thesis (1989).

[2] Burzlaff, J.: The soliton number of optical soliton bound states for two special families of input pulses. J. Phys. A21 (1988) 561 - 566

[3] Hasegawa, A.; Tappert, F.: Transmission of stationary nonlinear optical pulses in dispersive dielectric fibers. I. Anomalous dispersion. Appl. Phys. Lett. $\underline{23}$ (1973) 142 - 144

[4] Ifantis, E. K.; Siafarikas, P. D.; Kouris, C. B.: Conditions for solution of a linear first-order differential equation in the Hardy-Lebesgue space and applications: J. Math. Anal. Appl. $\underline{104}$ (1984) 454 - 466

[5] Ismail, M. E. H.; Muldoon, M. E.: Monotonicity of the zeros of a cross-product of Bessel functions. SIAM J. Math. Anal. $\underline{9}$ (1978) 759 - 767

[6] Kodama, Y.; Hasegawa, A.: Nonlinear pulse propagation in a monomode dielectric guide. IEEE J. Quant. Elect. $\underline{23}$ (1987) 510 - 524

[7] Mollenauer, L. F.; Stolen, R. H.; Gordon, J. P.: Experimental observation of picosecond pulse narrowing and solitons in optical fibres. Phys. Rev. Lett. $\underline{45}$ (1980) 1095 - 1098

[8] Mollenauer, L. F.; Smith, K.: Demonstration of soliton transmission over more than 4000 km in fiber with loss periodically compensated by Raman gain. Opt. Lett. $\underline{14}$ (1988) 674 - 678

[9] Mollenauer, L. F.; Stolen, R. H.: The soliton laser. Opt. Lett. $\underline{9}$ (1984) 13 - 15

[10] Muldoon, M. E.: Private communication

[11] Satsuma, J.; Yajima, N.: Initial value problems of one-dimensional self-modulation of nonlinear waves in dispersive media. Prog. Theor. Phys. Suppl. $\underline{55}$ (1974) 284 - 306

[12] Watson, G. N.: A treatise on the theory of Bessel functions. 2nd ed. Cambridge University Press 1944

[13] Zakharov, V. E.; Shabat, A. B.: Exact theory of two-dimensional self-focusing and one-dimensional self-modulation of waves in nonlinear media. Sov. Phys. JETP $\underline{34}$ (1972) 62 - 69

Jürgen Burzlaff
Dublin City University
Dublin 9, Ireland

MATHEMATICAL MODELS FOR THE DIFFUSION OF INNOVATIONS

V. CAPASSO

Department of Mathematics, University of Bari, Italy
and
SASIAM (School for Advanced Studies in Industrial and
Applied Mathematics)
Tecnopolis, Valenzano (Ba), Italy

M. ZONNO

University of Bari, Italy
(Computer Science curriculum)

1. Introduction

Models for the diffusion of innovations belong to the wider class of mathematical models for the spread of rumors and infections [1].

More recently a large interest has arisen in this kind of mathematical modelling to stimulate the penetration of a new product in the market. (See e.g. [7], [8] for an update review on the subject).

The present paper offers an outline of the fundamental ideas on which mathematical modelling is based upon, along with an original proposal of models with space structure borrowed from the authors' literature in epidemic theory [3]. Further research is under way, for stochastic modelling too.

Usually so called compartmental models have been used, according to which the total relevant population is divided into homogeneous classes which interact according to the specific structure of the system.

For example with respect to an innovation the total population may be divided into two main classes : uninformed (D) and adopters (A) which interact according to some information mechanism so that uninformed become themselves adopters. In this case a possible mathematical model may be

225

Hj. Wacker and W. Zulehner (eds.),
Proceedings of the Fourth European Conference on Mathematics in Industry, 225–233.
© 1991 B.G. Teubner Stuttgart and Kluwer Academic Publishers.

$$(1.1) \quad \begin{cases} \dfrac{d}{dt} \, D(t) = - \, g(t, A(t)) \, D(t) \\[3mm] \dfrac{d}{dt} \, A(t) = g(t, A(t)) \, D(t) \end{cases}$$

where $g(t, A(t))$ is the "force of infection" of uninformed individuals, due to the individuals $A(t)$, who are already adopters and are willing to transmit the information.
In this case the total population

$$\overline{A}(t) = A(t) + D(t) = \text{const}$$

so that system (1.1) can be reduced to

$$(1.2) \qquad \frac{d}{dt} \, A(t) = g(t, A(t)) \, (\, \overline{A} - A(t)).$$

The force of infection $g(t, z)$ can be of any form, suitably chosen according to the specific mechanism of transmission of the information. For simplicity it has often been chosen of the following linear form

$$(1.3) \qquad g(t, z) = a(t) + b(t) \, A(t)$$

where $a(t)$, usually called the innovation rate, describes the presence of external sources of information which induce the "infection" process (mass media such as radio, TV, posters, etc.); $b(t)z$ describes the infection process due to interpersonal contacts ($b(t)$ is called the innovation rate). The form (1.3) includes both possibilities.
Marketing actions can be taken into account by letting

$$a(t) = \tilde{a}(M(t)) \quad ; \quad b(t) = \tilde{b}(M(t))$$

where $M(t)$ describes the specific action (price, advertising, etc.). For example to include effects due to the price $p(t)$ of the new product we may assume

$$b(t) = \overline{b} \, e^{-kp(t)};$$

to include an advertising effort $M_A(t)$ we may assume

$$a(t) = \overline{a}_1 + \overline{a}_2 \, M_A(t) \quad ;$$

where $p(t)$ and $M_A(t)$ will be selected according to the specific action. It is clear that the price itself may vary as a function of the number of consumers [2]

$$p(t) = \frac{\alpha}{[A(t)]^\lambda} \quad .$$

Model (1.1) is usually called a binomial model, since the total population of potential adopters A has been divided in only two classes. Multinomial models include more than two classes; for example we may have

(D) uninformed   - the class of individuals who are not informed yet, but are susceptible to adoption;

(A) adopters   - the class of individuals who have already adopted the innovation, and are willing to spread the information;

(R) rejectors   - the class of individuals who have already adopted the innovation but are not willing any more to spread the information;

(L) disapprovers - the class of individuals who, having been informed, disapprove the innovation so that they will not adopt it, and will not spread the information.

The following picture (Fig. 1.1) illustrates a model proposed in [10] for the possible interactions among the different classes.

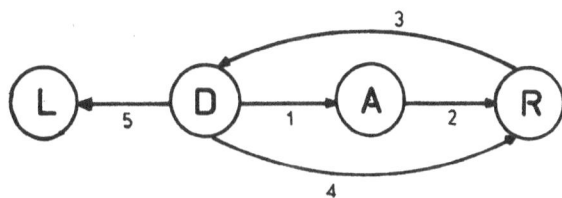

Fig. 1.1.

The general model illustrated in Fig. 1.1 is quite intractable from an analytic point of view. Computer simulations have been carried out in [10].

A much simpler situation is obtained if we assume that class (L) is empty and that no direct transfer is possible between class (D) and class (R). An example of this kind is the

classical Kermack - Mc Kendrick model for infectious
diseases [6] which is based on the following ODE system :

$$(1.4) \quad \begin{cases} \dfrac{d}{dt} D(t) = - \beta \, A(t) \, D(t) \\[2ex] \dfrac{d}{dt} A(t) = \beta \, A(t) \, D(t) - \alpha \, A(t) \\[2ex] \dfrac{d}{dt} R(t) = \alpha \, A(t) \end{cases}$$

This model applies for example to the adoption of a Color TV
as opposed to the BW TV since it can be assumed that no one
is a direct opposer of Color TV; imitation is much more
important than advertising; removal is only a consequence of
having been an adopter [10].
The Kermack - Mc Kendrick model has largely been analyzed in
connection with the spread of infectious deseases, hence we
refer to the literature for results (see e.g. [9]).
More interesting than the basic Kermack - Mc Kendrick model
is the case with vital dynamics which allows immigration in
the class of uninformed and migration of individuals from
the different classes, thus leading to the following
mathematical model

$$(1.5) \quad \begin{cases} \dfrac{d}{dt} D(t) = - \beta \, A(t) \, D(t) + \mu_1 - \mu_2 \, D(t) \\[2ex] \dfrac{d}{dt} A(t) = \beta \, A(t) \, D(t) - \alpha \, A(t) - \mu_3 \, A(t) \\[2ex] \dfrac{d}{dt} R(t) = \alpha \, A(t) - \mu_4 \, R(t) \end{cases}$$

The simplest case obviously is the one which the overall
population is constant (= 1) with

$$(1.6) \qquad \mu_1 = \mu_2 = \mu_3 = \mu_4 = \mu$$

In this case the analysis shows the possible existence of a (dynamic) nontrivial steady state D^*, A^* ($R^* = 1 - D^* - A^*$) which is globally asymptotically stable in

$$T := \left\{ (D, A) \in \mathbb{R}^2 \mid D + A \leq 1 \right\}$$

(see Fig. 2.1).

2. Models with space structure

The models presented in Section 1 can be modified to include heterogeneities.

Important features to be included in order to apply the models to real situations are the following: space structure; time structure; age structure; or any combination of them (see e.g. [1], [4], [9]).

Here we shall present a modification of model (1.5), (1.6) which includes the possibility of a discrete spatial structure.

This model describes the interaction between different geographical locations or different social groups, according to a Kendall - like mechanism [5].

The total population is divided into n groups, so that the ODE system describing the system is the following

$$
(2.1) \quad
\begin{cases}
\dfrac{d}{dt} D_i = - (K * A)_i D_i + \mu_i - \mu_i D_i \\[2ex]
\dfrac{d}{dt} A_i = (K * A)_i D_i - (\mu_i + \alpha_i) A_i \\[2ex]
\dfrac{d}{dt} R_i = \alpha_i A_i - \mu_i R_i
\end{cases}
$$

for $i = 1, \ldots, n$ where

$$(2.2) \qquad (K * A)_i = \sum_{j=1}^{n} \lambda_{ij} A_j \quad , \quad i = 1, \ldots, n$$

We may like to observe that (2.1), (2.2) can be seen as the discretization of a continuous space model.

Problem of existence and global asymptotic stability of equilibrium states for system (2.1), (2.2) have been faced in [3]. In particular conditions are given on the parameters such that a unique nontrivial steady state exists which is globally asymptotically stable in the interior of

$$\Omega := \left\{ z = (D_1, \ldots, D_n, A_1, \ldots, A_n) \in \mathbb{R}_+^{2n} \mid \sum_{i=1}^{2n} z_i \leq n \right\}$$

Figures 2.2 and 2.3 illustrate the behaviour of the system for different choices of the parameters.

REFERENCES

[1] Bailey, N.T.J., The Mathematical Theory of Infectious Diseases, 2 Edn. Griffin, London, 1975.

[2] Bass, F. M., A new product growth model for consumer durables. Management Sciences 15 (1969), 215-227.

[3] Beretta, E. and Capasso, V., Global stability results for a multigroup SIR epidemic model. In "Mathematical Ecology"(T. G. Hallam et al. Eds.). World Pub. Co., Singapore, 1988.

[4] Capasso, V., Mathematical Structures of Epidemic Systems, in preparation.

[5] Kendall, D. G., in discussion on Bartlett, Measles periodicity and community size. J. Roy. Stat. Soc. Sez. A 120 (1957), 48-70.

[6] Kermack, W. O. and Mc Kendrick, Contributions to the mathematical theory of epidemics. Proc. Roy. Soc. A 115 (1927), 700-721.

[7] Mahajan, V. and Peterson R. A., Models for Innovation Diffusion. Sage Publications, Beverly Hills, 1985.

[8] Mahajan, V. and Wind, Y. , (Editors), Innovation Diffusion Models of New Product Acceptance. Bellinger Pub. Co., Cambridge, Mass., 1986.

[9] Murray, J.D., Mathematical Biology, Springer – Verlag, Heidelberg, 1989.

[10] Sharif, M.N., and Ramanathan K., Polynomial innovation diffusion models. Technological Forecasting and Social Change 21 (1982), 301-323.

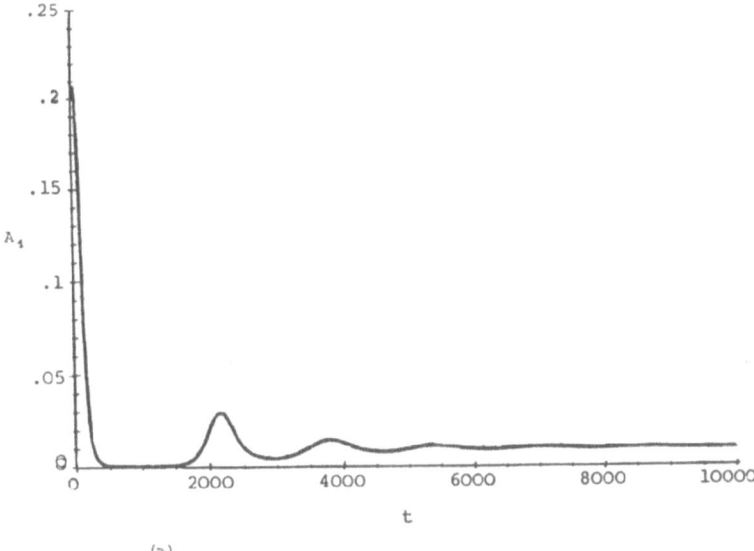

(a)

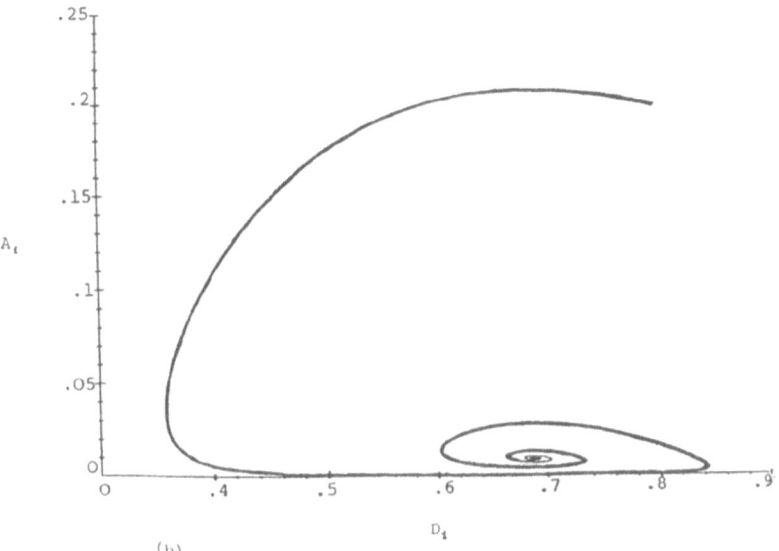

(b)

Fig. 2.1 λ_{ii} = 0.05 , λ_{ij} = 0   i, j = 1, 2   i $\neq$ j
μ_i = 10^{-3}, α_i = 30^{-1}, D_i^0 = 0.8, A_i^0 = 0.2,   i = 1, 2.

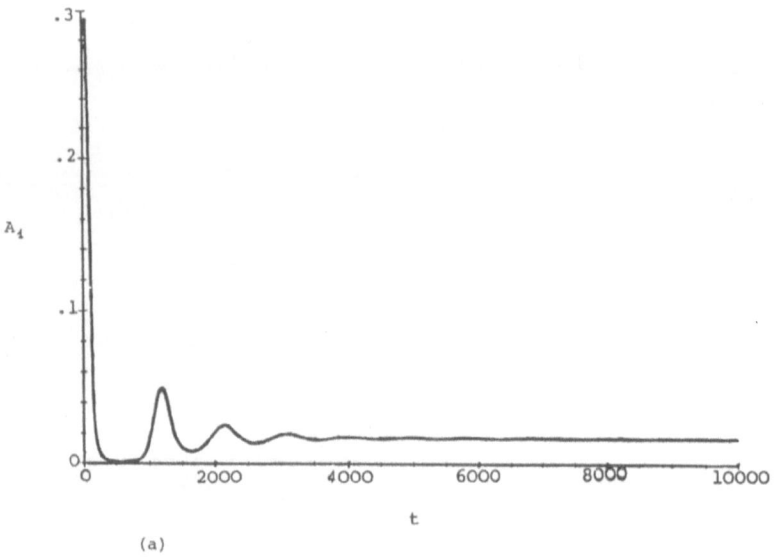

(a)

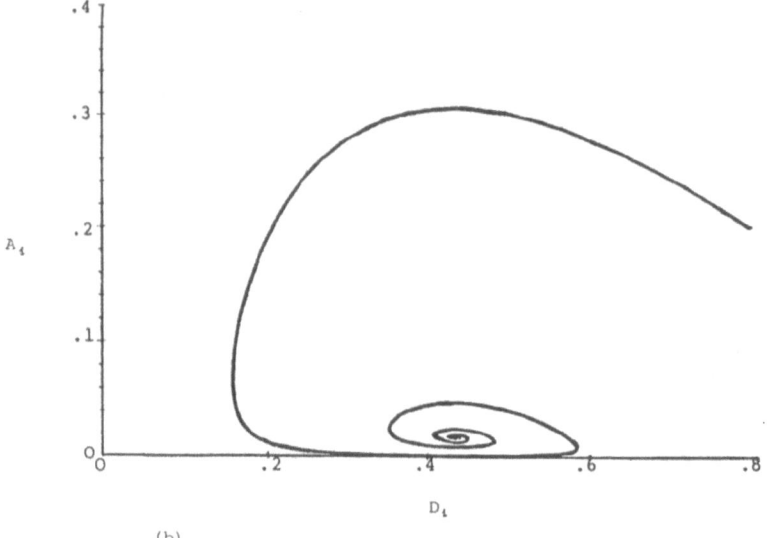

(b)

Fig. 2.2 $\lambda_{ii} = 0.05$ $\lambda_{ij} = 0.03,$ $i,j = 1,2$ $i \neq j$
$\mu_i = 10^{-3},$ $\alpha_i = 30^{-1},$ $D_i^0 = 0.8,$ $A_i^0 = 0.2,$ $i = 1,2.$

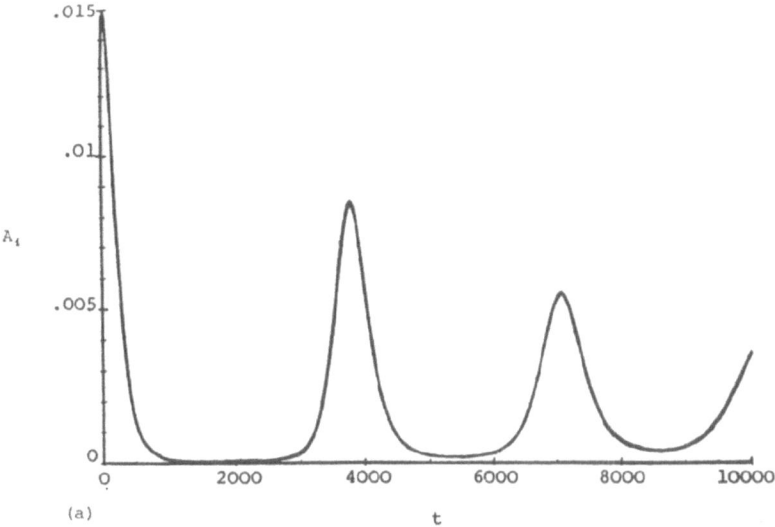

(a)

t

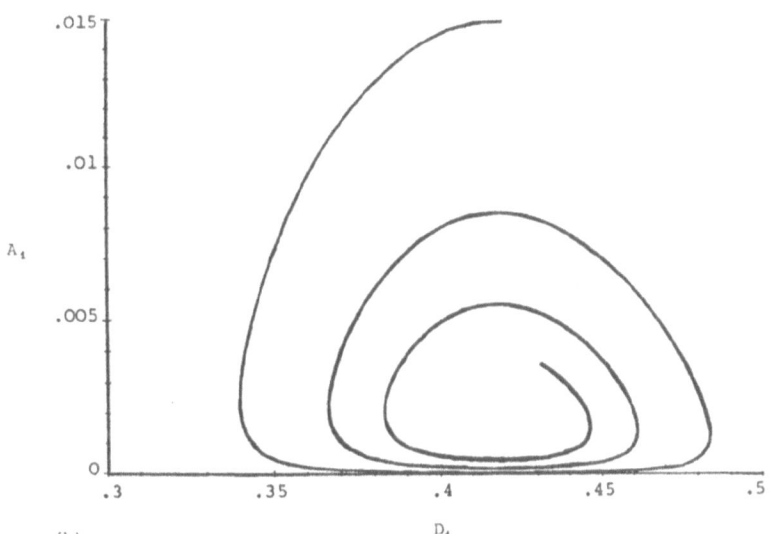

(b)

D_1

Fig. 2.3 $\lambda_{ii} = 0.05$, $\lambda_{ij} = 0.03$ $i, j = 1, 2$ $i \neq j$
      $\mu_i = 10^{-4}$, $\alpha_i = 30^{-1}$, $D_i^0 = 0.8$, $A_i^0 = 0.2$, $i = 1, 2$.

IMAGE   SEGMENTATION

BY

MINIMUM   INFORMATION   LOSS

R. CASELLI

SASIAM (School for Advanced Studies in Industrial and
        Applied Mathematics)
        Tecnopolis, Valenzano (Ba), Italy

B. FORTE

Department of Applied Mathematics, University of Waterloo,
Ontario, CANADA

ABSTRACT

The grey colour in a picture is a tool to convey information
about the subject. The transition from the original image to
its digitized version, by thresholding its grey levels,
yields a loss of information. From a " natural " set of
axiomatic properties a mathematical representation is first
derived for the information loss. This takes into account
the grey level histogram of the original picture and that of
its digitized thresholded version. This is indeed the only
relevant information one has when dealing with a global
point-dependent method for image representation by
thresholding. It is shown that by using a more sophisticated
measure of the information loss one can achieve substantial
improvements over previous results.

Possible areas of application of the above techniques are
Robotics, Computer Vision and Medicine when transmission and
storage of images can require to deal with a reduced number
of grey levels.

Hj. Wacker and W. Zulehner (eds.),
Proceedings of the Fourth European Conference on Mathematics in Industry, 235.
© 1991 B.G. Teubner Stuttgart and Kluwer Academic Publishers.

A BRIEF MATHEMATICAL ANALYSIS
OF
ELECTROCHEMICAL MACHINING PROBLEMS

Søren Christiansen, Lyngby, Copenhagen

Summary: Electrochemical machining (ECM) is a modern technological method for metal removal. For the solution of some two–dimensional electrochemical problems, formulated as moving boundary problems, we present a mathematical method in the form of a system N ordinary differential equations for describing the movement of N marker points on the boundary. The system has been analysed, in some cases, with respect to the linear stability of equilibrium solutions and to the nonlinear stability of non–equilibrium solutions. The mathematical model has features which are in accordance with those of the physical problem. This has also been demonstrated by actual computation of some examples.

1   Formulation of the moving boundary problem

Electrochemical machining (ECM) is a modern technological method of metal removal, based on Faraday's laws of electrolysis [16], which has found useful applications [18; p. 152]. In ECM the metal to be machined is made the positive electrode, the anode, in an electrolytic bath, while the tool is made the negative electrode, the cathode [18; p. 152], and metal is dissolved electrolytically from the anode [17; p. 472]. Much attention has been paid to the problems associated with the smoothing of the anodic surface, where a plane–faced cathode–tool is placed parallel to the anode––workpiece so that irregularities are preferentially removed by the effects of the current density over the surface [18; pp. 153–154]. The purpose of the present paper is to report on an analysis of a mathematical model for a moving boundary problem for ECM, with emphasis on saw–tooth instabilities [2], [5] (see also [3]). For a more detailed exposition, with many references, see [4].

We consider a two–dimensional ECM model-problem (for the smoothing of the anode) formulated in an xy–coordinate system, 2π–periodic in the x–direction, with y $= 0$ being the straight cathode, with $y = \eta(x,t) > 0$ being the (moving) anode, where t is the time. The xy–coordinate system here introduced follows the cathode. Between the electrodes a scalar potential $\Phi = \Phi(x,y)$, 2π–periodic in x, satisfies Laplace's equation $\Delta\Phi = 0$. On the electrodes Φ satisfies the conditions: $\Phi \equiv 0$ on $y = 0$ and $\Phi \equiv 1$ on $y = \eta$. (The cathode is grounded and the anode has a constant potential which is put equal to one.) On $y = \eta$ the potential Φ satisfies a kinematic free surface boundary condition (with electrochemical machining constant equal to one [9; p. 412]), which in turn can be reformulated as

237

Hj. Wacker and W. Zulehner (eds.),
Proceedings of the Fourth European Conference on Mathematics in Industry, 237–242.
© 1991 B.G. Teubner Stuttgart and Kluwer Academic Publishers.

$$\eta_t(x,t) = \Phi_y(x,\eta) + \frac{\Phi_x(x,\eta)^2}{\Phi_y(x,\eta)} - v \; ; \; 0 \leq x \leq 2\pi \; , \tag{1}$$

where the quantity v, which is called the feedrate, indicates the speed by which the cathode is moved forward. We assume here v to be a constant, which is at disposal. At t = 0 the elevation η is prescribed as $\eta(x,0) := \overset{o}{\eta}(x)$, where $\overset{o}{\eta}(x)$ is a given (single valued) function, and the mathematical problem is to determine $\eta(x,t)$ from a given $\overset{o}{\eta}(x)$.

The elevation $\eta(x,t) = v^{-1}$, with $\Phi(x,y) = vy$, is a solution of a free boundary problem with a straigth cathode. It is intuitively obvious that

$$\overset{\infty}{\eta}(x) := \eta(x,\infty) := \lim_{t \to \infty} \eta(x,t) \equiv v^{-1}. \tag{2}$$

2   Method of solution

When the initial boundary $y = \overset{o}{\eta}(x)$ is prescribed, we here choose to solve the moving boundary problem by a front-tracking method [6; § 4], [12] in that we represent the curve by an ordered set of marker points located on the curve, and the position of the marker points is represented by the distance from a reference line, viz. the cathode y = 0, cf. [12; § 2]. We shall represent $\eta(x,t)$ by means of the following "Eulerian" marker points: For $x := x_i \equiv i\, 2\pi/N$; $i = 1,2,\cdots,N$; the values $\{\eta_i\}_1^N$, with $\eta_i = \eta(x_i,t)$ are combined into one N-vector $\eta = [\eta_1,\eta_2,\cdots,\eta_N]^T$. The N functions $\eta_i(t) := \eta(x_i,t)$ represent the distances. In order to make possible the analytical calculations leading to the results (5), it is necesserary that the values $\{x_i\}_1^N$ are tractable fractions of π. This makes it necessary to use *even* values of N.

For a fixed value of t, with a given curve $y = \eta(x,t)$, the problem for Φ has a unique solution. This problem is solved approximately by replacing Φ by a sum with N terms (N *even*)

$$\Phi^{(N)}(x,y) = A_o y + A_n \cos nx \sinh ny + \sum_{j=1}^{n-1} (A_j \cos jx + B_j \sin jx) \sinh jy, \tag{3}$$

with n = N/2, each term being 2π-periodic in x, satisfying Laplace's equation and the condition $\Phi = 0$ on y = 0. Because N must be even, the term with sinnx sinhny is absent in (3), which therefore may look a little bit strange.

For the determination of the N unknown coefficients $\{A_j\}_0^n$ and $\{B_j\}_1^{n-1}$ we here only consider a *point-collocation method*, where we require, that $\Phi^{(N)} = 1$ at the marker-points, which leads to a system of linear algebraic equations for the coefficients. This is an easy and straightforward method for constructing an approximation of the form (3). Numerical experiments show that this approximation is satisfactory when the amplitude of η is small, while it is clearly unsatisfactory when the amplitude of η is large.

From $\Phi^{(N)}$ the partial derivatives $\partial\Phi^{(N)}/\partial x$ and $\partial\Phi^{(N)}/\partial y$ are determined analytically and evaluated at the marker points, whereby the right hand side of (1) is expressed as a function of η, so that the time derivative η_i' is a function F_i of η, viz. η_i' = $F_i(\eta)$, which is combined to a vector function $F = [F_1, F_2, \cdots, F_N]^T$. Therefore we have expressed η' in terms of η as a system of N autonomous, non linear, ordinary differential equations (ODEs)

$$\eta' = F(\eta) \quad , \tag{4}$$

which together with the initial vector $\overset{o}{\eta}$, constitutes an initial value problem from which $\eta(t)$ can be found by (numerical) integration.

3   Analysis of the system of ODEs

Because of the structure of the model–problem under consideration, it can be seen that when $\eta_1 = \eta_2 = \cdots = \eta_N =: \bar{\eta}$ then the Jacobi matrix $\underline{\underline{J}}$, with elements $J_{ij} := \partial F_i/\partial \eta_j$, is *circulant* [7] and symmetric. Therefore the eigenvectors can be expressed solely in terms of a sampling of trigonometric functions, and the eigenvalues can be determined solely from the first row. When the potential problem for Φ is solved by means of the point–collocation method, with an *even* number N of terms, it is possible, in the special case when $\{\eta_i\}_1^N$ are all equal to $\bar{\eta}$, by means of the language REDUCE [11], to find that the eigenvalues of $\underline{\underline{J}}$ are

$$\lambda^{(0)} = -\frac{1}{\bar{\eta}^2} \, , \tag{5a}$$

$$\lambda^{(k)} = -k\,\frac{\coth(k\bar{\eta})}{\bar{\eta}} \, ; \, k = 1, 2, \cdots, \frac{N}{2} \, . \tag{5b}$$

The eigenvector $e^{(N/2)}$ has the form of a saw–tooth: $[-1, 1, -1, \cdots, 1]^T$. Because $\lambda^{(N/2)}/\lambda^{(0)} = O(N)$ the system of ODEs is *stiff*, and because the eigenvalues are all negative the solutions of a corresponding system of linear ODEs are stable [1; § 4.2, Th. 1].

In order to see if the system (4) is contractive, an analysis is performed using concepts from nonlinear stability theory for systems of ordinary differential equations, and in particular the concept *logarithmic norm* of a matrix [8; p. 31], [15; p. 418]. If the logarithmic matrix norm of the Jacobi matrix, determined from the system of ODEs, is negative, then we are assured of the contractivity of the exact solution for the general nonlinear problem, in terms of the corresponding matrix norm.

For a given vector η the Jacobi matrix $\underline{\underline{J}}$ is determined approximately by numerical differentiation by a two-sided difference approximation [here using an IBM 3081, VM/CMS, FORTVS, double precision (i.e. using around 16 decimal digits)] where h is chosen to be 10^{-5}, in accordance with a machine epsilon around 10^{-16}, cf. [10; p. 286] and from $\underline{\underline{J}}$ the logarithmic norm is computed, and becomes a scalar function of the vector η. Numerical experiments show that when the amplitude of the boundary curve is not too large then the logarithmic norm of $\underline{\underline{J}}$ is negative, and the system is therefore contractive.

4   Stability of numerical solution

In order to get a better insight into the numerical process we first apply elementary integration formulas, where each method is characterized by a region of absolute stability. Let λ be an eigenvalue of $\underline{\underline{J}}$, evaluated at the equilibrium solution $\bar{\eta}$, and let h be the time step, then if λh is outside the region, and the numerical solution η contains a component of the corresponding eigenvector of $\underline{\underline{J}}$, then this component will be growing (not necessarily monotonically) with the number of time steps.

When N is sufficiently large, the eigenvalue $\lambda^{(N/2)}$, cf. (5), corresponding to the saw-tooth eigenvector $[-1,1,\cdots,1]^{T}$, will be the absolutely largest and therefore it will determine a maximal step length h_{max} from a consideration of stability. With all the eigenvalues being negative, cf. (5), the integration methods do *not* create saw-tooth instability, provided that the step length h satisfies $S \leq \lambda h \leq 0$, or

$$0 \leq h \leq h_{max} := N^{-1} \cdot 2|S|\bar{\eta} \tanh \left(\frac{N}{2}\,\bar{\eta}\right) , \tag{6}$$

where S depends upon the integration method used, and $\bar{\eta}$ tend to v^{-1}, cf. (2) and § 3. Therefore the requirement regarding stability put a bound on the admissible step-length, in that $\Delta t = 0(\Delta x)$. If the time step h is slightly larger than h_{max}, then the vector η will develop into $\bar{\eta}$, which −solely because of numerical reasons− will be polluted with a superimposed saw-tooth component stemming from the eigenvector $[-1,1,-1,\cdots,1]^{T}$. An accurate description of the moving boundary requires N to be sufficiently large, and technological applications requires $\bar{\eta}$ to be small compared to the x-period ($= 2\pi$). Suitable values may be N = 36 and $\bar{\eta} = 1/10$.

Actual numerical integrations have been carried out using various methods of integration, e.g. Adams-Bashforth-Moulton, 4th order [19; p. 458] : S = −1.28481626 (in PECE mode), cf. [14; pp. 106-107] stating S = −1.25 and S = −1.3. With N = 36 and $\bar{\eta} = 1/10$, this gives $h_{max} = 0.00676$. Given a curve $\eta(x,0)$, which is not too wavy, such that the point-collocation method is applicable and the system (4) is contractive, and with h smaller than h_{max}, then the result is a smoothing of the anode, as expected, and a straight anode is obtained as the equilibrium solution $\hat{\eta}$, cf. (2).

Because the maximal permissible timestep, (6), in practical cases, is so small, it is necessary to treat the system (4) using routines particularly suited to integrating stiff systems.

5 References

[1] Braun, M.: Differential Equations and Their Applications. Springer–Verlag; New York et al. 2nd Edition (1978) 13 + 518 pp. (Applied Mathematical Sciences 15).

[2] Christiansen, S.: Development of Saw–Tooth Instabilities on a Moving Boundary. Zeit. Angew. Math. Mech. (ZAMM) (Sonderheft, GAMM–Tagung, Regensburg 1984) 65 (1985) T373–T375.

[3] Christiansen, Søren: A stability analysis of a Eulerian method for some surface gravity wave problems. pp. 75–84 of: Ballmann, Josef & Rolf Jeltsch (Eds.): Nonlinear Hyperbolic Equations – Theory, Computation Methods, and Applications (Proceedings of the Second International Conference on Nonlinear Hyperbolic Problems, Aachen, FRG, March 14–18, 1988). Notes on Numerical Fluid Mechanics 24, Friedr. Vieweg & Sohn; Braunschweig, Wiesbaden. (1989) 10 + 718 pp.

[4] Christiansen, Søren: A Stability Analysis of a Eulerian Solution Method for Moving Boundary Problems in Electrochemical Machining. Jour. Computational Appl. Math., submitted.

[5] Christiansen, Søren: Application of integral equations to an electrochemical machining problem. Mathematisches Forschungsinstitut, Oberwolfach, BRD. Lecture 1980–12–09.

[6] Cryer, Colin W.: Numerical Methods for Free and Moving Boundary Problems. pp. 601–622 of [13].

[7] Davis, Philip J.: Circulant Matrices. John Wiley & Sons; New York et al. (1979) 15 + 250 pp.

[8] Dekker, K.; J.G. Verwer, J.G.: Stability of Runge–Kutta methods for stiff nonlinear differential equations. CWI Monograph 2. North–Holland; Amsterdam et al. (1984) 9 + 307 pp.

[9] Forsyth Jr., P.; Rasmussen, H.: Solution of Time Dependent Electrochemical Machining Problems by a Co–ordinate Transformation. Jour. Inst. Math. Applications 24 (1979) 411–424.

[10] Hassard, B.D.; Kazarinoff, N.D.; Wan, Y.–H.: Theory and Applications of Hopf Bifurcation. Cambridge University Press; Cambridge et al. (London Mathematical Society Lecture Note Series 41) (1981) 6 + 311 pp.

[11] Hearn, Anthony C. (Ed.): REDUCE, User's Manual, The Rand Corporation, Santa Monica, CA 90406, Version 3.1., (April 1984).

[12] Hyman, James M.: Numerical methods for tracking interfaces. Physica D 12 (1984) 396–407.

[13] Iserles, A.; Powell, M.J.D. (eds.): The State of the Art in Numerical Analysis. Proceedings of the joint IMA/SIAM conference held at the University of Birmingham 14–18 April 1986. Clarendon Press; Oxford (1987) 14 + 719 pp.

[14] Lambert, J.D.: Computational Methods in Ordinary Differential Equations. John Wiley & Sons; Chichester et al. (1973) 15 + 278 pp.

[15] Lambert, J.D.: Developments in Stability Theory for Ordinary Differential Equa-
      tion. pp. 409–431 of [13]

[16] McGeough, J.A.: Principles of Electrochemical Machining. Chapman and Hall;
      London. (1974) 15 + 255 pp.

[17] McGeough, J.A.: Free and moving boundary problems in electrochemical ma-
      chining and flame fronts. pp. 472–482 of Fasano, A. & M. Primicerio (Eds.): Free
      boundary problems: theory and applications. Pitman Advanced Publishing Prog-
      ram; Boston, London, Melbourne (1983) Vol. I, pp. 1–320, Research Notes in Ma-
      thematics 78, Vol. II, pp. 321–711, Research Notes in Mathematics 79.

[18] McGeough, J.A.: Unsolved moving boundary problems in electro–chemical
      machining. pp. 152–156 of Bossavit, A.; A. Damlamian & M. Fremond (Eds.):
      Free boundary problems: applications and theory. Pitman Advanced Publishing
      Program; Boston, London, Melbourne (1985) Vol. III, pp. 1–303, Research Notes
      in Mathematics 120, Vol. IV, pp. 305–613, Research Notes in Mathematics 121.

[19] Young, David M. ; Gregory, Robert Todd: A Survey of Numerical Mathematics.
      I+II. Addison–Wesley Publishing Company; Reading, Massachusetts (1973) 51 +
      63 + 1099 pp.

Laboratory of Applied Mathematical Physics
The Technical University of Denmark
DK–2800 Lyngby
Denmark

A MATEMATICAL MODEL FOR THE PRODUCTION OF BIMETALLIC STRIP

E. Comparini, C. Manni, Università di Firenze, Italy

Summary: *A mathematical model for the study of the thermal field in each component (bronze and steel) of a moving bimetallic strip is proposed. The problem of phase change in the bronze is considered in order to obtain an estimate of the width of the solidification zone.*

1. Introduction

The industrial process for the production of uniform bimetallic strips consists in pouring molten bronze onto a hot steel strip, which moves along a horizontal line (see [1]).

The steel strip is fed through the casting line at a prescribed steady speed, selected within a fixed range, in order to guarantee the quality of the product. Then it is quenched with oil jets from below in order to solidify the bronze. Heat is then removed from the oil by a heat exchanger.

Since we are in the presence of a steady state process at a known uniform time-rate, our purpose consists in determining the spatial temperature distribution in both metals, studying the relative influence of conduction and of convection on the temperature in each component of the bimetallic strip, and in providing an estimate of the width of the solidification zone.

Let us consider a longitudinal section of the bimetallic strip, which moves in the x direction at velocity V. The steel occupies the zone between $y=0$ and $y=\epsilon_1$, the bronze lies between $y=\epsilon_1$ and $y=\epsilon_1+\epsilon_2$. Indicating by θ_1, θ_2 the temperature in steel and bronze respectively, energy balance yields the equation

$$(1.1) \qquad K(y)\,\Delta\theta(x,y) - V\,\frac{\partial E}{\partial x}(x,y) = 0 \,,$$

where

$$(1.2) \qquad E = \begin{cases} \rho_1\ c_1\ (\theta-\theta_I) \,, & 0<y<\epsilon_1, \\ \rho_2[c_2\ \theta + \lambda\ H(\theta)] \,, & \epsilon_1<y<\epsilon_1+\epsilon_2 \,, \end{cases}$$

θ_I is the difference between the melting temperatures of steel and bronze (which is chosen equal to 0), $H(\theta) = 1$ in the liquid bronze and 0 in the solid bronze, and $K(y)$ assumes the values K_1, K_2, in the two components (the thermal conductivity K_i, the density ρ_i and the specific heat c_i have been considered constant in each component, and θ is less than the melting temperature of steel).

Moreover we impose boundary conditions on $x=0$ and $x=1$ (ϵ_1, $\epsilon_2<<1$), and conditions governing heat exchange on the surfaces $y=0$ and $y=\epsilon_1+\epsilon_2$:

$$(1.2) \qquad \theta(0,y) = \theta^0 \,, \quad \theta_x(1,y) = 0 \,,$$

243

Hj. Wacker and W. Zulehner (eds.),
Proceedings of the Fourth European Conference on Mathematics in Industry, 243–247.
© 1991 B.G. Teubner Stuttgart and Kluwer Academic Publishers.

(1.3) $K_1\theta_{1y}(x,0)=h_1[\theta_1(x,0)-T_1],$ $K_2\theta_{2y}(x,\epsilon_1+\epsilon_2)=h_2[T_2-\theta_2(x,\epsilon_1+\epsilon_2)],$

where T_1 and T_2 are constant temperatures, h_1 and h_2 are the surface heat transfer coefficients. Finally, on the interface between bronze and steel, continuity of temperature and flux is assumed:

(1.4) $\theta_1(x,\epsilon_1) = \theta_2(x,\epsilon_1),$ $K_1\theta_{1y}(x,\epsilon_1) = K_2\theta_{2y}(x,\epsilon_1).$

By means of the change of variables $\xi=xV$ one reduces (1.1) to

$$\frac{K}{V^2}\theta_{\xi\xi} + K\theta_{yy} - E_\xi = 0.$$

For sufficiently high values of the casting velocity we can neglect the conduction in the direction of the velocity and consider the parabolic problem

(1.6) $$\frac{\partial}{\partial y}\left(K(y)\frac{\partial\theta}{\partial y}\right) - V\frac{\partial E}{\partial x} = 0,$$

with initial and boundary conditions (1.2)-(1.4).

A proof of the convergence of the solution of (1.1) to the solution of (1.6) for $K/V^2 \rightarrow 0$ has been obtained in [2] in the case of the one-phase problem.

2. The problem for the averaged temperature

Taking advantage of the fact that ϵ_1, $\epsilon_2 \ll 1$, we can follow the methods of [3] for a model of a heat exchanger, and study the problem for the averaged temperature on each transversal section of the strip, in order to obtain a simpler scheme.

Let us define

(2.1) $w_1(x)=\frac{1}{\epsilon_1}\displaystyle\int_0^{\epsilon_1}\theta_1(x,y)\,dy,$ $w_2(x)=\frac{1}{\epsilon_2}\displaystyle\int_{\epsilon_1}^{\epsilon_1+\epsilon_2}\theta_2(x,y)\,dy,$

and analogously $\bar{E}_1(x),$ $\bar{E}_2(x).$

Integrating equation (1.6) w.r.t. y in the two zones, and using (1.3) we obtain

(2.2) $\frac{1}{\epsilon_1}[K_1\theta_y(x,\epsilon_1)-h_1(w_1(x)-T_1)]-V\bar{E}_1'(x)=0,$ $\frac{1}{\epsilon_2}[h_2(T_2-w_2(x))-K_2\theta_y(x,\epsilon_1)]-V\bar{E}_2'(x)=0.$

Now, let us divide the strip into three different zones:

a) $0 \leq x < \bar{x}$ is the zone where the bronze is liquid ($\bar{x}$ represents the point in which $\bar{E}_2=L=\rho_2\lambda$, corresponding approximately to the first point in which $w_2=0$).

b) $\bar{x} \leq x \leq \bar{\bar{x}}$ is the solidification zone.

c) $\bar{\bar{x}} < x \leq 1$ is the zone where the bronze is solid.

In the zone a), (see [4], sec. 4) the terms calculated

on the interface in equation (2.2) can be replaced by a radiation condition, with a suitable coefficient $h_0 = \dfrac{2K_1 K_2}{\epsilon_1 K_2 + \epsilon_2 K_1}$,

(2.3) $K_i \theta_{iy}(x, \epsilon_i) = h_0[w_2(x) - w_1(x)]$, i=1,2.

Considering non-dimensional variables $u_i = w_i/\theta^0$, $x = x/l$ (setting $T_i = T_i/\theta^0$), our problem reduces to a linear system of two ordinary differential equations, with constant coefficients, whose solution is

(2.5) $u_1(x) = \alpha_1 e^{\lambda_1 x} + \alpha_2 e^{\lambda_2 x} + u_1^{\,0}$,

$u_2(x) = \dfrac{\alpha_1}{\omega_1}(\lambda_1 + \omega_1 + \gamma_1) e^{\lambda_1 x} + \dfrac{\alpha_2}{\omega_1}(\lambda_2 + \omega_2 + \gamma_2) e^{\lambda_2 x} + u_2^{\,0}$,

with

$\lambda_{1,2} = -\tfrac{1}{2}(\omega_1 + \gamma_1 + \omega_2 + \gamma_2)\left\{1 \pm \sqrt{1 - 4\dfrac{\omega_1 \gamma_2 + \omega_2 \gamma_1 + \gamma_1 \gamma_2}{(\omega_1 + \gamma_1 + \omega_2 + \gamma_2)^2}}\right\}$,

$u_1^{\,0} = \dfrac{(\omega_2 + \gamma_2)\gamma_1 T_1 + \omega_1 \gamma_2 T_2}{(\omega_1 + \gamma_1)(\omega_2 + \gamma_2) - \omega_1 \omega_2}$, $u_2^{\,0} = \dfrac{\omega_1 + \gamma_1}{\omega_1} u_1^{\,0} - \dfrac{\gamma_1 T_1}{\omega_1}$, $\omega_i = \dfrac{h_0 l}{\rho_i c_i \epsilon_i V}$, $\gamma_i = \dfrac{h_i l}{\rho_i c_i \epsilon_i V}$,

and α_1, α_2 constants which can be determined by means of the condition at x=0.

Imposing $u_2(\bar{x}) = 0$, one can determine $\bar{x}$ and $u_1(\bar{x}) = \bar{u}_1$.

An estimate of the length of the solidification front can be obtained assuming that $0 < \bar{E}_2 < L$ in the zone b), that is, we impose $u_2(x) \equiv 0$ in $\bar{x} \leq x \leq \bar{\bar{x}}$, simplifying the boundary conditions accordingly. The resulting system can be immediately integrated, yielding

(2.6) $u_1(x) = \left(\bar{u}_1 - \dfrac{\gamma_1 T_1}{\omega_1 + \gamma_1}\right) \exp[-(\omega_1 + \gamma_1)(x - \bar{x})] + \dfrac{\gamma_1 T_1}{\omega_1 + \gamma_1}$,

$\dfrac{\bar{E}_2(x)}{\rho_2 c_2} = -\dfrac{\omega_2}{\omega_1 + \gamma_1}\left(\bar{u}_1 - \dfrac{\gamma_1 T_1}{\omega_1 + \gamma_1}\right)\{\exp[-(\omega_1 + \gamma_1)(x - \bar{x})] - 1\}$

$+ \left(\dfrac{\omega_2 \gamma_1 T_1}{\omega_1 + \gamma_1} + \gamma_2 T_2\right)(x - \bar{x}) + \dfrac{L}{\rho_2 c_2}$,

The equation $\bar{E}_2(\bar{\bar{x}}) = 0$ has one solution which can be estimated by

(2.7) $\dfrac{L/\rho_2 c_2}{-\gamma_2 T_2 - \dfrac{\omega_2 \gamma_1 T_1}{\omega_1 + \gamma_1}} \leq \bar{\bar{x}} - \bar{x} \leq \dfrac{L/\rho_2 c_2 + \dfrac{\omega_2}{\omega_1 + \gamma_1}\left(\bar{u}_1 - \dfrac{\gamma_1 T_1}{\omega_1 + \gamma_1}\right)}{-\gamma_2 T_2 - \dfrac{\omega_2 \gamma_1 T_1}{\omega_1 + \gamma_1}}$.

Numerical computations show that in the zone b) the liquid bronze has temperature very close to 0, so that a more refined approximation of the solidification zone can be obtained considering a one-phase Stefan problem for the temperature in the solid bronze (the origin of the x, y axes has been shifted to the point $(\bar{x}, \epsilon_1)$, determined above), with a parabolic front $y = \alpha \sqrt{x} = s(x)$:

(2.8) $K_2 \, \theta_{2yy}(x,y) = \rho_2 c_2 \, V \, \theta_{2x}(x,y),$ in $D_{\bar{x}} = \{0 \leq x, \, 0 < y < \epsilon_2\},$

$\theta_2(x,0) = w_0, \quad \theta_2(x,s(x)) = 0, \quad L \, V s'(x) = -K_2 \, \theta_{2y}(x,s(x)),$

α and w_0 being constant to be determined.

The solution of problem (2.8) is $\theta(x,y) = f\left(\frac{y}{2\sqrt{x}}\right),$

$$f(z) = w_0 + \frac{VL}{K_2} \, \alpha \, e^{\frac{\alpha^2}{4b}} \sqrt{b} \int_0^{z/\sqrt{b}} e^{-\eta^2} \, d\eta \, ,$$

with $b = \dfrac{K_2}{\rho_2 c_2 V}$, and $w_0 = -\dfrac{VL}{K_2} \, \alpha \, e^{\frac{\alpha^2}{4b}} \sqrt{b} \displaystyle\int_0^{\alpha/2\sqrt{b}} e^{-\eta^2} \, d\eta \, .$

Now we impose the energy balance in the solidification zone: integrating (2.1) in $\{0 \leq x \leq x_0 \, , \, 0 \leq y \leq s(x)\}$, where x_0 is such that $s(x_0) = \epsilon_2$, and imposing the bronze-steel interface boundary condition

(2.9) $K_i \, \theta_{iy}(x,0) = h_0(w_0 - w_1(x)) \, ,$

we have (using non-dimensional variables as above and setting $u_0 = w_0/\theta^0$)

(2.10) $u_2\left(\dfrac{x_0}{l}\right) = w_2 \displaystyle\int_0^{x_0/l} u_1(x) \, dx - w_2 \, u_0 \, \dfrac{x_0}{l} + \dfrac{L}{\rho_2 c_2 \theta^0} \, .$

In this equation, $u_1(x)$ is obtained solving the first of (2.2), with (2.9):

(2.11) $u_1(x) = (\bar{u}_1 - \hat{u}) \exp[-(\omega_1 + \gamma_1) \, x] + \hat{u} \, , \quad \hat{u} = \dfrac{\gamma_1 \, T_1}{\omega_1 + \gamma_1} + \dfrac{\omega_1 \, u_0}{\omega_1 + \gamma_1} \, .$

Now, equation (2.10), with condition $x_0 = \dfrac{\epsilon_2^2}{\alpha^2}$, gives a unique value for the constant α .

An explicit expression for α can be obtained considering that, in practical cases,

$\alpha \ll 2\sqrt{b} = 2 \sqrt{\dfrac{K_2}{\rho_2 c_2 \, V}}$,   which implies $w_2(x_0) \simeq -\dfrac{VL}{K_2} \dfrac{\alpha^2}{4}$, and $w_0 \simeq -\dfrac{VL}{K_2} \dfrac{\alpha^2}{2}$,

from which (2.10) becomes

(2.12) $A_0 \, V^2 \, \alpha^4 + A_1 \, V \, \alpha^2 + A_2 = 0,$

where $A_0 = \dfrac{L}{2K_2}\left\{\dfrac{1}{2} + \dfrac{\omega_1 \omega_2}{(\omega_1 + \gamma_1)^2}\right\}, \quad A_2 = \dfrac{\epsilon_2^2}{l} V \dfrac{\omega_2 \gamma_1 T_1}{\omega_1 + \gamma_1} \, ,$

$A_1 = \dfrac{L}{\rho_2 c_2} + \dfrac{L \, V \, \omega_2 \epsilon_2^2}{2K_2 l} \dfrac{\gamma_1}{\omega_1 + \gamma_1} + \dfrac{\omega_2 \bar{u}_1 \theta^0}{\omega_1 + \gamma_1} - \dfrac{\omega_2 \gamma_1 T_1}{(\omega_1 + \gamma_1)^2} \, .$

It can be seen from estimate (2.7) and from the solution

of (2.12) that the width of the solidification zone depends linearly on the tickness of the strip and on the casting velocity. If however $h_i = O(\epsilon_i)$, the width does not depend on ϵ_i, in agreement with [5].

For more details and numerical computations see [6].

3. References

[1]   H.J.Sewell, M.J.Sewell, Mathematical modelling of the Quench Region in the continuous casting of Bimetallic Strip, Math. Eng. in Industry, 1,(4),(1987), 289-312.

[2]   J.F. Rodriguez, L. Santos, Asymptotic convergence in one phase continuous casting Stefan problem with high extraction velocity, I.M.A. J. Appl. Math. (1988).

[3]   A.Fasano, M.Primicerio, S.Grochmal, Su una classe di fenomeni diffusivi con scambio di materia o calore tra mezzi in moto relativo, Atti Sem. Univ. Modena, (1989).

[4]   E. Comparini, C. Manni, On a model for the temperature distribution in moving bimetallic strips, to appear on Meccanica.

[5]   A. Fasano, M. Primicerio, Phase change without sharp interfaces, Mathematical Models for Phase change Problems, ISNM 88, Rodriges ed., Birkhäuser Verlag, Basel, (1989).

[6]   E. Comparini, C. Manni, An approximation of the thermal field in a continuous casting process of a thin metal layer, to appear.

Computing guaranteed error bounds for problems in renewal theory

Hans–Jürgen Dobner

1 Introduction

Equations of renewal type are linear Volterra integral equations having the special form

$$(1) \qquad x(s) = g(s) + \int_0^s k(s-t)x(t)dt, \ 0 \le s \le a, \ a \in (0, \infty),$$

where the kernel and inhomogeneity are continuous. We are searching solutions in the real Banach Space $B = C[0, a]$ which has the usual maximum norm.

Apart from special cases equation (1) must be solved numerically, where various quadrature formulae (see. [7]) are applied. These discretization methods have a simple structure and are easy employable but having the disadvantage that there is a lack of tools for assessing precise the approximation error. Even when a standard error analysis gives explicit bounds the error is often ignored because these theoretical estimates are not easy to use and lead to an unrealistic overestimation; therefore an exact error analysis is left undone.

Equations of type (1) play for example an important role in pharmokinetics, describing there the interaction between drug and receptors (cf. [3]). In biomathematics there is need for reliable high accurate algorithms, so that traditional approaches frequently fail. Therefore in this paper a new kind of algorithm is described, avoiding the disadvantages of conventional numerics. Due to Kaucher (cf. [4]) this algorithms, where close mathematically guaranteed error bounds are part of the output, are called E–Methods. The letter E arises from the following quality properties, typical for an E–Method:

- Enclosing the solution within a ball of a small diameter.

- Existence verification of this solution.

2 Mathematical foundations

There are two sorts of errors in numerical procedures. The first one arises from roundoff- and the second one from approximation errors. If constructing safe bounds these errors must be estimated and controlled during the whole computation.

The roundoff errors are included by a precise computerarithmetic in connection with interval analysis.

A precise formulation of floating point arithmetic can be shortly chara·terized by the requirement:
No floating point number between the exact result and the rounded result of elementary floating point operations.

This arithmetic with maximum accuracy is realized with the help of monotone, antisymmetric roundings (cf. [5], [6])) and available for contemporary scientific data types on many computer systems. It is impossible to compute reliable error bounds without a precise formulation and implementation of the arithmetic.

249

Hj. Wacker and W. Zulehner (eds.),
Proceedings of the Fourth European Conference on Mathematics in Industry, 249–253.
© 1991 B.G. Teubner Stuttgart and Kluwer Academic Publishers.

The set of all real intervals

(2) $$[\alpha] = [\underline{\alpha}, \overline{\alpha}] = \{x \in \mathbb{R} \mid \underline{\alpha} \leq x \leq \overline{\alpha}\},$$

is denoted with $\mathbb{IR}$. In $\mathbb{IR}$ the elementary operations are defined according to

(3) $$[\alpha] * [\beta] := \{x \mid x = \alpha * \beta, \ \alpha \in [\alpha], \ \beta \in [\beta]\}, \ * \in \{+, -, \cdot, /\}$$

with nonvanishing denominator in case of division.

These constructs are supported by higher programming languages which allow the use of new concepts such as
- arithmetic operations with controlled rounding including the exact scalar product for avoiding cancellation
- interval arithmetic
- general function and operator concept
- higher data types.

PASCAL–SC and FORTRAN– SC, extensions of PASCAL and FORTRAN are examples for such programming languages.

The second kind of error occurs when approximating the solution of (1) in a finite dimensional space. Therefore the original problem formulated in the infinite dimensional space B must be made capable of a direct encoding into a computer in such a way that the unavoidable approximation errors are enclosed automatically. This goal is reached by producing a function tube inside of which the unknown solution is guaranteed to exist. The set is described by its boundaries. Let $b_i = b_i(s)$ denote a generating system of B, then with a fixed integer n we consider functions of the form

(4) $$[f](s) = [f] = [\alpha_0]b_0(s) + \ldots + [\alpha_n]b_n(s), \ \text{with} \ [\alpha_i] \in \mathbb{IR}$$

The set of all such interval or set functions is denoted with I_n; note that the elements of I_n are closed, bounded and convex. For many purposes the monomials $b_i(s) = s^i$, are an appropriate choice for a generating system.

Definition 1

The space I_n together with the approximated operations $[+], [-], [\cdot], [/], [\int_o^s]$ is called a functoid, where this operations are defined such that I_n is algebraically closed, that is

(5) $$[f][*][g] \in I_n, \ \text{for} \ [f], [g] \in I_n, \ * \in \{+, -, \cdot, /\}$$

(6) $$[\int_o^s][f] \in I_n, \ \text{for} \ [f] \in I_n.$$

A functoid for continous real valued functions $f(s,t)$, $0 \leq t \leq s \leq a$ contains elements of the form

$$[f](s,t) = [f] = \sum_{i=0}^{n} \sum_{j=0}^{n} [\alpha_{ij}]b_i(s)b_j(t) \, , [\alpha_{ij}] \in \mathbb{IR}$$

and will be indicated with $I_{n \times n}$.

Definition 2
$[f] \in I_n$ is called an inclusion approximation of $f \in B$ iff

(7) $$f(s) \in [f](s), \quad 0 \le s \le a,$$

holds.

An inclusion approximation for two dimensional real valued continous functions $f(s,t)$, $0 \le t \le s \le a$, or operators $T : B \to B$ are explained analogously.

Remark 1
An inclusion approximation $[f]$ contains all functions $f \in B$, for which $f \subseteq [f]$ in the graph sense is true. Furthermore we have for all $f \in [f], g \in [g]$ the relation $f * g \in [f][*][g]$.

An inclusion approximation for a given element of B can be obtained by computing with E-Methods the corresponding Taylor– or Fourier formulae where the remainder term is enclosed automatically, thus obtaining a function of type (4). The computation of inclusion approximations and the functoid operations are available as standard routines.

3 Inclusion Methods

Now our presented tools are connected yielding a selfvalidating method for equations of type (1), where we use the notations and abbreviations as introduced before.

Theorem 1
If we iterate in I_n according to
$$[x_0] := [g]$$

(8) $$[x_{i+1}] := [g][+][\int_0^s][k](s-t)[x_i](t)dt,$$

and the inclusion condition
(9) $$[x_{i+1}] \subseteq [x_i]$$

can be achieved for nonvoid iterates $[x_i]$, $[x_{i+1}]$, then there exists a solution x of (1) and additionally the error estimation $x \in [x_{i+1}]$ is validated.

Proof
From (8) and the property of inclusion approximations we can conclude with (9) and Schauders fixed point theorem that the compact operator $g(s) + \int_0^s k(s-t)x(t)dt$ has a fixed point in the set $[x_{i+1}]$.

Remark 2
From (9) there follows for all $\tilde{x} \in [x_{i+1}]$ the error estimation

(10) $$\frac{\|x - \tilde{x}\|}{\|x\|} \le \frac{\delta[x_{i+1}]}{\|[x_{i+1}]\| - \delta[x_{i+1}]},$$

here δ denotes the diameter and $\|[x_{i+1}]\|$ the maximnorm for sets, defined as follows

$$\|[x_{i+1}]\| = \sup_{\tilde{x} \in [x_{i+1}]} \|\tilde{x}\| \quad .$$

Remark 3

It is guaranteed by computational means alone that the unknown solution x exists and lies in a function tube bounded by two elements $\underline{x}_{i+1}$, $\overline{x}_{i+1}$ of I_n. Note that all relations have to be interpreted pointwise in the graph sense.

Remark 4

If the diameter of the solution containing set is too large it can be decreased by a subsequent residual correction step.

Remark 5

In applications, one uses monotone basis systems, i.e. a system where all elements $b_i(s)$ have the same monotonicity property, so that the inclusion condition (9) is proved for elements of type (4), when (9) holds for the corresponding interval coefficients. The latter test can be done easily by computational means.

4 Numerical examples

- *Cancer chemotherapy*

$$x(s) = 1 - \int_0^s \exp^{-t^2} dt + \int_0^s \exp^{-(s-t)^2} x(t)dt$$

E–Method: guaranteed accuracy of 8 correct digits.

- *Life distribution of a machine component*

$$x(s) = 1 - 0.5s^3 + 1.35s^8 + \int_0^s \exp^{-\lambda(s-t)} \frac{\lambda^k(s-t)^{k-1}}{(k-1)!} x(t)dt$$

E–Method: guaranteed accuracy of 11 correct digits.

References

[1] Alefeld,G.; Herzberger,J.: Introduction to interval computing. New York, Academic Press 1983

[2] Dobner, H.–J.: Contributions to Computational Analysis. Bull. Austral. Math. Soc. Vol. 41(1990),231–235

[3] Feldmann, U.; Schneider,B.: A General Approach to Multicompartment Analysis and Models for the Pharmacodynamics. Berlin, Heidelberg, New York. Lecture Notes in Biomathematics 11(1976),243–277

[4] Kaucher,E.; Miranker, W.L.: Self–Validating Numerics for Function Space Problems, Academic Press, New York 1984

[5] Kießling, J.; Lowes, M.; Paulik,A.: Genaue Rechnerarithmetik, B.G. Teubner, Stuttgart 1988

[6] Kulisch, U.: Grundlagen des Numerischen Rechnerns. Bibliographisches Institut, Mannheim 1976

[7] Linz, P.: Analytical and Numerical Methods for Volterra Equations. SIAM Studies in Applied Mathematics 7, Philadelphia 1985.

Dr. Hans–Jürgen Dobner
Mathematisches Institut II
Universität Karlsruhe
Kaiserstr.12
D 7500 Karlsruhe
FRG

Threshold Accepting Algorithms For 0-1 Knapsack Problems

Gunter Dueck and Jens Wirsching

1. Introduction

In [1] Drexl presented a Simulated Annealing (SA) algorithm for multi-constraint 0-1 knapsack problems (MCKP). Drexl studied 57 different MCKP's which have been published in the literature. In [2], the new optimization heuristic Threshold Accepting (TA) has been introduced. It is demonstrated in [2] that TA yields better results than SA for Traveling Salesman problems. In this paper we give a suited TA algorithm for MCKP's and report the computational results of test runs for the same set of 57 MCKP's. Since we were able to use the very same computer as Drexl (an IBM 3090-200 VF), we are able to make quite a fair comparison between the results with SA and TA.

We shall see that TA is faster and gives considerably better results. This fact substantiates the claim of [2] that TA seems to be superior to SA.

In the end of the paper we shall present a large application of the TA algorithm to a problem in marketing research which we were able to solve for the Gesellschaft für Konsumforschung (GfK), Nürnberg.

We give a short description of a MCKP:

$$\text{Maximize} \quad z \;=\; \sum_{j=1}^{n} c_j x_j$$

$$\text{subject to} \quad \sum_{j=1}^{n} a_{ij} x_j \;\leq\; b_i \qquad \text{for } i \;=\; 1, \dots, m$$

$$\text{and} \quad x_j \in \{\, 0\,,\, 1\,\} \qquad \text{for } j \;=\; 1, \dots, n \;.$$

We assume that all c_j, a_{ij}, and b_j are nonnegative.

The algorithms TA and SA we discuss here work as follows. First an initial configuration is chosen, i. e., an initial setting of the x-variables. For the case of MCKP's we study here the all zero sequence is always a feasible initial setting for these variables.

Each step of the two algorithms consists of a slight change of the old configuration into a new one. One compares the "qualities" of the two configurations, i. e. the two values of the objective function for the two configurations. Then a decision is made if the new configuration is "acceptable". If the new configuration is acceptable it serves as the "old" configuration for the next step. If it is not acceptable, the algorithm proceeds with a new change of the old configuration.

The two algorithms differ in their decision rule to determine whether a configuration is acceptable or not.

Hj. Wacker and W. Zulehner (eds.),
Proceedings of the Fourth European Conference on Mathematics in Industry, 255–262.
© 1991 B.G. Teubner Stuttgart and Kluwer Academic Publishers.

Kirkpatrick et al. [6] introduced the concept of "annealing" and combined it with the well-known Monte-Carlo algorithm by Metropolis et al. [7] which originally was used to numerically perform averages over large systems from Statistical Mechanics. The idea of SA runs as follows.

SA Algorithm for maximization
    choose an initial configuration
    choose an initial temperature $T > 0$
    Opt: choose a new configuration which is a stochastic small
           perturbation of the old configuration
           compute $\Delta E := $ quality(new configuration)- quality(old configuration)
           IF $\Delta E > 0$
               THEN old configuration := new configuration
               ELSE with probability $\exp(\Delta E/T)$
                           old configuration := new configuration
           IF a long time no increase in quality or too many iterations
               THEN lower temperature T
           IF the temperature is too low to promise further improvements
               THEN stop
    GOTO Opt

The method called **_Threshold Accepting_** (TA) which is studied in [2], is formally very similar to SA.

TA Algorithm for maximization
    choose an initial configuration
    choose an initial THRESHOLD $T > 0$
    Opt: choose a new configuration which is a stochastic small
           perturbation of the old configuration
           compute $\Delta E := $ quality(new configuration)- quality(old configuration)
           IF $\Delta E > -T$
               THEN old configuration := new configuration
           IF a long time no increase in quality or too many iterations
               THEN lower THRESHOLD T
           IF the threshold is too low to promise further improvements
               THEN stop
    GOTO Opt

TA accepts **every** new configuration which is **not much worse** than the old one (SA accepts worse solutions only with rather small probabilities) .

In the next Section we shall present a special TA algorithm for MCKP's. This algorithm is definitely more successful than Drexl's SA algorithm and seems to be also superior to many other heuristics which have been designed for MCKP's, cf. [1].

2. The TA Algorithms

The following TA algorithm is designed formally very similar to Drexl's SA algorithm. It is essentially an exchange algorithm. For the design of a local search algorithm we have to define what a "small perturbation" of a current solution should mean. The simplest idea: One changes single x-varibles from zero to one or from one to zero. However, the restriction to such trivial changes does not lead to successful algorithms. In MCKP's, a change of an x-variable in a feasible solution from one to zero gives always a new feasible solution.

On the other hand, a change of a single x-variable from zero to one produces mostly infeasible solutions. Thus, a simple "drop or add" algorithm yields very poor results. A much better approach is to study 2-exchanges, i.e. to choose two unequal x-variables and to exchange their values. Like Drexl we use here a combined form of simple changes and 2-exchanges. In [1], Drexl mentions: "The use of deterministic exchange algorithms for solving combinatorial optimization problems has the main drawback that - heavily depending on chosen starting solutions - the algorithms quite often only find bad local optima." The following version of our TA approach gives a proof to the contrary. Already in [2], we observed that deterministic TA versions perform equally well as stochastic versions. Hence, we designed also deterministic TA algorithms for the MCKP. We do not choose the indices h and k in the above exchange step at random. We examine the index pairs (h,k) in a prescribed order. There is a single randomness still in our algorithm: This is essentially the choice of the starting configuration. Hence, we can say: In the following TA algorithm "we choose randomly an initial configuration" and proceed completely deterministically.

Deterministic TA Algorithm for the MCKP

```
logical change
real threshold, φ
x_i : = 0 ∀ i
z : = 0
change : = .true.
compute a starting value for threshold
initialize the array permut of size 3n with three randomly created permutations of the numbers 1 to n
WHILE change DO
BEGIN
    change : = .false.
    FOR i : = 1 TO 3n DO
    BEGIN
        h : = permut(i)
        IF (x_h = 0) THEN
        BEGIN
            IF (a change of x_h from 0 to 1 reaches a feasible solution) THEN
            BEGIN
                x_h : = 1
                z : = z + c_h x_h
            END
        END
        IF (x_h = 1) THEN
        BEGIN
            IF (c_h < threshold) THEN
            BEGIN
                x_h : = 0
                z : = z - c_h x_h
                change : = .true.
            END
            ELSE BEGIN
                search the maximal k in the neighborhood of h satisfying (x_k = 0) and
                (c_h - c_k < threshold)
                IF (there exists such a k) AND (the changes of x_h and x_k reaches a feasible solution)
                THEN
                BEGIN
                    x_h : = 0
                    x_k : = 1
                    z : = z + c_k x_k - c_h x_h
                    change : = .true.
                END
            END
        END
    END
    threshold : = min(threshold · φ, threshold - 2)
END DO
```

Initially, we select three random permutations of the integers 1 to n. The algorithm is partitioned into "rounds". One round consists of 3*n exchange trials with the choice of the index h in the selected order. The algorithms stops if there has been no change during one round. After every round, the threshold is lowered by a multiplication by a constant factor φ. At least, however, the threshold is lowered by 2. We used φ in the range from .95 to .6 . The computational results reported of in the next Section have been obtained with the factor .8 . We lower the threshold at least by 2 to avoid unnecessary many loops in the end of the algorithm.

There is one severe and very important modification in the TA version we stated above: Once the index h is chosen we **do not** choose the index k by random or in the same order given by the initial random permutation. We make the following improvement:

We reorder the whole MCKP, i.e. we reorder the columns of the MCKP such that the coefficients c_i are sorted in a decreasing order. Then, for every index h chosen for an exchange trial, we choose k in the neighborhood of h, i.e. we choose k to be in the range from h-l, h-l+1, ..., h-1, h+1, ..., h+l for some number l.

Also in [2], it is observed that for Traveling Salesman problems such a neighborhood concept is very successful. We found out that for all real world Traveling Salesman problems the consideration of about 10 neighbors is completely satisfactory. Nevertheless, we were very astonished how successful the same concept works here. It turned out that the number l which defines the size of the "neighborhood" of an index h can be chosen in the order of 5 to 10 -- without affecting the quality of the computational results.

Thus, in the final version, we make exchange trials with indices h chosen in the initial order and with indices k chosen in the very close "neighborhood" of h in increasing order.

This simple idea guarantees that the number of exchange steps in our TA version grows only **linearly in the problem size** given by the number n of x-variables.

Hence, this modification is very important for the running time of the algorithm for large problem sizes. For the 57 test problems we study here, the effect is not very significant because these problems have a problem size only up to 105.

3. Computational Results

We give a tables of the results on the same 57 problems studied in [1]. The problems can be found in the papers [3] to [5] and [8] to [12]. In order to have comparable results we made the same approach "best of ten" as in [1]. From ten runs on the same problem we took the best result and computed the gap to the optimal solution. In a further column, we report the average gap to the optimum (10 runs), and we give eventually the number of times the optimum has been found during the ten runs.

Compared with the results obtained SA in [1], the TA results are definitely better. Drexl reported that SA solves 23 of the 57 problems to the optimum, TA solves 48. Moreover, the TA algorithm is on the average faster than Drexl's SA algorithm.

The TA algorithm runs particularly well on randomly generated problems. The Wei Shih problems are of that kind. The TA algorithm solves "nearly all" of those problems to the optimum.

The TA algorithm solved also many of the other (harder) problems. Here, Drexl's results indicate that the SA approach tends to find only the optimum for the smaller problems.

Problem	m	n	Gap to optimal solution (%)	Average gap to optimal solution (%)	number of found optima	CPU time in milliseconds
Shih 1	5	30	0	0	10	9.4
2			0	0.09	2	8.1
3			0	0.20	9	9.6
4			0	0.07	9	6.4
5			0	0	10	6.6
6		40	0	0.19	2	11.9
7			0	0.42	5	12.8
8			0	0.08	7	13.9
9			0	0	10	9.7
10		50	0	0.40	4	15.0
11			0	0.46	2	13.4
12			0	0.34	4	15.7
13			0	0.96	4	15.9
14		60	0	0.43	2	22.6
15			0	0.08	8	18.7
16			0	0.23	3	22.5
17			0	0.16	1	27.8
18		70	0.16	0.33	-	29.0
19			0	0.34	4	24.5
20			0	0.15	4	27.9
21			0	0.34	2	26.9
22		80	0	0.39	4	33.3
23			0	0.51	2	31.2
24			0	0.20	4	32.5
25			0	0.24	2	33.1
26		90	0	0.45	3	45.6
27			0	0.50	2	40.7
28			0	0.63	2	43.1
29			0	0.75	2	40.6
30			0	0.27	1	45.6
Peterson 1	10	6	0	14.74	6	0.9
2		10	0	2.54	4	4.1
3		15	0	0.15	8	7.5
4		20	0	0.38	3	7.9
5		28	0.08	0.34	-	11.9
6	5	39	0.45	4.03	-	18.7
7		50	0.11	1.11	-	25.7
Hansen, Plateau 1	4	28	1.11	1.94	-	12.8
2		35	0	4.94	2	15.4
Weingartner, Ness 1	2	28	0	0.44	7	12.6
2			0	0.02	8	10.7
3			0.06	1.34	-	7.8
4			0	0.59	8	10.2
5			0	2.15	4	8.8
6			0	0.27	1	10.6
7		105	0	0.05	2	88.1
8			0	1.52	1	47.4
Senju, Toyoda 1	30	60	0	0.36	1	26.5
2			0	0.17	1	41.1
Fleisher	10	20	0.79	1.87	-	3.8
Pb 1	4	27	0.97	2.25	-	10.8
2		34	0	2.50	2	13.6
3	2	19	0	3.72	2	6.5
4		29	0	5.52	2	11.6
5	10	20	0.79	1.89	-	5.1
6	30	40	0	2.14	5	20.1
7		37	0	1.18	1	17.0

Table 1. Results on 57 test problems from literature (TA with 2-exchange)

If one looks at the average gap to the optimum, it is seen that the results of the TA algorithm are very stable. Thus, for real world problems, one could possibly be content just with a single run rather with "the best of ten". There is one exception in the table: The problem Peterson 1. Here, the final solution for our algorithm is either the value 2400 or 3800. TA very often finds the optimum. However, if the optimum is missed, then the second best solution is very much worse. Peterson 1 is very ill-conditioned in this sense.

4. An Application of TA to Marketing Research

The G&I Forschungsgemeinschaft für Marketing, Nürnberg, is responsible for the marketing research for the GfK Gesellschaft für Konsumforschung, Nürnberg. The G&I reported to us a strong need for solutions of large 0-1 integer **quadratic** knapsack problems.

We describe here one of these problems which has a typical structure. There are many refinements of these problems which are usually a few times larger than that one we explain now.

For market analysis, the G&I inquires every week some market-relevant information from a panel of 5000 households. These 5000 households form a representative sample, i.e. the statistical properties of this sample reflect the statistical properties of the totality of the German households.

For any reasons, every month approximately 100 households are not further willing (able etc.) to report to the G&I. The G&I has to substitute these "losses" by adding 100 new households to the panel to rebuild a new representative model. For this purpose the G&I disposes of a certain pool of households who are willing to report for the G&I (approximately 500 households).

Thus, the following problem arises:

One is given 4900 "old" households and 500 possible "new" households. Select 100 households out of the reserve pool such that the insertion of the elected 100 households into the panel of the 4900 old households structures again a representative sample.

More formally:

To every household $H_i \in \{H_1, \dots, H_{500}\}$ in the reserve of the 500 households there is assigned a 0-1 vector $a_{i1}, \dots, a_{i40}$ of length 40 representing the relevant information for the panel structure. For instance,

$$a_{i29} = 1 \quad \text{resp.} \quad = 0$$

may contain the information: "H_i is resp. is not situated in the state Bavaria within Germany" or: "H_i's net income per month exceeds resp. does not exceed 3000 DM".

If we select 100 households $H_{j_1 k}, \dots, H_{j_{100} k}$ out of the reserve, the equation

$$\sum_{i=1}^{100} a_{j_i k} = b_k$$

means: b_k of the 100 households have the attribute k. To select 100 from 500 households appropriately in the G&I problem requires that for every $k = 1, \dots, 40$ a condition like that above has to be satisfied. We can now present a formal statement of the problem:

"Solve":

$$\sum_{i=1}^{500} a_{ik} x_i \; = \; b_k \qquad \text{for} \; k \, = \, 1, \dots , 40$$

$$\text{s.t.} \; \sum_{i=1}^{500} x_i \; = \; 100$$
$$\text{and} \quad x_i \; \in \; \{\,0\,,\,1\,\} \qquad \text{for} \; i \, = \, 1, \dots , 500$$

($x_i = 1$ or 0 means: household i is selected or not)
Of course, this system has generally no solution.
We reformulate:

Solve:

$$\sum_{i=1}^{500} a_{ik} x_i \; \text{"as near as possible to} \; b_k\text{"} \qquad \text{for} \; k \, = \, 1, \dots , 40$$

$$\text{s.t.} \; \sum_{i=1}^{500} x_i \; = \; 100$$
$$\text{and} \quad x_i \; \in \; \{\,0\,,\,1\,\} \qquad \text{for} \; i \, = \, 1, \dots , 500$$

and finally:

$$\text{minimize} \; \sum_{k=1}^{40} \left(\sum_{i=1}^{500} a_{ik} x_i \, - \, b_k \right)^2$$

$$\text{s.t.} \; \sum_{i=1}^{500} x_i \; = \; 100$$
$$\text{and} \quad x_i \; \in \; \{\,0\,,\,1\,\} \qquad \text{for} \; i \, = \, 1, \dots , 500$$

We were able to give with a TA algorithm a routine for this problem which was about 1000 times faster than the "traditional" method used by the G&I and which gave much better results. Furthermore it is very easy to implement a different "taste function" resp. objective function in the above problem. The reason is that for some k the requirement "near to b_k" has to be stronger than for some other k. Here, TA is very flexible and allows quick solutions for a very large class of objective functions. A TA algorithm for this G&I problem implemented in FORTRAN runs approximately half a second on an IBM 3090-200 VF.

5. *Acknowledgements*

It is a pleasure to thank for encouraging support: W. Kepper, management consultant of the G&I, draw our attention to the marketing research problem we discussed above, and we profitted from his friendly assistance during our work. Angela Neumann implemented the first prototype APL2-program for the G&I problem. Having her first TA version we knew that we were on the right way. We thank M. Broder, W. Höfling, and Dr. W. Adlwarth from the G&I for the perfect and uncomplicated co-operation and for their open-mindedness to our mathematical research.

References

[1]   Drexl, A.: *A simulated annealing approach to the multiconstraint zero-one knapsack problem.*
      Z. Computing, <u>40</u> (1988) 1 - 8

[2]   Dueck, G.; Scheuer, T.: *Threshold Accepting: A general purpose algorithm appearing superior
      to simulated annealing.* Technical Report 88.10.011, IBM Heidelberg Scientific Center

[3]   Fleisher, J.: Sigmap Newsletter <u>20</u> (1976)

[4]   Fréville, A.; Plateau, G.: *Méthodes heuristiques performantes pour les problèmes en variables
      0-1 à plusieurs contraintes en inégalité* Publication ANO-91, Université des Sciences et
      Techniques de Lille (1982)

[5]   Fréville, A.; Plateau, G.: *Hard 0-1 multiknapsack test problems for size reduction methods.*
      Prepublication informatique <u>72</u> Université Paris-Nord (1987)

[6]   Kirkpatrick, S.; Gelatt, C.D.; Vecchi, M.P. *Optimization by simulated annealing.* Science,
      <u>220</u> (1983) 671 - 680

[7]   Metropolis, N.; Rosenbluth, A.; Rosenbluth, M.; Teller, A.; Teller, E.: *Equation of state cal-
      culation by fast computing machines.* Journ. Chem. Phys. <u>21</u> (1953) 1087 - 1092

[8]   Peterson, C.C.: *Computational experience with variants of the Balas algorithm applied to the
      selection of R and D projects.* Management Science <u>13</u> (1967) 736 - 750

[9]   Plateau, G.: *Reduction de la taille des problèmes linéaires en variables 0-1.* Publ. 71 du Lab.
      de Calcu de l'Université des Sciences et Techniques de Lille 1 (1976)

[10] Shih, W.: *A branch and bound method for the multiconstaint 0-1 knapsack problem.* Journ.
      of the Operational Research Society <u>30</u> (1979) 369 - 378

[11] Toyoda, Y.: *A simplified algorithm for obtaining approximate solutions to 0-1 programming
      problems.* Management Science <u>21</u> (1975) 1417 - 1427

[12] Weingartner, H.M.; Ness, D.N.: *Methods for the solution of the multidimensional 0-1
      knapsack problem.* Operations Research <u>15</u> (1967) 83 - 103

Gunter Dueck and Jens Wirsching
IBM Germany, Heidelberg Scientific Center
Tiergartenstraße 15
D-6900 Heidelberg
EARN: DUECK at DHDIBM1
Phone: 06221-404-0

DIGITAL TOPOLOGY IN DOCUMENT PROCESSING

Ulrich Eckhardt

Gerd Maderlechner

Abstract. Since thirty years it is popular in automatic document processing to use thinning algorithms for extracting topological features from binary images. In spite of this fact, however, a large number of fundamental questions concerning the basic algorithms of digital topology are not well understood. The aim of this paper is to report on some results obtained by the authors which are relevant for the design of algorithms for document processing.

1. Introduction. Black-and-white pictures or *binary pictures* constitute a special class of pictures having great practical importance. Most of man-made pictures are of this type, e.g. office documents, technical drawings and diagrams, but also works of art like woodcuts. These pictures have a very simple structure: They can be interpreted mathematically as subsets of the plane since they can be described uniquely by the set of all black points contained in them. So they are very simple models for more complex pictures and they are very well suited for studying fundamental mathematical questions related to picture processing.

Usually binary pictures are represented in the computer in discretized form, i.e. they are given as subsets of the *digital plane* which is the set of all points in the plane having integer coordinates. These subsets are also termed *digital sets*. A fundamental problem arising in this context is how to model topological and geometrical properties of the Euclidean plane in terms of the digital plane. Since topological properties are very fundamental and important in applications (e.g. automated testing of printed circuits) we concentrate here exclusively on them.

In picture processing, very large amounts of data are typical. If a standard office document is discretized with usual resolution of 200 dots per inch, 4 Mbits of raw data are obtained. In contrast, the useful information of such a document is much less than 2 500 letters or some 20 kbit which means only 0.5 %

263

Hj. Wacker and W. Zulehner (eds.),
Proceedings of the Fourth European Conference on Mathematics in Industry, 263–267.
© 1991 *B.G. Teubner Stuttgart and Kluwer Academic Publishers.*

of the initial data amount. In order to achieve tolerable processing times, parallel processing becomes mandatory. Specifically, parallel data reduction is necessary. When the topological properties of the sets contained in a picture are of importance, the reduction should be performed in such a way, that the
"topology" is preserved. Minsky and Papert |3|, however, proved
that a certain model for a parallel computing machine, a so-called perceptron, can only decide "trivial" topological predicates.
Rosenfeld |4| showed that the decision whether a specific point
of a digital set is important for the topology of this set, is
a trivial predicate in this sense (see Theorem 1 below). This
latter result is the mathematical justification for the so-called
thinning methods which were introduced in 1959 by Sherman |5|.

2. The Topology of the Digital Plane. Since it is not possible
to topologize the digital plane in a nontrivial and practically
relevant way, usually a graph structure is introduced. Two
points in a digital set S with integer coordinates (i,j) and
(k,ℓ) are *(8-) neighbors* if

$$\max \left(|i - k|, \ |j - \ell| \right) \leq 1.$$

Two points in the complement ¢S of S with integer coordinates
(i,j) and (k,ℓ) are *(4- or direct) neighbors* if

$$|i - k| + |j - \ell| \leq 1.$$

Two points in S (or ¢S, respectively) are *connected*, if there
exists a sequence of points in S (or ¢S) such that each two successive points in this sequence are neighbors. The equivalence
classes with respect to this relation are termed *connection components* of S (or ¢S). For a justification of this definition see
|4| or the book of Voss |6|.

For a given point P in the digital plane with coordinates (i,j)
the *neighborhood* N(P) of P is the set of all points with coordinates (k,ℓ) such that

$$\max \left(|i - k|, \ |j - \ell| \right) = 1.$$

According to Rosenfeld |4| a point P of a digital set S is ter-
med *simple*, if S and S - {P} have the same number of (8-) connection components and if ¢S and ¢S ∪ {P} also have the same
number of (4-) connection components. Consequently, S and
S - {P} have identical "topological" properties in the sense in-

dicated above. Rosenfeld |6| proved:

Theorem 1: A point P of a digital set S is simple if and only if
N(P) ∩ S contains exactly one (8-) connection component and
N(P) ∩ ∁S contains exactly one (4-) component which is (4-) con-
nected to P.

As a consequence, it can be decided on the basis of local information only, whether a specific point is essential for the topology or not. The idea of thinning consists in removing (sequentially or, more preferably, in parallel) simple points from a
digital set until an irreducible set is obtained. Usually the
latter will have a very simple structure so that its topological
properties can be derived quite efficiently by means of a sequential method.

In a series of publications (see |2| for the statement of re-
sults and further references) the authors investigated the following topics:
- Characterization of irreducible sets obtained by thinning,
- Determination of practical conditions guaranteeing that the
  irreducible sets are "not too complex",
- Finding a thinning method having the properties of well-defi-
  nedness and invariance with respect to motions of the digital
  plane,
- Parallelization of the method and investigation of its properties.

3. Irreducible Sets. Arcelli |1| gave the first example of a
nontrivial irreducible set containing *interior points*, i.e.
points in S whose direct neighbors also belong to S. It can be
shown that Arcelli's set is in a certain sense the only prototype of a nontrivial irreducible set containing interior points.
Define the digital set

$$S_n = \{(i,j) \mid i,j \geq 0,\ i + j \leq n\} \cup$$

$$\cup \{(i,j) \mid i,j > 0,\ i + j = n + 2\}.$$

Then the *Arcelli set* $A_n^{(r,s)}$ (r, s = 0,1) is constructed by reflecting S_n at the horizontal line $(t, -\frac{r}{2})$ (t real) and subsequently reflecting the result at the vertical line $(-\frac{s}{2}, t)$.
The set given in Arcelli's paper |1| is $A_2^{(0,0)}$:

<u>Theorem 2</u>: If an irreducible set contains interior points, then
it contains a translate of an Arcelli set.

(For the proof see |2|).

A *1-hole* of a digital set S is a point in ₵S whose direct neighbors belong to S. The following Theorem holds:

<u>Theorem 3</u>: If a digital set contains no 1-holes then its (8-)
connected subsets containing exclusively interior points have at
most twelve elements. There are only eleven possible types of
such subsets.

In |2| a catalogue of possible configurations is given. The
last Theorem offers a very simple possibility for preprocessing
a digital picture so that large clusters of interior points in
the reduced set are not possible. Since usually 1-holes are removed anyway (as they are regarded as 1-pixel errors), the Theorem explains why such clusters were never observed in "realistic" situations.

4. <u>Perfect Points</u>. In order to make the thinning process well-
defined and invariant with respect to motions of the digital
plane and moreover, to make it implementable in parallel, the
concept of a *perfect point* was introduced. A point P of a digital set is termed perfect if there exists an interior point in
N(P) such that all direct neighbors of P which are not neighbors
of this interior point, belong to ₵S. The following Theorem
holds |2|:

<u>Theorem 4</u>: When all simple and perfect points are removed from
a digital set simultaneously then the following holds:
- The reduced set contains exactly as many connection components as the original set and the same holds for the comple-
  ments. Hence, the original set and the reduced set have the
  same "topology",
- The process is well-defined and invariant with respect to motions of the digital plane.
- *End points* (i.e. points P in S with $|N(P) \cap S| = 1$) are

(automatically) preserved from elimination.

In the thinning methods published in the literature, the last assertion of the Theorem is forced to be true by explicitly requiring it to hold (see |2| for a rigorous topological justification of this requirement).

It can be shown that simple and perfect points exhibit a very close relation to the structure of the boundary of a digital set |2| so that it becomes possible to implement the parallel algorithm described here by using boundary information only.

The method described here was implemented and tested on a large number of different binary images and the results were compared with those of other methods. Some typical results are published in |2|.

References

1. Arcelli, C.: Pattern thinning by contour tracing. Computer Graphics and Image Processing 17 (1981) 130-144

2. Eckhardt, U.; Maderlechner, G.: Thinning of Binary Images. Hamburger Beiträge zur Angewandten Mathematik, Reihe B, Bericht 11, April 1989

3. Minsky, M. L.; Papert, S.: Perceptrons. An Introduction to Computational Geometry. Cambridge: The MIT Press 1969

4. Rosenfeld, A.: Digital topology. Amer. Math. Monthly 86 (1979) 621-630

5. Sherman, H.: A quasi-topological method for the recognition of line patterns. Information Processing, Proc. UNESCO Conf. 1959

6. Voss, K.: Theoretische Grundlagen der digitalen Bildverarbeitung. Berlin: Akademie-Verlag 1988

Ulrich Eckhardt
Institut für
Angewandte Mathematik
der Universität Hamburg
Bundesstrasse 55
D-2000 Hamburg 13
Germany

Gerd Maderlechner
SIEMENS AG
Corporate Research
and Development
Otto-Hahn-Ring 6
D-8000 München 83
Germany

MODELLING THE RHEOLOGY OF A COAL-WATER SLURRY

A. Fasano, M. Primicerio, Università di Firenze, Italy

1. Introduction

A coal-water slurry (CWS) is a suspension of coal particles (up to 70 % in weight) in water. A small amount of an appropriate chemical substance is added in order to fluidize the mixture which becomes stable against sedimentation and can be pumped through pipelines.

A CWS has a peculiar rheological behaviour: when stirred in batch or circulated through a loop, its apparent viscosity first decreases and then increases dramatically after a sufficiently long time [1], [2].

In some previous papers [3] we developed a macroscopic model providing a qualitative description of the final stage (deterioration). The basic idea was to link the rate of change of rheological parameters to the power dissipated by internal friction.

Such a conjecture was confirmed by experiments. Indeed, plotting rheological parameters versus the total energy dissipated (i. e. assuming energy as a new time scale), the curves corresponding to different experiments with different nominal shear rate are basically coincident in the long run. The comparison between fig. 1 and fig. 2 illustrates this concept.

Thus, the following step is to describe the influence of the microscopic dynamics on the evolution of the macroscopic rheological parameters. Here we suggest a model which is focussed on some particular aspect and is aimed at the interpretation of the experimental results obtained in the test loop of Eniricerche. It will be refined when a new series of data will be available. We wish to thank E. Carniani and D. Ercolani (Snamprogetti) and S. Meli

Hj. Wacker and W. Zulehner (eds.),
Proceedings of the Fourth European Conference on Mathematics in Industry, 269–274.
© 1991 *B.G. Teubner Stuttgart and Kluwer Academic Publishers.*

(Eniricerche) for their cooperation and for a number of interesting and stimulating discussions.

2. Basic ideas of the model.

A crucial point is to model the dynamics of the fluidizing agent. Basically we can say that it can exist in two states:

(i)   dissolved in water (with concentration A),

(ii)   adsorbed on coal particles.

However not all of the substance adsorbed exerts a fluidizing action. There can be a fraction of it which is not efficient, depending on the way it is linked to the particle.

We shall denote by B the concentration of the dispersant adsorbed in the active way and by C the concentration of the inert fraction.

Apparently, for reasons not yet completely understood, internal friction can cause the transition $B \to C$.

Also a temperature increase seems to produce a similar effect (or the direct transition $A \to C$), but we will not consider this point for the sake of simplicity.

On the other hand a transition $A \to B$ can occur to replace the dispersant becoming inert. The dynamics of the three "populations" can be described by the system

$$(2.1) \qquad \dot{A} = -\lambda A (B_0 - B) ,$$

$$(2.2) \qquad \dot{B} = \lambda A (B_0 - B) - f(W) B ,$$

$$(2.3) \qquad \dot{C} = f(W) B ,$$

where $f(W)$ is a function of the power dissipated.

The system (2.1), (2.2) (the evolution of C is obtained as a consequence and is irrelevant) can be studied once initial conditions are prescribed: $A(0) = A_0$, $B(0) = B_0$. First consider the case $A_0 = 0$, implying $A(t) \equiv 0$ so that the system is reduced to the single equation

(2.4) $\dot{B} = -f(W) \ B$.

    We assume that the rheological properties of the CWS
depend on B. If the evolution of B as a function of a new
"time variable"   E   has to be independent on the nominal
shear rate, it will be necessarily

(2.5) $f(W) = F_0 \ W$,

F_0 being a positive constant, so that

(2.6) $B = B_0 \ e^{-F_0 E}$

which is to be satisfied irrespectively of the partucular
rheological model for the slurry.

3. The general model.

    The experimental data we want to interpret present the
CWS as a Bingham fluid. If τ is the stress and $\dot{\gamma}$ is the
shear rate, the τ-$\dot{\gamma}$ relationship for $\tau > \tau_0$ is

(3.1) $\tau = \tau_0 + \eta_B \ \dot{\gamma}$

where τ_0 is the yield stress, η_B the Bingham viscosity.

    We assume that:

(i) τ_0 is a decreasing function of B . Here we use the simple
relation

(3.2) $\tau_0 = \tau_m \ B_0/B$,

τ_m being the initial value of the yield stress; of course
(3.2), and the whole model, makes sense until $\tau_0/\tau_m \simeq 2 \div 3$.

    Together with (2.6) the above relationship gives

          $\tau_0/\tau_m = e^{F_0 E}$

which fits the final branch of the experimental curve
(E>2 . 10^5 J/Kg) for $F_0 = 2.3 \div 2.4$. 10^{-6} MKS units (fig. 2).

(ii) η_B is related to the dispersion degree of the slurry,
reaching a minimum η_B^∞ for the maximum dispersion. Assuming

that dispersion is caused by the energy dissipated and
arguing as in Sec. 2 we write

$$(3.3) \qquad \dot{\eta}_B = -C\,W\,(\eta_B - \eta_B^\infty) + r\,(\eta_B^0 - \eta_B) \ ,$$

$$(3.4) \qquad \eta_B(0) = \eta_B^0 \ ,$$

where the second term accounts for the recombination effect,
tending to reproduce the initial situation, which however is
a very slow mechanism and will be neglected.

Setting $M = \eta_B/\eta_B^\infty$, $\quad M_0 = \eta_B^0/\eta_B^\infty$, we have

$$(3.5) \qquad M = 1 + (M_0 - 1)\,\exp[-C\,E]$$

which is in agreement with the experiments (fig. 2), taking
$C \simeq 1.5 \ . \ 10^{-5}$ MKS units.

Coming back to the complete problem and setting $\alpha = A/A_0$,
$\beta = B/B_0$, $\bar{t} = F_0\tau_m\dot{\gamma}t$, we obtain

$$(3.6) \qquad \frac{d\alpha}{d\bar{t}} = -\epsilon_1\,\alpha\,(1-\beta)$$

$$(3.7) \qquad \frac{d\beta}{d\bar{t}} = \epsilon_2\,\alpha\,(1-\beta) - 1 - \epsilon_3\,\beta\,\{1+(M_0-1)\,e^{-Ct}\}$$

with $\epsilon_1 = \dfrac{\lambda\,B_0}{F_0\,\tau_m\,\dot{\gamma}}$, $\qquad \epsilon_2 = \epsilon_1\dfrac{A_0}{B_0}$, $\qquad \epsilon_3 = \dfrac{\eta_B^\infty\,\dot{\gamma}}{\tau_m}$.

If A_0 is of the same order of B_0 (the fluidizer in
excess is a non-negligible quantity), then $\epsilon_1, \epsilon_2 \gg 1$ and we
have essentially three stages:

(i) β decreases linearly with $\bar{t}$ as long as $\epsilon_2\bar{t} \ll 1+\epsilon_3 M_0$.
However, this time interval is very small and it cannot be
observed in reality.

(ii) The reaction $A \to B$ dominates and β is slowly
decreasing while α tends to zero and η_B tends to η_B^∞ .
Assuming a form $1-\beta=c_1(1+c_2\bar{t})$ one finds that $c_1 \ll 1$ results
from the balance between $B \to C$ and $A \to B$ and c_2 accounts
for the competition between $B \to C$ and the decay of η_B .

(iii)The third phase starts when the first term in (3.7)
becomes negligible and the quantity in braces in the third

term is essentially 1. Then we are back to the case we studied in Sec. 2 and the evolution of τ_0 depends on E only.

Essentially, this means that curves τ_0–E corresponding to different nominal shear rates actually coincide in the third stage, while a dependence on $\dot{\gamma}$ is still present in the second stage, again in agreement with experiments (see fig. 2).

References.

[1] ERCOLANI, D. et al.: Shear Degradation of Concentrated Coal-Water Slurries in Pipeline Flow. Proc. of 13th Int. Conf. on Slurry Techn., Denver, April 1988.

[2] ERCOLANI, D. et al.: Effect of mechanical and thermal energy on CWS rheology. Proc. 11th Int. Conf. Hydr. Transp. of Solids, Stratford, October 1988.

[3] PRIMICERIO, M.: Dynamics of "slurries". Proc. of the Second European Symp. on Math. in Industry - ESMI II, March 1-7, 1987 Oberwolfach - Kluwer. Ac. Publ. - Ed. by H. Neunzert, Kaiserlautern.

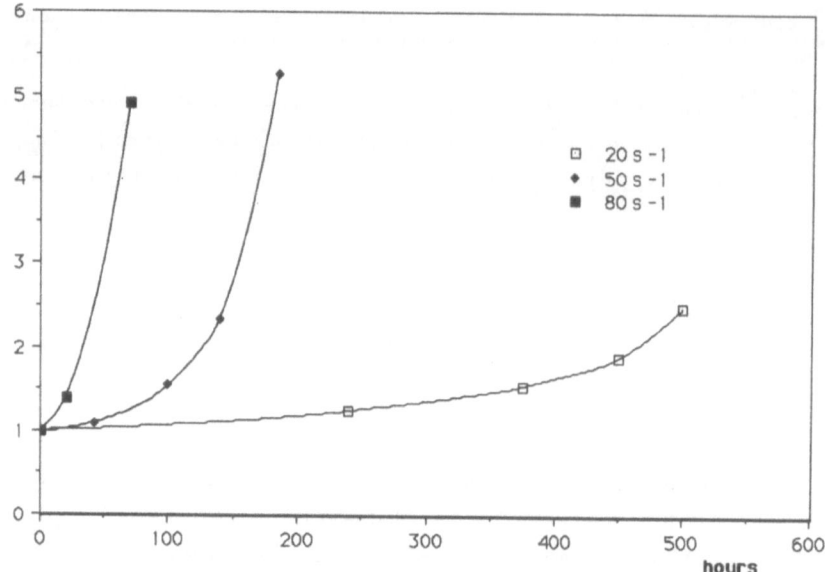

Fig1. Relative threshold stress vs. time for different values of the nominal shear rate.
(Courtesy of Eniricerche)

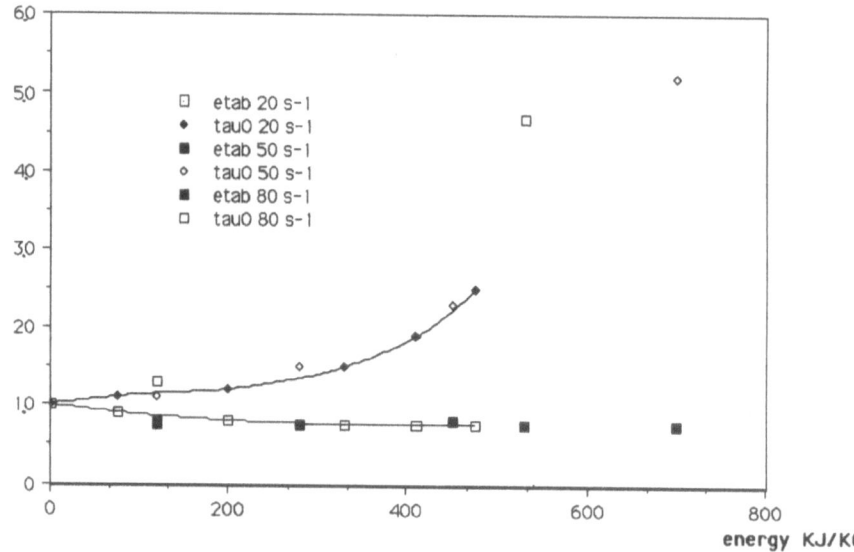

Fig 2. Same data and Bingham viscosity vs. total energy dissipated in the flow.
(Courtesy of Eniricerche)

NONORTHOGONAL EXPANSIONS OF SIGNALS AND SOME OF THEIR APPLICATIONS

By   Hans G.Feichtinger and K.Gröchenig

Hans G.Feichtinger
Institut für Mathematik
Universität Wien
Strudlhofgasse 4
A-1090 WIEN , AUSTRIA

Karlheinz Gröchenig
Department of Mathematics, U-9
University of Connecticut
STORRS CT 06269
U S A

Classical orthogonal expansions (such as Hermite functions or Fourier series representations) lack important properties, such as locality. Changes in a piece of a given multivariate function may produce unpredictable changes of many of the coefficients. A new class of orthogonal and nonorthogonal coherent series expansions has found much interest in the last few years, both in the theoretical and the applied community. The key to this development was the invention of orthogonal wavelets by Y.Meyer in 1985. These are expansions of the form $f = \sum_{i \in I} \lambda_i \cdot \pi(x_i) g_0$, indicating that the building blocks $g_i := \pi(x_i) g_0$ are generated in a coherent way by certain transformations $\pi(x_i)$ out of a continuous group of transformations $\pi(x)$ from a 'template' g_0.

Important special cases are the case of affine wavelets (translations and dilatons) and the Heisenberg group (translations and multiplication with complex exponentials). There are different variants of such expansions. Orthogonal expansions and frames are described in much detail in a recent survey by C.Heil and D.Walnut (SIAM Review 34 (1989), 628-666 and the references given there). A more flexible approach, allowing much more freedom in the choice of the template and also the proof of stability results for these expansions have been developed on a group theoretic basis in a series of papers by the authors. Besides this all these expansions enjoy the following properties: Good locality, smoothness of the building blocks, linear dependence of the (sometimes non-unique) coefficients, which can be obtained in a constructive way. There are upcoming applications in fields such as the analysis of transient signals, data compression, image analysis and pattern recognition. Our main reference describing the general theory of atomic decompositions is the joint paper in J.Funct.Anal. 86 (1989), 307-340.

A COPY OF THE FULL REPORT CAN BE OBTAINED FROM THE AUTHORS.

Hj. Wacker and W. Zulehner (eds.),
Proceedings of the Fourth European Conference on Mathematics in Industry, 275.
© 1991 *B.G. Teubner Stuttgart and Kluwer Academic Publishers.*

TWO-PHASE FLOW METHODOLOGY AND MODELLING

A.D. Fitt, Faculty of Mathematics, Southampton University, Hampshire, England

1. Introduction

The problem of how to treat flows in which there is more than one phase present has long been a difficult one. Recently interest has increased in such flow problems, notably in the nuclear power and defence industries where multiphase mixtures are commonplace. By 'multiphase' here we mean flows where the multiconnectedness or complexity and unsteadiness of the interphase boundaries makes phase 'tracking' impractical.

In the past a popular way of proceeding has been to average the equations in some way, introducing some form of void fraction function which indicates how much of each phase is present. Unfortunately the progress from generalized single - component equations to 'two - phase' equations has proved to be littered with problems (witness the controversy in [6]). These difficulties frequently arise from insufficient generality of the basic equations of motion, failures in the averaging procedure, or a neglect of important quantities. Very often in the past the result has been a set of conservation laws which possesses complex characteristics. The consequences of this are severe. The Cauchy problem becomes ill-posed, and any attempt at a numerical solution to such sets of equations is likely to suffer from exponential instability and other problems. Additionally, some of the best methods for shock capturing (for example the method of random choice [2]) and other robust implicit schemes (for example the implicit MacCormack method as described in [7]) cannot now be used at all, as they all rely on the existence of real characteristics. From many angles therefore the appearance of complex characteristics is highly undesirable.

Solutions to these shortcomings of the equations have varied in their ingenuity and physical reality, ranging from the completely artificial where terms in the equations giving rise to complex characteristics are ignored) to the more considered but still dubious piecemeal addition of extra terms to certain equations. What is really required however is a more general approach to the modelling of the problem, carried out in such a way that the assumptions which have been made can be clearly appreciated and justified.

In some ways the 'moral' of the present study [1] is a pessimistic one. The theory tells us that two - phase flow is essentially a MODELLING problem, and the modelling will change for different flows. It is too much to hope that a single set of simple equations will suffice for all two phase flows. Instead we must proceed from a GENERAL set of equations, and use modelling skills, combined with suitable small parameter analysis to indicate which terms should be included in the final equations of motion which we wish to solve to determine the flow. If we are unable to accomplish this, then the fault is with the modelling (for example, insufficient knowledge of turbulent effects) and not the basic equations themselves. It is also worth stating that most of the initial development of the equations of motion is very similar to that carried out in [3]. Both ensemble and cross-sectional averaging procedures are used before the gas/particulate flow regime is examined.

---

[1] A fuller version of this paper will be submitted to Int. J. Multiphase Flow

Hj. Wacker and W. Zulehner (eds.),
Proceedings of the Fourth European Conference on Mathematics in Industry, 277–281.
© 1991 *B.G. Teubner Stuttgart and Kluwer Academic Publishers.*

2. General Equations for Two-Phase Flow and their One-Dimensional Interpretation

Because of the anticipated sensitivity of the modelling, we begin consideration of two-phase flow equations from a suitably general viewpoint. This entails proposing equations for each phase separately, before averaging. Using standard theory, the equations of motion and associated jump conditions for each phase may be written

$$(1) \qquad \frac{\partial}{\partial t}(\rho\Sigma) + \nabla.(\rho q\Sigma - J) = \rho f, \quad f_i = [(\rho\Sigma(q - q_i) + Q).\hat{n}_i]_1^2.$$

Here ρ represents the phase density, q_i the interfacial velocity, f_i the interfacial source density term and $\hat{n}_i$ the unit interface normal. The divergence term is to be suitably interpreted depending on whether its operand is a vector or a scalar. We then develop the equations of motion for each phase by suitably choosing the variables in (1).

Accepting now that to use the above equations separately for each phase would involve an unmanagable amount of interface tracking, we must now consider how to average the equations. First, we employ a (fully three-dimensional) standard ensemble averaging process where the average of a quantity f is given by

$$\bar{f} = \int_M f(\pmb{x}, t; \omega) dm(\omega)$$

where M is the total probability space and the measure $m(\omega)$ represents the probability of observing state ω. Since more than one phase is present, we cannot assume that the derivative of $\bar{f}$ will be equal to the average of the derivative of f. Instead, phase indicator functions must be introduced (see [3]). Using these, and various other properties of the averaging operator, it is possible to propose general three-dimensional averaged equations. It is also possible to include the effects of fluctuations such as turbulence, but for the purposes of the present study these are ignored. At this juncture, we could start to specialize to the particular regime of gas/particulate flow, but first we simplify further by employing quasi one-dimensional assumptions and area averaging. For any variable $\bar{f}$ which has already been ensemble averaged we define

$$< \bar{f}(\pmb{x}, t) >= \frac{1}{A(x)} \int \int_{A(x)} \bar{f}(\pmb{x}, t) dA$$

where $A(x)$ represents the cross - sectional area of the channel at ordinate x. Again, standard averaging results may be used (see, for example [4]), and the cross-sectional average of the general conservation law becomes

$$< \bar{\Sigma}\bar{\alpha}_k\bar{\rho}_k >_t + \frac{1}{A}\oint_C \frac{\bar{\alpha}_k\bar{\rho}_k\bar{q}_k(\bar{\Sigma}.\pmb{n})}{N}ds + \frac{1}{A}[A < \bar{\Sigma}\bar{\alpha}_k\bar{\rho}_k\bar{q}_k >]_x.\pmb{e}_x =$$

$$\frac{1}{A}\oint_C \frac{\bar{\alpha}_k\bar{\pmb{J}}_k.\pmb{n}}{N}ds + \frac{1}{A}[A < \bar{\alpha}_k\bar{\pmb{J}}_k >]_x.\pmb{e}_x + < \bar{f}_k > + < \bar{j}_k > + < \bar{\Sigma}_{ki}\bar{\Gamma}_k >.$$

Now choosing Σ, J and f as usual and resolving the momentum equation in the axial direction gives six equations of motion, three for each phase.

3. The Modelling of Gas/Particulate Two-Phase Flows

We must now consider how to model specific terms in the equations of motion. This modelling must reflect the exact nature of the flow which we are considering; if we do not make the correct assumptions we cannot expect to end up with a hyperbolic system.

Firstly, we assume that $A(x) = $ constant and neglect all purely algebraic source terms since these cannot affect the hyperbolicity of the system. Next, we assume that phase 1 is a (compressible) gas phase, and phase 2 an (incompressible) solid phase which is dispersed in the form of 'blobs' within the continuous phase 1. The gas phase is assumed to be a Newtonian viscous fluid, and the particles of the solid phase are regarded as being composed of highly viscous fluid, but with no displacements inside any parts of phase 2 so that the stress tensor has only pressure components. Further, the particles are assumed to be so dispersed that there can be no stress created between neighbouring solid particles, and collisions are neglected. Then *within each solid particle* the pressure p_2 is constant, and depends on the pressure in phase 1. To model the interfacial momentum and work terms it is normal to separate out the pure interfacial terms from the mean flow terms. This introduces the interfacial pressures p_{1i} and p_{2i} which we assume are related to the bulk pressures p_1 and p_2 via some inviscid flow calculation. The incompressibility of the solid phase requires that $p_2 = p_{1i} = p_{2i}$, and to relate p_1 to these pressures we use the fact that for inviscid flow streaming with velocity V past a sphere of radius a it is easily shown (see, for example [7]) that the pressure on the surface of the sphere will have the form

$$p\mid_{r=a} = p_\infty - \frac{1}{2}\rho_\infty V^2 F(\theta)$$

where p_∞ and ρ_∞ are the pressure and density respectively far away from the sphere and θ is the azimuthal angle. This leads to the introduction of an 'interfacial pressure coefficient' C_s which remains when all the θ-dependence has been averaged out.

The bulk added interfacial terms, representing many of the forces which have been averaged out, must also be considered. A careful non-dimensionalization is needed here in general to decide which of these terms must be retained. In the present study, only the virtual mass and drag terms are important. Following [1] we use a standard virtual mass term and assume that the drag is proportional to the square of the relative velocity. The jump conditions allow the interfacial terms relevant to one phase to be inferred from the other.

It remains to consider the gas phase energy equation. Assuming no heat flux and a standard relation between the interfacial work and interfacial momentum transfer, and closing the system with the standard perfect gas law, the final 'working equations' are

$$(\alpha_1\rho_1)_t + (\alpha_1\rho_1 u_1)_x = 0$$

$$(\alpha_2)_t + (\alpha_2 u_2)_x = 0$$

$$(\alpha_1\rho_1 u_1)_t + (\alpha_1\rho_1 u_1^2 C_{u1})_x + \alpha_1 p_{1x} = -C_s\rho_1(u_1 - u_2)^2\alpha_{1x}+$$
$$C_{vm}\alpha_2\rho_1[(u_{1t} + u_1 u_{1x}) - (u_{2t} + u_2 u_{2x})]$$

$$(\alpha_2\rho_2 u_2)_t + (\alpha_2\rho_2 u_2^2 C_{u2})_x + \alpha_2 p_{1x} = \alpha_2[C_s\rho_1(u_1 - u_2)^2]_x-$$
$$C_{vm}\alpha_2\rho_1[(u_{1t} + u_1 u_{1x}) - (u_{2t} + u_2 u_{2x})]$$

$$(\alpha_1\rho_1(e_1 + u_1^2/2))_t + (\alpha_1\rho_1 u_1 C_{e1}(e_1 + u_1^2/2))_x + (p_1 u_1\alpha_1)_x + p_1\alpha_{1t} =$$
$$C_s\rho_1(u_1 - u_2)^2\alpha_{1t} + u_2 C_{vm}\alpha_2\rho_1[(u_{1t} + u_1 u_{1x}) - (u_{2t} + u_2 u_{2x})].$$

Here subscripts refer to phase number or differentiation, α is the void fraction, ($\alpha_1 + \alpha_2 = 1$), ρ represents density, u velocity, p pressure, e is internal energy, C_s and C_{vm} are respectively the drag and virtual mass coefficients, and C_{u1}, C_{u2} and C_{e1} are profile parameters in the momentum and energy equations.

4. Hyperbolicity Analysis of the Equations

We must now analyze the hyperbolicity of the working model. It should be emphasized that in order to accomplish this task in any reasonable length of time, a symbolic manipulator (MAPLE was used in the calculations reported below) is essential. Indeed, it is fair to say that before the advent of such algebraic computation systems the calculations below would have been well-nigh impossible to perform. Before dealing specifically with the gas/particulate flow case, we note a number of other special cases.

4.1 Case 1: Constant-pressure model, no added terms

The model with $p_1 = p_2 = p_{1i} = p_{2i}, C_{u1} = C_{u2} = C_{e1} = 1$ and $C_{vm} = C_s = 0$ has long been used for the modelling of many different two-phase flows. The hyperbolicity result here (see [5]) is that the eigenvalues are $\lambda = u_1$ and $\lambda = yc + u_1$ where y is a root of the quartic equation

$$y^4 - 2Vy^3 + y^2(V^2 - 1 - q) + 2Vy - V^2 = 0$$

and $c^2 = \gamma p/\rho, V = (u_2 - u_1)/c, q = \alpha_2\rho_1/(\alpha_1\rho_2)$. Four real roots exist if and only if $V^2 > (1 + q^{\frac{1}{3}})^3$. Hence the system is totally hyperbolic only for large enough relative speeds.

4.2 Case 2: Bubbly Flow

One model for bubbly flow, where we assume that phase 1 is an inviscid fluid, and phase 2 is composed of incompressible gas bubbles is similar to the working system, but requires no energy equation. With $\rho_1 = $ constant, we find that there are two zero eigenvalues, and the other two satisfy a quadratic equation. We can analyze this by setting $\epsilon = \rho_2/\rho_1 \ll 1$. Then the leading order quadratic exhibits no dependence on C_{u2}. For typical values of say $C_{vm} = 1/2, C_{u1} = 1$, we find that the condition for real roots is

$$-4\alpha_1 C_s^2 + 4\alpha_1^3 C_s - 6\alpha_1 C_s + 2C_s + 2\alpha_1^2 - 3\alpha_1 + 1 < 0,$$

so that for $C_s = 0$ the roots are real so long as $1/2 < \alpha_1 < 1$. For non-zero values of C_s, the range of values which α_1 may take INCREASES. In other words the *interfacial pressure term helps the hyperbolicity*. Physically, it seems very reasonable that the modelling should set a limit on the size of α_2.

4.3 Case 3: Gas/Particulate flow with added terms

Now we consider the working equations. A MAPLE calculation for the full system yields a very large determinant with nearly a thousand terms. Suppose however we decide to ignore the effects of profile parameters for the present and set $C_{u1} = C_{u2} = C_{e1} = 1$. The main difference between the present case and the bubbly flow considered above is that here the density ratio is large, as we assume that the density of the solid is much greater than that of the gas. setting $\epsilon = \rho_1/\rho_2 \ll 1$, we find that to first order the eigenvalues are given by $\lambda = u_1, u_2, u_2$ and the roots of a quadratic with discriminant

$$C_{vm}^2(\gamma - 1)^2(u_1 - u_2)^2\alpha_2^2 + 4c^2\alpha_1(\alpha_1 C_{vm} + \alpha_1 - C_{vm}).$$

Hence the conclusions here are totally different from bubbly flow: As usual the hyperbolicity of the system is guaranteed for large enough relative velocities, but otherwise we have hyperbolicity only when

$$\alpha_1 > \frac{C_{vm}}{1 + C_{vm}}.$$

Thus in gas/particulate flow, the interfacial pressure term does not help the situation: C_s does not appear to lowest order. Moreover, the extra virtual mass term can have a *bad* effect on the hyperbolicity, since for $C_{vm} = 0$ the system is hyperbolic, but for $C_{vm} = 1/2$ for example we require $\alpha_1 > 1/3$. Clearly further analysis is needed of the gas/particulate flow case. One way of proceeding is to note that for all cases considered so far, the biggest threat to hyperbolicity seems to arise for low relative velocities. Accordingly we analyse the case where $u_1 = u_2$ and $\epsilon = r$ say is not nessecarily small. Now the eigenvalues are given by $\lambda = u_1$ (three times) and the roots of a quadratic with discriminant

$$4\alpha_1 c^2 (\alpha_1 (rC_{vm} - C_{vm} - 1) + C_{vm})(\alpha_1^2 (r - 1) - r\alpha_1 + rC_{vm}).$$

Certainly for $C_{vm} = 0$ the system is hyperbolic, but for $C_{vm} = 1/2$ for example (a reasonable value) the hyperbolicity requirement (for $r < 2/3$) is

$$\alpha_1 > \frac{1}{3 - r}.$$

5. Conclusions and Possibilities for Further Work

In this short study there has only been time to indicate how to proceed and to analyze some simple cases. The conclusion for gas/particulate flow is that more terms are needed in the model. Both profile parameters and fluctuation terms have been totally ignored. Also, for internal ballistics flows of the type which we ultimately wish to study, interparticulate stresses and collisions may be important. Clearly, much work remains.

References

[1] Batchelor, G.K. (1979) 'An Introduction to Fluid Dynamics' C.U.P.

[2] Chorin, A. (1976) 'Random Choice Solutions of Hyperbolic Systems' J. Comput. Phys. **22** pp. 517-536

[3] Drew, D.A. & Wood, R.T. (1985) 'Overview and Taxonomy of Models and Methods for Workshop on Two-Phase Flow Fundamentals' National Bureau of Standards, Gaithersburg, Maryland.

[4] Fitt, A.D. (1987) 'A Rigorous Derivation of a Set of Two-Dimensional Equation for Gas/Particulate Flow in a Channel' CIT Maths and Ballistics Report No. MB 1/87

[5] Fitt, A.D. (1988) 'Mixed Hyperbolic-Elliptic Systems in Industrial Problems' Proc. ECMI88 Strathclyde, Teubner.

[6] Gough, P.S. & Zwarts, F.J. (1979) 'Modeling Heterogeneous Two-Phase Reacting Flow' AIAA J. **17** pp. 17-25

[7] MacCormack, R.W. (1982) 'A Numerical Method for Solving the Equations of Compressible Viscous Flow' AIAA J. **20** pp. 1275-1281

MAXIMUM EXPECTED INFORMATION (MEI) DISCRETIZATION METHOD

FOR SPATIAL DATA ANALYSIS (*)

B. FORTE - Department of Applied Mathematics, University of
              Waterloo, Ontario, Canada.

R. MININNI - SASIAM/Tecnopolis Csata Novus Ortus.

ABSTRACT

The objective of this study is to establish the degree of
stochastic independence of the two random variables coordinates
X and Y on the basis of their pictorial representation.

Theoretically the method that it has been used, called Maximum
Expected Information (MEI) Discretization, is based on the principle of maximum entropy as a criterion for optimality of expected information. In fact, by means of this principle, the equivalence between the classical definition of stochastic independence of X and Y and a definition of stochastic independence based
on the existence, for each $m, n \in N$, of a $m*n$ rectangular equiprobable discretization of the range Q of (X,Y) can be proved.

The method generates $m*n$ equiprobable discretizations of Q and
allows to detect the dependence of X and Y from the graphical
representation of each discretization if at least one of these
is not rectangular. Moreover, the method computes in practical
way the discrepancy between all possible (m,n) equiprobable discretizations and rectangular equiprobable discretizations of Q,
as a measure of the degree of independence of X and Y.

Two possible applications of the above mentioned method are one
related to the town-planning, where the spatial pattern of retail
establishments in urban areas has been analysed, and one in the
biomedical area where the distribution of a disease with respect
to the age of the patients has been studied. Another application
of great importance, is related to the simulation of space-time
stochastic system, such as the statistical analysis of pseudo-
random number generators.

(*) A SASIAM project supported by IBM-Italy

Hj. Wacker and W. Zulehner (eds.),
Proceedings of the Fourth European Conference on Mathematics in Industry, 283.
© 1991 *B.G. Teubner Stuttgart and Kluwer Academic Publishers.*

Automatic mesh generation for 3D domains with application in fluid mechanics, structures, electrotechnics and aerospace problems.

P.L. George, F. Hecht, E. Saltel
INRIA, B.P. 105, 78153 LE CHESNAY CEDEX, France

Finite element simulation requires the generation of an adequat mesh recovering the domain of interest. Domains, assumed of arbitrary shape, are generally described via the data of their boundaries (a list of faces due to CAD-CAM systems).

To solve this problem 3 methods are currently investigated:

- a structured partition of coarse mesh of the domain based upon mapping [1,2];
- a generation of interior elements following a frontal approach [3];
- a Delaunay's philosophy to obtain the tesselation [4,5].

The method, we are interested in, can be seen as one of the 3rd type and is based upon an updating process of an existing mesh whose elements are only 3-simplices. Briefly, we suggest to extract from the data of the given boundary the connected set of points, then we mesh with some simplices a box containing this set and update this mesh by considering a new point : when all points are introduced the mesh of the box is obtained which contains all the given points.

Unfortunaly, this mesh **does not contain,** in general, even in 2D, the edges (or the faces) of the specified boundary as shown in the following figure :

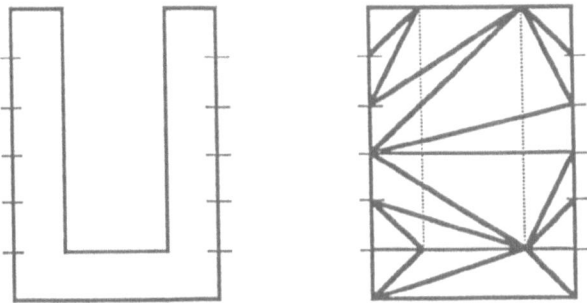

Figure 1: hedgehog effect

285

Hj. Wacker and W. Zulehner (eds.),
Proceedings of the Fourth European Conference on Mathematics in Industry, 285–289.
© 1991 *B.G. Teubner Stuttgart and Kluwer Academic Publishers.*

The aim of this paper is to propose a method which, from a given mesh, derives a mesh fitting exactly the boundary and then to enrich the resulting mesh with internal points to obtain a suitable final mesh.

So two problems are to be investigated : that due to this specification of the boundary and that connected to a suitable creation of internal points.

The first crucial result of the paper can be summarized as the following theorem:

Theorem : *Let's assume :*

i) T, a given set of tetraedrons meshing a 3D domain,
ii) all the given points are points of T,
iii) the boundary of T is not crossed,

then we can derive T' an equivalent mesh which fits exactly the boundary.

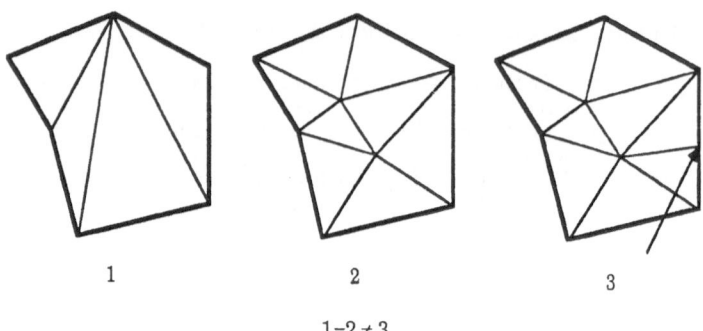

1   2   3

1=2 ≠ 3

Figure 2: example of equivalent and no equivalent meshes

To prove this theorem, we proceed in 2 steps. First, we consider the case of missing edges and then the case of missing faces. The method is based upon the 3 following operations:

1) swapping edges in faces;
2) swapping faces in edges;
3) adding points in some tetraedrons.

So, assuming the existence of the specified boundary in the mesh of the above box, we can recognize the inside and the outside of the domain. The theory of graphs is then used to suppress all exterior elements.

The second point to be examined is that relative to the creation of suitable internal points. When specified boundary is regenerated, we obtain a mesh of the domain whose points are those of the boundary and, in general, some extra points required by this process of regeneration. Such mesh is not adapted in the view of finite element computation, so internal points are to be properly created.

An analysis of the shape of elements is performed which is used to define if an internal points is needed in the examined element , its location is computed according to the properties to be satisfied (distance, shape and density). Isotropic or anisotropic analysis can be envisaged.

The main application of such method is to use it in a mesh generator of previously seen class. As briefly mentionned, the final mesh consistes of well balanced tetrahedrons and can be used for finite element computation. The generator has been designed with the special aim to be applied for arbitrary shape, so various applications can be envisaged. In particular, we emphasize that creation of mesh for exterior of domains (aerospace problems), can be easily obtained.

As illustration of capabilities of the proposed method, we only provide 4 examples extracted from the more than 100 domains tested [6] to validate the corresponding program. This one has been developped in Apollo workstations using UNIX system and currently run in Cray, IBM and Vax computers.

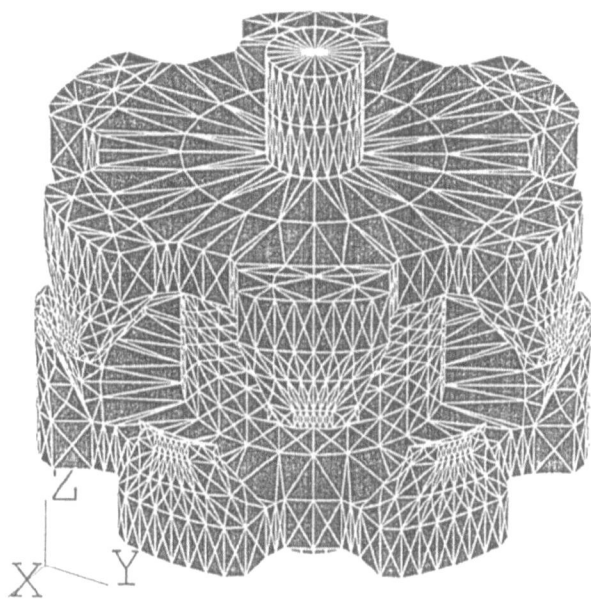

Figure 3: Rotor of alternator

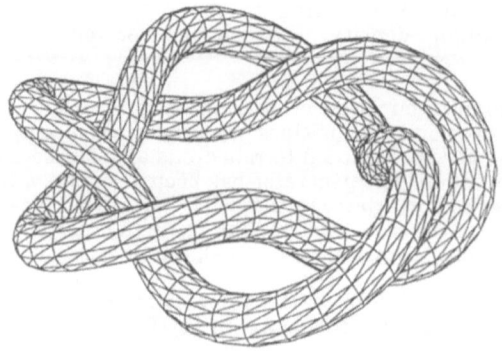

Figure 4: Basic example

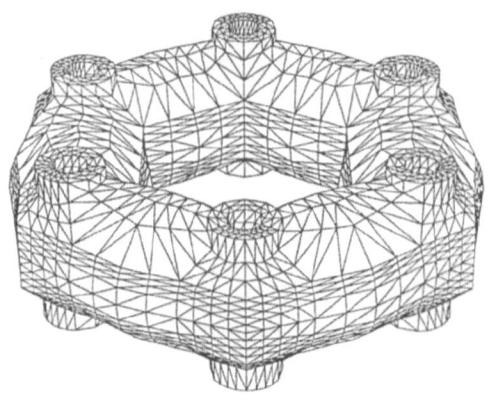

Figure 5: Part of homocinetic junction

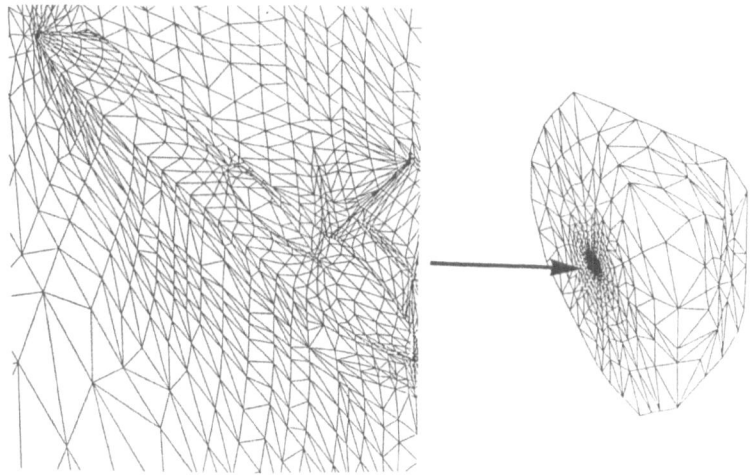

Figures 6: mesh around an half aircraft (courtesy of AMD-BA)

In conclusion, we claim that such method could be powerful for solving the problem of mesh in 3D geometry using mesh generator based on points only (say Voronoi's method for instance). Anyway, a large collection of searchers are actually working in such direction, which seems to be very promising.

Elements of bibliography:

[1] **J.F. Thomson, Z.Warsi, C.W. Master:** Numerical grid generation foundation and applications, North Holland 1985.

[2] **P.L. George:** MODULEF: Génération automatique de maillages, 2ième édition, INRIA éditeur 1988.

[3] **R.Löhner, P. Parikh:** Generation of 3-D unstructured grids by advancing-front method, AIAA 26th Aerospace Sciences Meeting, January,11-14 1988 Reno Nevada USA.

[4] **T.J Baker:** Generation of tetrahedral meshes around complete aircraft, Second Conference on Nunerical grid generation in computationnal fluid dynamics, December 5-9 1988, Miami Beach Florida USA.

[5] **F. Hermeline :** Triangulation automatique d'un polyèdre en dimension N. RAIRO, Numerical Analysis Vol 16 n°3 page 211-242,1982.

[6] **P.L. George, F. Hecht, E. Saltel :** Automatic mesh generator with specified boundary. To appear in Comp. Meth. in Appl. Mech. and Eng., 1989

Quality Control of Artificial Fabrics

Peter Hackh
Arbeitsgruppe Technomathematik
Universität Kaiserslautern
D-6750 Kaiserslautern

Abstract:

One problem which arises in different branches of industry is to find an objective measure for the irregularities of tissues. Up to now the valuation of these irregularities often depends on somebodies personal visual impression.

In order to find an objective procedure one has to resort to measurements of the thickness of the fabric resp. transmission rates of a laser beam. The idea is to compare the distribution given by such a record with a uniform distribution 'given' by the mean thickness resp. the mean transition rate; we introduce numbers which measure the deviation from this uniform distribution.

In an abstract setting we have to look for suitable distance concepts for probability measures μ, ν; we propose two such distance concepts:

the Bounded Lipschitz Distance ρ and the Interval Discrepancy D,

$$\rho(\mu,\nu) = \sup\left\{ |\int fd(\mu-\nu)| \ : \ |f(x)-f(y)| \leq |x-y|, \ f \text{ bounded}\right\} ,$$

$$D(\mu,\nu) = \sup|(\mu-\nu)(I)| .$$
$$\text{I interval}$$

There exist very easy and fast algorithms in order to calculate these numbers even for the case of, say, 5000 or 10000 measurement values; so you don't need the Simplex Scheme or something like this.

After some investigations it turned out that the Bounded Lipschitz Distance is less appropriate compared with the Discrepancy, which seems to work in many situations.

Hj. Wacker and W. Zulehner (eds.),
Proceedings of the Fourth European Conference on Mathematics in Industry, 291.
© 1991 *B.G. Teubner Stuttgart and Kluwer Academic Publishers.*

Developing a Unidimensional Simulation Model for Producing Silicon in an Electric Furnace

Svenn A. Halvorsen, James H. Downing, Anders Schei
Elkem a/s

Summary: Within Elkem a/s two different approaches were followed in the early stages of developing a model for the Si-process. In Kristiansand, Norway, we chose to keep close to a rather fundamental formulation. Serious numerical difficulties occurred, however, during implementation, and the model was not finished. In Niagara Falls, USA, a simple and direct approach was followed, leading to a complete simple model. The model was then significantly improved by implementing time integration, reaction rate formulations etc based on the more fundamental formulations developed in Kristiansand.

In the final model difficult gas dynamic computations have been avoided at the cost of solving a nonlinear equation (with one unknown) for each segment in the model. Fast, irreversible chemical reactions rates caused serious numerical convergence problems also for this model, but in this case the problem could be solved by implementing discontinuity handling for the time integration procedure. For the original Kristiansand model the numerical problems remain open, implying that academic research is wanted.

1   Introduction

In a silicon smelting furnace quartz (SiO_2) and carbon react to produce silicon metal and CO gas. The silicon yield for a reasonably well run furnace is about 85%, which means that 15% of the silicon leaves the furnace as microsilica in the offgas (typical particle size less than 1 μm). Increasing the silicon yield has a huge economic potential, especially for Elkem which supplies some 20% of the world's total demand.

2   Silicon smelting

In a silicon smelting furnace four condensed (solid/liquid) species are present in reasonable amounts: SiO_2, C, SiC and Si. The gas consists mainly of the two components SiO and CO. The chemical reactions running in this system can (formally) be described by a linear combination of the three reactions

(a) $Si + SiO_2 = 2\ SiO$

(b) $2\ C + SiO = SiC + CO$

(c) $SiC + SiO = 2\ Si + CO$

The reactions will combine to the overall reaction

(d) $SiO_2 + (1+x)\ C \rightarrow x\ Si + (1+x)\ CO + (1-x)\ SiO$

where x is the Si-recovery for the furnace. After leaving the furnace top CO will burn to CO_2 and SiO to SiO_2 (microsilica particles).

Silicon will be produced in the furnace hearth, where the temperature is above 2000 °C. The necessary heat is supplied from an

293

Hj. Wacker and W. Zulehner (eds.),
Proceedings of the Fourth European Conference on Mathematics in Industry, 293–298.
© 1991 B.G. Teubner Stuttgart and Kluwer Academic Publishers.

electric arc. The gas produced will roughly consist of 50% CO and 50% SiO according to the thermodynamics for the system [1]. On its way to the furnace top SiO can be captured. Carbon will react to silicon carbide in the upper part of the furnace according to (b), while silicon carbide will react according to (c) at high temperatures close to the furnace hearth.

In the upper part of the furnace where the temperature is in the range 1000–1800 °C, excess SiO will "condense" (reaction (a) backwards) while releasing large amounts of heat. The charge will be heated until reaching a steady state in which the furnace top is hot enough to let part of the SiO escape.

The reactions involved are irreversible in the sense that the reaction rates depend on whether the reaction runs forward or backward. SiO-"condensation" is for instance fast, while reaction in the forward direction is negligibly slow in the upper parts of the furnace.

3   Process modelling

3.1   Rigorous model

Since 1984 we have been engaged in developing a unidimensional model for the metallurgical aspects (including heat balance) of the Si-process at Elkem's R&D Center in Kristiansand, Norway. We decided to concentrate on details in the upper part of the furnace. Thus the model is distributed in this region, while the hearth is described by a point model (see figure 1). Our aim was to keep as close as possible to the fundamentals with just enough simplifying to make the computer implementation fairly straightforward. A general preliminary model was presented for the ECMI II meeting in Oberwolfach [2].

In the upper parts (furnace shaft) hyperbolic transport equations were formulated both for the concentration of condensed species (downwards motion), for the concentration of the two gas species (upwards motion), for the enthalpy of condensed materials, and for the gas enthalpy. Chemical reaction rates were included

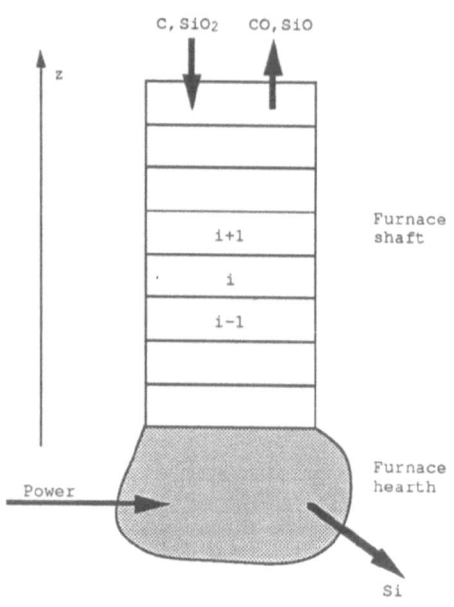

Figure 1.   Furnace hearth and shaft partitioned into segments

as highly nonlinear source terms. The system of partial differential equations was coupled to a corresponding set of ordinary differential equations describing the furnace hearth.

The line method was adopted, discretizing the partial differential system by control volume integration and applying an A-stable method for solving the resulting ODE-system (system of ordinary differential equations). As described in [2] serious numerical problems occurred after implementing just one reaction (SiO-"condensation"). The ODE-solver often got convergence problems when solving the internal system of nonlinear equations by a quasi-Newtonian method. Hence the time steps were (drastically) reduced.

Further work reduced, but did not eliminate the problem. The preliminary model could be run only with comparatively short time steps. Further, a stabilizing procedure applying implicit (backward) Euler and very short time steps, was called whenever the computer code detected convergence troubles. Presently we believe that the problem can be solved by modifying the solver for the internal system of nonlinear equations. This may turn out to be a difficult task, which so far has not been pursued by us. A general solver is wanted from a user's point of view, but analysis may reveal that a problem dependent solver is much better. We feel that academic research is needed to clarify and solve the problem.

3.2 Directly formulated model

In Elkem's R&D Center in Niagara Falls, USA, a model which incorporated mix movement, heat transfer, and chemical reactions including "condensation" and "vaporization" of SiO, was programmed directly using a forward Euler type integration method. The hearth was modeled rather coarsely and the shaft more detailed like the rigorous model.

In the directly formulated model the chemical reactions were implemented sequentially: When gas enters a segment in the furnace, the reaction with highest precedence is first considered. The gas composition is then updated to account for this reaction before the next reaction is considered, etc. This approach can produce reasonable results provided a proper (realistic) reaction precedence is chosen. Interpreting computed results must be done carefully as the reaction precedence can cause artificial effects.

A brief comparison between the rigorous and the direct modelling approach is given in table 1.

3.3 Combined model

The directly formulated model produced a reasonable simulation of the silicon process, while the rigorous approach had met serious numerical difficulties. We therefore decided to base further development on the former model and implement features from the rigorous model gradually.

The modelling work done in Kristiansand enabled a straightforward implementation of a more efficient time integration procedure.

DIRECTLY PROGRAMMED MODEL	RIGOROUS MODEL
- Complete simple model implemented	- Sophisticated model formulations - Prototype not finished
- Direct programming : Hearth and a shaft sectioned into several segments - Explicit (Euler type) method for the time integration	- Shaft modeled with partial differential equations, hearth with ordinary differential equations. - Get a system of ordinary differential equations by discretizing the shaft - Implement a stiff solver (subroutine package) for solving the ordinary differential equations.
- Model and solution method are integrated (Variables are updated along with the computations)	- Model and solution method are separated (separate FORTRAN routines)
- No gas dynamics	- Includes gas dynamics (gas pressures and gas velocity) - Difficult nonlinear computations (serious numerical problems)
- Sequential reaction computations - Side effect: Kinetics depend on 1) Descriptions / formulas 2) Order of computations	- Parallel reactions - Reaction rates depend on computed gas pressures, temperature etc.

Table 1. A comparison of the two preliminary models

We then proceeded by implementing improved chemical kinetic relations based on deviation from equilibrium. The reaction rates are functions of the state variables in a shaft segment (local temperature, concentrations (moles/m^3) of condensed species) and the local SiO and CO gas pressures. Assuming one atm total pressure, the SiO and CO pressures can be expressed by the local SiO/CO ratio. Applying upwind differences on the stationary gas transport equations, then gives

(1) $V_{SiO-i} = V_{SiO-(i-1)} + \sum a_k R_k(x_i, y_i) \Delta z_i$

(2) $V_{CO-i} = V_{CO-(i-1)} + \sum b_k R_k(x_i, y_i) \Delta z_i$

where

$\quad V_{SiO-i}$ = SiO flux (moles/m^2 s) leaving shaft segment no i[*]

$\quad V_{CO-i}$ = CO flux (moles/m^2 s) leaving shaft segment no i

$\quad R_k(x_i, y_i)$ = Reaction rate for reaction no k in shaft segment no i

$\quad x_i$ = model state variables in segment no i

$\quad y_i$ = SiO/CO ratio in segment no i

$\quad a_k, b_k$ = Stoichiometric coefficients (negative for reactants, positive for products)

$\quad \Delta z_i$ = Segment height for segment no i

---

[*]   i=0 for the furnace hearth

According to the upwind difference scheme, the gas composition leaving a segment is equal to the gas composition in this segment, implying:

$$(3) \qquad y_i = \frac{V_{SiO-(i-1)} + \sum a_k R_k(x_i, y_i)\, \Delta z_i}{V_{CO-(i-1)} + \sum b_k R_k(x_i, y_i)\, \Delta z_i}$$

Given the state variables, the non-linear equation (3) must be solved for each model segment in order to compute the reaction rates and the gas fluxes. Fortunately the equations can be solved sequentially, starting with the (slightly modified) equation for the furnace hearth, and proceeding segment by segment upwards. Computing the gas dynamics has thus been avoided at the cost of solving a non-linear equation for each model segment. This is the main difference between the combined model and the rigorous one.

The model now performed excellently on test cases, even for very high rates of the SiO-"condensation" (which had caused serious troubles for the rigorous model). Further testing revealed, however, that some combinations of parameters for the reaction rates caused troubles. In such cases the program would often perform well for a while, and then suddenly start getting convergence problems.

The problem was solved by implementing discontinuity handling for the (irreversible) chemical reactions: For all model segments switches are set according to the direction of each reaction. During the computations for a time step, forward or backward reaction rate formulations are applied *according to the switch settings*. The equilibrium expressions are evaluated after completing the time step. If the equilibrium had been passed, the time for crossing the equilibrium is computed, and the appropriate switches are changed. The time integration can then continue from the time for crossing the equilibrium.

Our current computer code now performs reasonably well. Some convergence problems still remain, probably caused by problem formulations giving rapid variation in some elements of the Jacobian. We expect that these problems can be gradually eliminated by minor model reformulations during the continuous further development of our model.

Discontinuity handling had been implemented for the rigorous model [2]. Consequently, the convergence problem for this model remains open.

4   Conclusions

A unidimensional dynamic model for simulating the Si-process has been successfully developed and implemented. Our experience confirms (the probably well-known guide-line) that a simple model should be established before trying to implement a more sophisticated one.

In order to avoid convergence problems and short time steps for an implicit ODE solver, *discontinuity handling* is necessary for dynamic systems involving fast, irreversible chemical reactions .

Academic research is wanted to establish improved nonlinear solvers for the system of equations encountered when solving (highly) nonlinear ODEs applying an implicit method.

5   References

[1]   Schei A., Larsen K. : "A Stoichiometric Model of the Ferrosilicon Process", Proceedings of the 39th Electric Furnace Conference, Houston, 1981.

[2]   Halvorsen, S.A. : "Dynamic Model of a Metallurgical Shaft Reactor with Irreversible Kinetics and Moving Lower Boundary", Proc. of the Second European Symposium on Mathematics in Industry, ESMI II, March 1987 Oberwolfach, Teubner 1988.

Acknowledgement
The authors wish to thank Prof. Syvert P. Nørsett (The Norwegian Institute of Technology) and Dr. Per G. Thomsen (Technical University of Denmark) for making their ODE solver available and for helpful discussions, especially advice on discontinuity handling.

Address of the authors: Svenn Anton Halvorsen and Anders Schei
                                Elkem a/s, R & D Center
                                P.O. Box 40 Vaagsbygd
                                N-4602 KRISTIANSAND, Norway

                                James H. Downing
                                Elkem Metals Company, Technology Center
                                P.O. Box 1344
                                Niagara Falls, NY 14302, USA

Numerical Study Of The Nonlinear Barotropic Instability Of Rossby Wave Motion

By

P F Hodnett and W M O'Brien

ABSTRACT

Using β-plane geometry the non-linear barotropic, non-divergent vorticity equation is solved numerically over a long time period to test for stability Rossby's original wave solution of the equation. This equation represents a simple model of the dynamics of the atmosphere. The numerical results show that the initial growth rate of the solution (when unstable) is correctly predicted by the linearized analysis of the problem conducted by Lorenz (1972) but that eventually the solution settles into a bounded oscillating state (which contrasts with the unbounded growth predicted by linear theory). This eventual bounded oscillating behaviour is shown to occur for all values of the basic wave amplitude, A which extends the result of the _weakly_ nonlinear analysis of the problem by Loesch (1978) and Deininger and Loesch (1982) who showed similar behaviour for values of A _slightly_ _in_ _access_ of A_C (the critical amplitude for instability determined by linear theory). Significantly, the numerical results show that A_C is not an accurate indicator of stability/instability for the nonlinear solution.

1.   Introduction

Rossby [5] waves occur as solutions of equations which are idealized models of the real atmosphere but are important since

Hj. Wacker and W. Zulehner (eds.),
Proceedings of the Fourth European Conference on Mathematics in Industry, 299–307.

they represent slow, large scale motions similar to what is observed in the atmosphere in the development of some weather systems. The stability or otherwise of Rossby wave motion is then of considerable interest in relation to the extent to which future states of the atmosphere are predictable. Lorenz [4] established that Rossby waves can be barotropically unstable (resulting from exchanges of kinetic energy). Employing linear perturbation analysis he demonstrated that Rossby's [5] wave solution of the barotropic, non-divergent vorticity equation (representing an east-west propagating planetary wave embedded in a constant westerly flow) is unstable provided the wave amplitude, A is sufficiently large or its (zonal) wavenumber, k_o is sufficiently high. Loesch [3] extended the linear stability study of Lorenz [4] into the weakly non-linear regime by retaining elements of the non-linear term in the non-divergent barotropic vorticity equation (neglected by Lorenz in his linear analysis) under the constraint that the amplitude of the basic wave, A exceeds the critical value for instability, A_C by a small increment, Δ . For such values of A he found that the perturbation initially grows exponentially as predicted by Lorenz [4] but eventually settles into a bounded oscillating state. Some of the analysis and conclusions in the Loesch [3] paper were later corrected by Deininger and Loesch [1].

The purpose of the present investigation is to extend the linear stability study of Lorenz [4] and the weakly non-linear stability analysis of Loesch [3] and Deininger and Loesch [1] to the fully nonlinear regime by solving numerically the non-divergent barotropic vorticity equation with all nonlinear terms retained. The objective is (a) to compare the growth rate of the nonlinear solution (when _unstable_) with the linearized solution of Lorenz [4], (b) to follow the evolution of the nonlinear solution with time so as to identify if the solution eventually oscillates for all values of the wave amplitude A as Loesch [3] has shown occurs when A slightly exceeds A_c, (c) to determine whether the critical amplitude, A_c yielded by the linearized analysis of Lorenz [4] is an accurate predictor of stability/instability for the nonlinear solution.

2.   Governing Equation

The atmosphere is modelled here by the shallow water equations which assume that the atmosphere is a thin layer of homogeneous, incompressible fluid where viscous effects are largely negligible. A further approximation, following Rossby [5], is the use of the beta-plane where a tangent plane to the earth is drawn at a mid-latitude position, ϕ_0 and the equations are solved in the resulting Cartesian space rather than in the real spherical geometry of the earth. When the atmosphere is assumed to have a rigid lid (i.e. the height of the atmosphere is constant) the governing equations are the non-divergent shallow water equations and do not admit gravity-inertia wave solutions but do admit wave solutions called Rossby waves following Rossby [5]. These equations yield a single equation called the non-divergent barotropic vorticity equation which is

(2.1) $$\frac{\partial \zeta}{\partial t} = \frac{\partial \Psi'}{\partial y}\frac{\partial \zeta}{\partial x} - \frac{\partial \Psi}{\partial x}\left(\frac{\partial \zeta}{\partial y} + \beta\right)$$

where t is time, x and y are distances in the eastward and northward directions, Ψ is the streamfunction of the flow so that u, v the eastward and northward components of the velocity are $u = -\dfrac{\partial \psi}{\partial y}$, $v = \dfrac{\partial \psi}{\partial x}$. The vorticity relative to the earth

is $\zeta = \nabla^2 \psi$ so that the absolute vorticity is $\zeta + f$ where in

equation (2.1) the Coriolis parameter $f = 2\Omega\sin\phi$ has been

approximated by $f \approx f_o + \beta y$ where $f_o = 2\Omega\sin\phi_o$,

$\beta = 2\Omega\cos\phi_o/R$ and Ω is the earth's rotation rate, R is the

earth's radius and ϕ_o is a mid-latitude value of the latitude

ϕ. Also assumed in the derivation of equation (2.1) is that

the surface of the earth is flat. Equation (2.1) has a

solution (the original Rossby [5] wave solution) of the form

(2.2) $\psi_o = -Uy + A \sin k_o(x-Ct)$,

where U, A, k_o, C are constants and the wave speed, C is given by

(2.3) $C = U - \beta/k_o^2$.

It is the stability of the solution (2.2) to small initial

perturbations that we study here.

2.1 Previous Work

In his _linear_ analysis Lorenz [4] wrote $\psi = \psi_o + \psi'$ where
ψ' is an _initially_ _small_ perturbation to the basic wave

solution ψ_o (as given in (2.2)) and assumed that squares and

higher powers of ψ' are always negligible. This procedure

yields a _linear_ equation for ψ' which Lorenz solved by looking

for solutions of the form

(2.1.1) $\psi' = U L \displaystyle\sum_{n=-\infty}^{\infty} X_n \exp[i(nk_ox + ly - \sigma t)]$,

where L is a typical length, X_n are non-dimensional constants,

l is real and σ may be either real or complex and must be

complex for ψ' to represent a perturbation which grows

without bound in time. Fuller details of the procedure are

also presented in Hodnett and O'Brien [2].

Lorenz [4] imposed cyclic boundary conditions (as we do here) in both the x and y directions by requiring that $\psi(t,x,y) = \psi(t, x + 2\pi L, y + 2\pi L)$ where $2\pi L$ is the length of a circle of latitude at the midlatitude $\phi_0 = 45°N$ i.e. $L = R/\sqrt{2}$ where R is the earth's radius. The cyclic condition in the x direction then requires that if exactly N (an integer) primary waves cover the cyclic distance $2\pi L$ then the wave number, k_0 of the primary wave is given by $k_0 = \sqrt{2} N/R$. Lorenz [4] took N = 6 which corresponds to typical atmospheric observations. The cyclic condition in the y direction requires that $l = \sqrt{2} M/R$ where M is a positive integer. Lorenz showed that ψ' represents an amplifying solution <u>only</u> if M < N so that for N = 6 there are amplifying solutions when M has any of the values 1,2,3,4,5. Lorenz also showed that for most purposes ψ', given by (2.1.1), is sufficiently accurately represented by summing the series in (2.1.1) only over the values n = -1, 0, 1.

To allow comparison with later numerical results an averaged value of ψ' represented by H is plotted against time, t in Fig 1 (indicated by dashed curves) for the case N = 6, M = 3. The quantity $H = \underset{i}{\Sigma} \underset{j}{\Sigma} (|\psi'| / ULJ)$ where $\underset{i}{\Sigma} \underset{j}{\Sigma}$ represents summation over the grid points in the region $0 \leq x \leq 2\pi L$, $0 \leq y \leq 2\pi L$ and J is the number of grid points. In Figure 1, H is plotted against time, t for a range of values of the amplitude, A of the primary wave i.e. $A/A_c = 1.06, 1.28, 1.49, 1.75, 2.13, 2.74$, where A_c represents the critical amplitude for instability as determined by Lorenz, and show a set of curves growing exponentially with time.

In his <u>weakly</u> non-linear stability analysis Loesch [3]
(later corrected by Deininger and Loesch [1]) studied the nonlinear barotropic vorticity equation when the perturbation
stream function ψ' lies in the neighbourhood of the <u>neutrally</u>
<u>stable</u> perturbation, ψ'_N , given by the linear stability
analysis of Lorenz [4] i.e. when $A = A_C$. He expanded the
nonlinear term (in the governing equation for ψ') about the
<u>neutrally</u> stable perturbation and used the method of multiple
time scales to trace the time evolution of the perturbation ψ'
in the neighbourhood of the neutrally stable perturbation. His
results (valid when the basic wave amplitude A exceeds the
critical value A_C by a <u>small</u> increment Δ) show the perturbation
stream function, ψ', initially growing exponentially with time
similar to the linear analysis of Lorenz [4] until for large
values of time ψ' begins to oscillate in a bounded mode under
the influence of the nonlinear term which Lorenz [4] neglected.

3.   **Presentation and Interpretation of Numerical Results**

In the numerical study reported here the non-divergent
barotropic vorticity equation (2.1) is rewritten in the form

(3.1) $\dfrac{\partial \zeta}{\partial t}$ = $J(\zeta + \beta y, \psi)$,

where $J(a,b) \equiv \dfrac{\partial a}{\partial x} \dfrac{\partial b}{\partial y} - \dfrac{\partial b}{\partial x} \dfrac{\partial a}{\partial y}$. This equation is integrated
integrated numerically for a given initial streamfunction ψ^0 at
time, $t = 0$. To allow direct comparison with the linear
stability results of Lorenz [4], equation (3.1) is integrated
numerically with the initial value of the streamfunction, ψ^0
given by the sum of ψ_0 as given by (2.2) at $t = 0$ and ψ' as
given by (2.1.1) at $t = 0$ (summed over $n = -1,0,1$). Full
details of the numerical procedure and the computation of

results are presented in Hodnett and O'Brien [2]. The growth

of the non-linear numerical solution is characterised by

plotting $H = \sum_i \sum_j [(|\psi - \psi_0'|)/U\, L\, J]$ against time, t where ψ

is the nonlinear numerical solution and ψ_0' is the basic Rossby

wave solution given by (2.2) and as before $\sum_i \sum_j$ denotes

summation over the numerical grid points and J is the number of

grid points. In Figure 1, for the case N = 6, M = 3 the

quantity H is plotted against t (solid curves denote the

numerical solution) for A/A_c = 2.74, 2.13, 1.75, 1.49, 1.28,

1.06, 1, 0.85, 0.11. This allows a direct comparison with the

linear results of Lorenz [4] (dashed curves) and the following

interesting facts are noted. The linear results of Lorenz [4]

initially give a good description of the behaviour of the

system but as time progresses the linear solution begins to

grossly overestimate the growth of the perturbation. Also the

critical amplitude for instability, A_c (determined by linear

theory) appears not to be a good indicator of

stability/instability for the nonlinear solution since for

example the curve A/A_c = 0.85 (stable by linear theory) is seen

to grow to approximately three times its initial value over the

first 400 hours before declining again over the period 400 to

800 hours. Other results, not given here, presented in Hodnett

and O'Brien [2] confirm that A_c is not a good indicator of

stability/instability for the nonlinear solution. The large

time oscillation of the solution as predicted by Loesch [3],

Deininger and Loesch [1] for A/A_c slightly greater than one is

apparent for all values of A/A_c with a stronger oscillation for

large values of A/A_c and a weaker oscillation for small values

of A/A_c.

A range of checks on the reliability of the numerical solutions were performed to ensure that what appears to be instability is not really due to computational error. These are reported in detail in Hodnett and O'Brien [2] and for example when the <u>initial</u> value of the perturbation $\psi'|_{t=0}$ is given the value zero there is no <u>growth</u> in the perturbation (for $A/A_c = 0.85$) over 800 hours.

ACKNOWLEDGEMENT

The National Board for Science and Technology is thanked for their financial support of this research.

REFERENCES

[1] Deininger, R.C.,; Loesch, A.Z. : A note on Rossby wave instability at finite amplitude. J. Atmos. Sci. <u>39</u> (1982) 688-690.

[2] Hodnett, P.F.; O'Brien, W.M. : On the nonlinear barotropic instability of Rossby wave motion. Beitr. Phys. Atmosph. <u>62/2</u> (1989) 90 - 98.

[3] Loesch, A.Z. : Finite amplitude stability of Rossby wave flow. J. Atmos. Sci. <u>35</u> (1978) 929-939.

[4] Lorenz, E.N. : Barotropic instability of Rossby wave motion. J. Atmos. Sci. <u>29</u> (1972) 258-264.

[5] Rossby, C.G. : Relation between variations in the intensity of the zonal circulation of the atmosphere and the displacement of the semi-permanent centres of action. J. Marine Res. <u>2</u> (1939) 39-55.

Department of Mathematics
University of Limerick
Limerick, Ireland

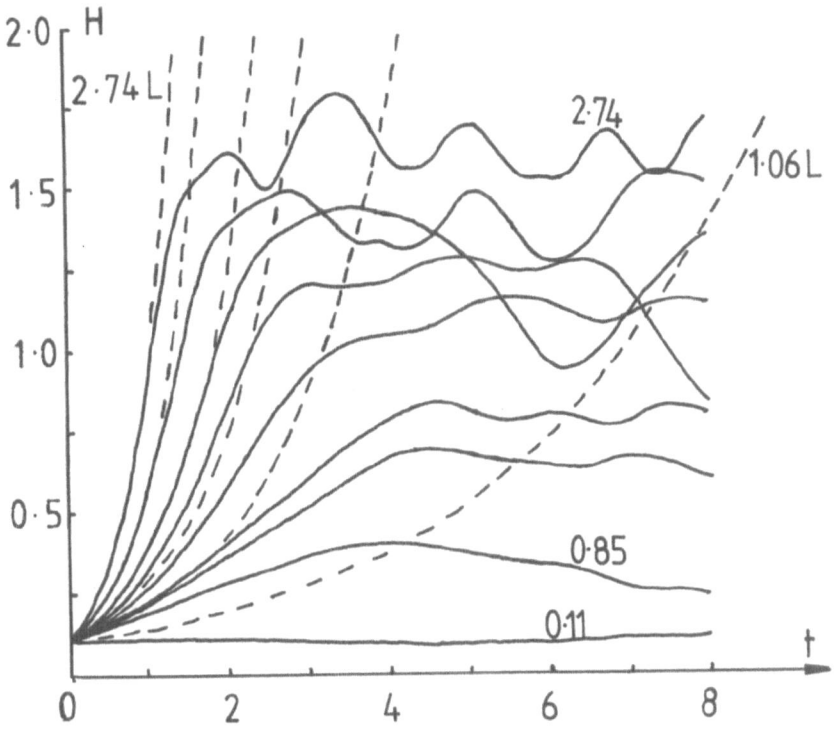

Figure 1. H (which is a measure of the perturbation field averaged over the grid points of the region) as computed numerically is shown (solid curves) plotted against time, t (measured in hours in units of 100), for the case N=6 and M=3, for a range of values of the amplitude, A i.e. A/A_C = 2.74, 2.13, 1.75, 1.49, 1.28, 1.06, 1, 0.85, 0.11. For comparison the linear results of Lorenz (1972) are superimposed and are indicated by the dashed curves for A/A_C = 2.74, 2.13, 1.75, 1.49, 1.28, 1.06.

The method of region analysis and its application for optimal control problems of hydroelectric power plants *

Hoàng Xuân Phú

Abstract

The *method of region analysis* is presented, by which one can solve some classes of optimal control problems with state constraints. In particular, problems with one state and one control variable are investigated in which the performance index and the state equation are linear in the control variable. As practical example, a problem of control of a hydroelectric power plant is considered.

1   Introduction

For solving optimal control problems with state constraints one can use Pontryagin's maximum principle as represented in [4]. This is a necessary optimality condition which provides the local behavior of optimal solutions, resulting in a series of two-point boundary-value problems. In consequence of the state constraints, the possible number of boundary-value problems is very large or maybe even infinite, and the location of the boundary points is generally unknown.

In order to overcome the above difficulties in solving optimal control problem with state constraints, we developed the *method of region analysis* whose procedure depends on the concrete problem class. In the second section of this paper, it is shown how a control problem with one state and one control variable in which the performance index and the state equation are linear in the control variable can be solved by the method of region analysis.

In the third section, a control problem of a hydroelectric power plant is considered. For this problem, the method of region analysis can deliver quickly some qualitative results.

---

*This paper was written when the author was invited by Prof. Dr. R. Bulirsch to work at the Munich University of Technology with the assistance of the Alexander von Humboldt Foundation.

309

Hj. Wacker and W. Zulehner (eds.),
Proceedings of the Fourth European Conference on Mathematics in Industry, 309–313.
© 1991 *B.G. Teubner Stuttgart and Kluwer Academic Publishers.*

2   The method of region analysis

We investigate the following problem: Determine the control function u which minimizes the cost functional

$$J = \int_{t_0}^{t_f} L(t, x(t), u(t))dt \ , \ L(t, \xi, v) = L_1(t, \xi) + L_2(t, \xi)v, \tag{1}$$

subject to the constraints

$$\dot{x}(t) = f(t, x(t), u(t)) \ , \ f(t, \xi, v) = f_1(t, \xi) + f_2(t, \xi)v,$$
$$\beta_1 \le u(t) \le \beta_2 \ , \ \alpha_1 \le x(t) \le \alpha_2 \ , \ x(t_0) = x_0 \ , \ x(t_f) = x_f. \tag{2}$$

Let us assume

$$\beta_1 < \beta_2 \ , \ \alpha_1 < \alpha_2 \text{ and } f_2(t, \xi) > 0 \text{ for all } (t, \xi) \in [t_0, t_f] \times [\alpha_1, \alpha_2]. \tag{3}$$

Here α_i and β_i are constant. But it means no loss of generality. If $f_2(t, \xi) < 0$ for all (t, ξ) we also get a similar result.

The *main idea* of the *method of region analysis* (for solving the problem (1)-(3)) consists in *analyzing* the *state region*

$$G := \{(t, \xi)|t_0 \le t \le t_f \ , \ \alpha_1 \le \xi \le \alpha_2\}$$

by the function h

$$h(t, \xi) := L_{1\xi} - \frac{L_2 f_{1\xi}}{f_2} - \frac{1}{f_2^2}[(L_{2\xi}f_2 - L_2 f_{2\xi})f_1 + L_{2t}f_2 - L_2 f_{2t}]|_{(t,\xi)} \tag{4}$$

(all functions on the right side depend on (t, ξ)). That means, the sign of h on G has to be determined. In consequence of this, we get subregions of G in which h is always positive or negative or equal to zero. By this resulting structure of G, the form of optimal processes can be determined or estimated. Of course, the structure of G may be very complicated. But in this paper, let us assume that there exists a *continuous* function η with the property

$$h(t, \xi) \begin{cases} > 0 & \text{in } G^+ := \{(t, \xi) \in G | \xi > \eta(t)\} \\ = 0 & \text{in } G^0 := \{(t, \xi) \in G | \xi = \eta(t)\} \\ < 0 & \text{in } G^- := \{(t, \xi) \in G | \xi < \eta(t)\}. \end{cases} \tag{5}$$

I.e., the state region G is divided into three subregions and h is positive in G^+, negative in G^- and equal to zero in G^0. Of course, some of these subregions may be empty. For example, if h is negative in whole G we can say that (5) holds for any function η with $\eta(t) > \alpha_2$ for all t in $[t_0, t_f]$. Define now a function η^* by

$$\eta^*(t) := \begin{cases} \alpha_2 & \text{if } \eta(t) > \alpha_2 \\ \eta(t) & \text{if } \alpha_1 \le \eta(t) \le \alpha_2 \\ \alpha_1 & \text{if } \eta(t) < \alpha_1. \end{cases} \tag{6}$$

First, let us consider the case with

$$f(t, \eta^*(t), \beta_1) < \dot{\eta}^*(t) < f(t, \eta^*(t), \beta_2) \text{ a.e. in } [t_0, t_f]. \tag{7}$$

Theorem 1. *Suppose (3) and (5)-(7). If (x^*, u^*) is an optimal process of the problem (1)-(2) with*

$$x_0 \leq \eta^*(t_0) \text{ and } x_f \leq \eta^*(t_f) \tag{8}$$

then there exist z_1 and z_2 such that $u^*(t) = \beta_2$ for $t \in (t_0, z_1)$, $x^*(t) = \eta^*(t)$ for $t \in (z_1, z_2)$ and $u^*(t) = \beta_1$ for $t \in (z_2, t_f)$.

Note that (t_0, z_1) or (z_1, z_2) or (z_2, t_f) may be empty. This theorem was proved in [7].

What does it say? The optimal trajectory has to approach as rapidly as possible to the graph of η^*. Then it has to keep running on this graph, except the last period of time when it has to move to the final point (t_f, x_f). The last assertion holds also if $x_0 > \eta^*(t_0)$ or $x_f > \eta^*(t_f)$ (i.e., if (8) is not fulfilled).

Of course, an admissible trajectory can only keep running on the graph of η^* under the assumption (7). What happens in the other case? Then the optimal trajectory has to try to remain nearby this graph. Theorem 2 formulates this property for a special case. For that we need an *attainable bound* E and a *security bound* S defined as follows:

$$E(t_0) = x_0 \ , \ \dot{E}(t) = \begin{cases} 0 & \text{if } E(t) = \alpha_2 \text{ and } f(t, \alpha_2, \beta_2) \geq 0 \\ f(t, E(t), \beta_2) & \text{otherwise,} \end{cases} \tag{9}$$

$$S(t_f) = x_f \ , \ \dot{S}(t) = \begin{cases} 0 & \text{if } S(t) = \alpha_2 \text{ and } f(t, \alpha_2, \beta_1) \leq 0 \\ f(t, S(t), \beta_1) & \text{otherwise.} \end{cases} \tag{10}$$

Theorem 2. *Suppose (3), (5)-(6) and $G = G^-$, i.e., $\eta^*(t) = \alpha_2$ for all $t \in [t_0, t_f]$. If (x^*, u^*) is an optimal process of (1)-(2) then $x^*(t) = \min\{E(t), S(t)\}$ for all t.*

This theorem was proved in [6]. An equivalent formulation of it is: If $M(t) := \{\xi |$ there exists an admissible process (x, u) of (1)-(2) with $x(t) = \xi\}$ and if (x^*, u^*) is an optimal process of (1)-(2) then $x^*(t) = \max\{\xi | \xi \in M(t)\}$ for all t.

More about the method of region analysis can be found in [6], [7] and other papers. For our purpose in the next section we only need the above two theorems.

3   Optimal control of a hydroelectric power plant

The control of a hydroelectric power plant is an important practical problem which was investigated by many authors, see for example [1], [2], [3], [5]... In this section, let us consider a power plant, the water changing in the reservoir of which is described by $\dot{x} = b - w - s$. Here, x is the water volume, b is the natural influx and w is the flow through the turbines. s denotes the spilling water, which is *unregulated*, i.e., s only depends on the water volume in such a way: There exists an $\alpha_0 \in [\alpha_1, \alpha_2]$ such that

$$s(\xi) = 0 \text{ if } \xi \in [\alpha_1, \alpha_0] \text{ and } s(\xi) > 0 \text{ if } \xi \in (\alpha_0, \alpha_2]. \tag{11}$$

If l is the *head* (i.e., the height difference between the water surface of the reservoir and the turbines) and ω is the efficiency of the plant, the energy produced at the time t is given by $c w(x(t), w(t))l(x(t))w(t)$ (c is a constant). For the sake of simplicity we assume that the efficiency is constant. Then, with the so-called *valuation function* a the *total rated energy* in the time interval $[t_0, t_f]$ can be represented by $\int_{t_0}^{t_f} c w a(t)l(x(t))w(t)dt$. Our aim is to maximize this rated energy. In order to get the problem type (1)-(2) with (3), we introduce a new control function $u := -w$. Our control problem is now: Determine the control function u which minimizes the cost functional

$$\int_{t_0}^{t_f} c w a(t)l(x(t))u(t)dt \tag{12}$$

subject to the constraints

$$\dot{x}(t) = b(t) - s(x(t)) + u(t) \ , \ x(t_0) = x_0 \ , \ x(t_f) = x_f,$$
$$-\beta^+ \leq u(t) \leq -\beta^- \ , \ \alpha_1 \leq x(t) \leq \alpha_2. \tag{13}$$

Here, α_1 is the minimal and α_2 the maximal allowed level of the reservoir volume, β^- is the minimal and β^+ the maximal flow through, x_0 is the start and x_f is the final volume.

For solving (12)-(13) by the method of region analysis, we have to determine the function h. It follows from (4)

$$h(t, \xi) = c w [a(t)l'(\xi)(r(\xi) - b(t)) - \dot{a}(t)l(\xi)] \tag{14}$$

where $r(\xi) := l(\xi)s'(\xi)/l'(\xi) + s(\xi)$. Let us consider only the case with *constant a*. In the following, we need some properties of the head and the spilling function which hold practically: l is strictly increasing and strictly concave, s is increasing and convex. For that reason, r is increasing and therefore, there exists a function η with (5). η is continuous because the natural influx b is continuous. Moreover, because of (11) and b is positive, it holds

$$\eta(t) > \alpha_0 \text{ for all } t \in [t_0, t_f]. \tag{15}$$

First case: If the function η^* defined by (6) satisfies (7) (that holds if the variation of the natural influx b is not very large), then the form of the optimal control is given by Theorem 1 (since (15) it is normal if assume $x_0 \leq \eta(t_0)$ and $x_f \leq \eta(t_f)$). That means, in the first period $[t_0, z_1)$, the flow through has to be equal to the minimal value β^- so that the water volume can reach as rapidly as possible the level of η^*. Then it has to be chosen so that the water volume remains on the level of η^*, except in the last period $[z_2, t_f]$ when the flow through is equal to the maximal value β^+ so that x^* can reach as rapidly as possible the final level x_f. In this case the optimal solution has an interesting property: (11) and (15) imply that $s(x^*(t)) > 0$ for all $t \in (z_1, z_2)$ if $\alpha_2 > \alpha_0$. Hence, there is spilling water in most of the time if $[t_0, t_f]$ is sufficiently large. That means that in order to get maximal electrical energy a part of water is "wasted".

Second case: Let us consider the case where η^* does not satisfy (7) and $\alpha_2 = \alpha_0$ (i.e., the spilling water is not allowed). Here we have $G = G^-$. Hence, Theorem 2 shows that $x^* := min\{E, S\}$ is the only optimal state function if it is admissible, i.e., if $x^*(t_0) = x_0$, $x^*(t_f) = x_f$ and $x^*(t) \geq \alpha_1$ for all $t \in [t_0, t_f]$. The optimal control $u^*(t) = -w^*(t)$ is equal to $-\beta^-$ if $E(t) < S(t)$ and equal to $-\beta^+$ if $E(t) > S(t)$.

Otherwise, $u^*(t) = -w^*(t) = -b(t)$ almost everywhere in $\{t|S(t) = E(t)\}$. What does this result say? The power plant has to be regulated so that the water volume is always *as great as possible*. The optimal flow through is equal to the minimal value β^- if the water volume is less than the maximal allowed level α_2 and it is equal to the natural influx b if the water volume is equal to α_2. But if the natural influx is greater than the maximal flow through β^+ in the next period, the optimal flow through has to be switched to β^+ in time so that there is no inundation and x^* is equal to α_2 if the natural influx b is contained in $[\beta^-, \beta^+]$ again. Of course, in the last period, the flow through is equal β^+ so that the water volume reaches the final level in the shortes time.

More results about optimal control of hydroelectric power plants by the method of region analysis can be found in [8], [9] ...

References

[1] Bauer, W.; Buchinger, S.; Wacker, Hj.: *Einsatz mathematischer Methoden bei der Hydroenergiegewinnung*, ZAMM 63 (1983) 227-243

[2] Duong, P. C.:*Aproximate synthesis of controls of a cascade of water reservoirs*, Inf. in Appl. Math., Computation Centre of the Acad. of Sci. of the USSR, Moscow 1983

[3] Gfrerer, H.:*Optimization of hydro energy storage plant problems by variational methods*, Z. Oper. Res. Ser. B 28 (1984) 87-101

[4] Ioffe, A. D.; Tichomirov, V. M.: *Theorie der Extremalaufgaben*, VEB Deutscher Verlag der Wissenschaften, Berlin (GDR) 1979

[5] Menshikov, J. S.; Menshikova, O. R.:*Methods for optimal control and differential games in control problems for cascades of water reservoirs*, Inf. in Appl. Math., Computation Centre of the Acad. of Sci. of the USSR, Moscow 1983

[6] Phú, H. X.:*Some necesssary conditions for optimality for a class of optimal control problems which are linear in the control variable*, Systems Control Lett. 8 (1987) 261-271

[7] Phú, H. X.:*A method for solving a class of optimal control problems which are linear in the control variable*, Systems Control Lett. 8 (1987) 273-280

[8] Phú, H. X.:*On optimal control of a hydroelectric power plant*, Systems Control Lett. 8 (1987) 281-288

[9] Phú, H. X.:*Optimal control of a hydroelectric power plant with unregulated spilling water*, Systems Control Lett. 10 (1988) 131-139

Address: Institute of Mathematics, P.O.Box 631 Bo Ho, 10000 Hanoi, Vietnam

STOCHASTIC MODELLING OF THE DIELECTRIC RELAXATION IN CONDENSED MATTER

A. Janicki, A. Weron, and K. Weron, Wroclaw

Abstract:A simple stochastic model describing the dielectric relaxation function is proposed. It is based on modelling of inherent disorder which is associated with all dielectric materials.

A study of relaxation phenomena in condensed matter is of
great technological importance.For instance,it may be important
for optical fibers because of a gradual increase of the ·dielec-
tric loss, for insulators employed in such diverse applications
as power transmission cables and xerographic films, cf.[1,2].

The dielectric response function or the relaxation function
f(t) as the response of polarization P(t) to an electric field E
can be defined in the following way

$$P(t) = \varepsilon \int f(u)E(t-u)du.$$

The present uderstanding of the microscopic dynamic response is
based on the Debye's work. His approach depends on an examina-
tion of the dynamics of the response of a charged dipole embedded
in a viscous medium and gave the relaxation function in the exponential form. Experimental evidences collected in last years [2]
show that the relaxation in many materials follows a universal pa
ttern of the stretched exponential form

$$f(t) = \exp(-t/u)^{\alpha} , \quad 0 < \alpha < 1.$$

In this paper we consider any orientationally polarizable sy
stem as containing an inherent degree of disorder in which should
arise distributions of relaxation times [3] or distributions of
activation energies.Consequently, one can apply stochastic model
ling based on limit theorems.It gives stretched exponential form
as a result of Levy's limit theorem and the fact that both relaxa
tion times and activation energies are non-negative. Thus comple
tely asymmetric α-stable distributions with $0 < \alpha < 1$ describe
relaxation behaviour in condensed matter.

References

[1] K.L.Ngai, *Comm. on Solid State Phys.*9(1980),141.
[2] A.K.Jonscher, *Dielectric Relaxation in Solids*, London 1983.
[3] K.Weron, *Acta Phys.Polon.A* 70 (1986), 579\

Institute of Mathematics
Technical University of Wroclaw
50-370 Wroclaw, Poland

Hj. Wacker and W. Zulehner (eds.),
Proceedings of the Fourth European Conference on Mathematics in Industry, 315.
© 1991 B.G. Teubner Stuttgart and Kluwer Academic Publishers.

A CALCULATION AND CONTOUR PLOTTING SYSTEM FOR FLAME TEMPERATURES

Simon B Jones

ICI Chemicals & Polymers Ltd, Research & Technology Department,
P O Box 8, The Heath, Runcorn, Cheshire WA7 4QD, U.K.

Summary: Details are given of a computer program called FLAME,
which plots contours of flame temperatures for a fuel-oxidant-
inert system on a triangular diagram. This program can be used
to estimate the flammability envelope for such a mixture. The
computational parts of the program are based on the STANJAN program which was developed at Stanford University for calculations
of equilibrium chemical reaction conditions. The calculations
can be carried out under either adiabatic or explosive detonation conditions. They assume ideal gases, pure condensed
species and equilibrium thermodynamics. The computer code also
includes a contouring algorithm which is specially designed to
handle the sharp-cornered contours which occur in some flame
temperature diagrams. Three examples of computed flammability
diagrams are given.

Keywords: Flammability, Contouring, Triangular diagrams, Equilibrium thermodynamics.

1   Introduction

The development of the FLAME program arose from a request
from the Explosion Hazards team of ICI Chemicals & Polymers Ltd.
Their interest is in determining the range of compositions for
which a particular mixture of chemicals is flammable. Many mixtures can be characterised as fuel-oxidant-inert systems, so
that the flammable region can be plotted on a triangular
diagram.

Determining flammable regions experimentally requires a
substantial amount of data, which takes a long time to obtain.
Flammability limits are not always needed to a high level of
accuracy, because often plant designs involve a mixture which is
either well inside the flammable region or well away from it. It
was seen that if the flammability limits could be estimated to a
reasonable degree of accuracy by theoretical means, then the

Hj. Wacker and W. Zulehner (eds.),
Proceedings of the Fourth European Conference on Mathematics in Industry, 317–322.
© 1991 B.G. Teubner Stuttgart and Kluwer Academic Publishers.

pressure on experimental work could be reduced, as only a few
experimental data points might be needed to confirm a theoretical conclusion.

Underlying the theoretical calculation of flammability
limits is the hypothesis that the flame temperature is approximately the same at all points on the boundary of the flammable
region. This is a hypothesis which has been confirmed by a
considerable amount of experimental evidence, but it remains an
approximation. Thus the flammability envelopes which are determined using the FLAME program are also only approximate. However, results from FLAME will always be backed up by
experimentation.

2   Theory

The FLAME program is based on the STANJAN package which was
developed at Stanford University. STANJAN is a general purpose
program for chemical equilibrium thermodynamics of ideal gases
and pure condensed phases.

The basis of chemical equilibrium calculations is the minimisation of the Gibbs function for the system, given a number of
constraints.

The Gibbs function for ideal mixtures is usually written
as:

(1.1) $G = \sum_i \mu_i n_i$

where G is the Gibbs function, μ_i is the chemical potential
for species i, and n_i is the number of moles of species i.

Equation (1.1) can be transformed [4] into:

(1.2) $G = \sum_j \chi_j N_j$

where χ_j is the element potential for element j and N_j is
the number of moles of element j.

This summation is over elements rather than species. For
any given system at equilibrium, the element potential is characteristic of an element throughout the system, regardless of
the species and phase in which it is found. Thus if we minimise

the Gibbs function in equation (1.2), we have many fewer variables to minimise over - one for each element, rather than for
each species as in equation (1.1). STANJAN uses this more
efficient element potential method and can be run on microcomputers, while the chemical potential method is usually only
practicable on mainframes.

The minimisation of G is always subject to constraints on
thermodynamic variables. Therefore we need to relate χ_j to these
variables. We have

(1.3) $\mu_i = \sum_j a_{ij}\chi_j$

where a_{ij} is the number of atoms of element j in species i.

Now, for ideal species, we have the following [1]:

(1.4) $\mu_i(T,P,\underline{x}) = \mu_i^*(T,P) + RT \ln x_i$

(1.5) $x_i(T,P,\underline{\chi}) = \exp\left\{\frac{1}{RT}\left(\sum_j a_{ij}\chi_j - \mu_i^*(T,P)\right)\right\}$

where x_i is the mole fraction of species i and $\underline{x}$ is the
vector $(x_1, x_2, ...)$

The value of μ_i^* can be related to its value at a reference
pressure P_o (usually atmospheric pressure) by

(1.6) $\mu_i^*(T,P) = \mu_i^o(T) + RT \ln(P/P_o)$ (gas phase)

(1.7) $\mu_i^*(T,P) = \mu_i^o(T) + \int_{P_o}^{P} v\,dP$ (condensed phase)

where v is the molar volume of the condensed phase and

(1.8) $\mu_i^o(T) = H_i^o(T) - S_i^o(T)T$

where $H_i^o(T)$ and $S_i^o(T)$ are the enthalpy and entropy of species
i at the reference pressure. These quantities are tabulated
against temperature in standard thermodynamic tables such as
JANNAF [3].

Thus we can use the χ_j and two thermodynamic quantities from
T, P, H, S, U, V (U is internal energy and V is volume) as our independent variables in the minimisation.

A flame temperature is achieved with two Gibbs function
minimisations. We are given the starting species (usually a
fuel, an oxidant and an inert) and the initial P and T.

STEP 1: Fix P and T and minimise to find the initial H, U, S, V.
STEP 2: Fix two of H, U, S, V and minimise to find the others,
including T, which is the flame temperature. In the FLAME program, one of two options is used: either adiabatic (H and P
fixed) or explosive detonation (U and V fixed).

The contouring routine in the FLAME program is based on the
routine J06GFF from the Graphical Supplement to the NAG library,
modified to draw contours on a triangular diagram rather than a
rectangle and to handle the very sharp corners which are found
in flame temperature contours.

3   Examples

In this section, the results of applying the FLAME program
to the computation of flammability diagrams for three systems
are presented. The results are compared with experimental flammability curves.

The thermodynamic data required for these runs have been
taken from the files of JANNAF data supplied with the STANJAN
program from Stanford University.

The data and plots for the three example systems is shown
in table 3.1, in which T_L^{calc} and T_U^{calc} are the flame temperatures
at the lower and upper flash points as calculated by FLAME. T_L^{exp}
and T_U^{exp} are the corresponding values from experiment.

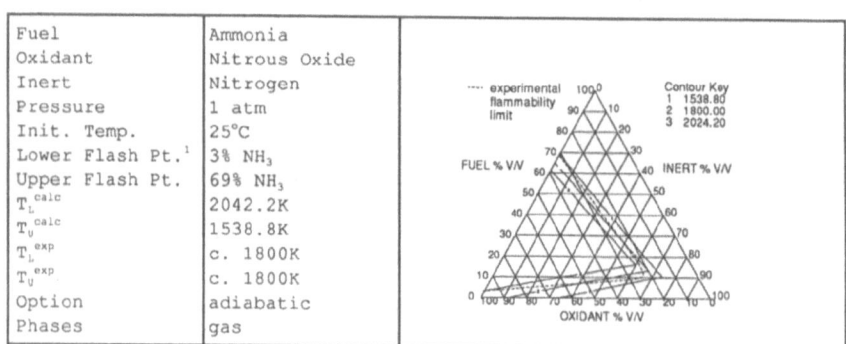

Fuel	Ammonia	
Oxidant	Nitrous Oxide	
Inert	Nitrogen	
Pressure	1 atm	
Init. Temp.	25°C	
Lower Flash Pt.[1]	3% NH₃	
Upper Flash Pt.	69% NH₃	
T_L^{calc}	2042.2K	
T_U^{calc}	1538.8K	
T_L^{exp}	c. 1800K	
T_U^{exp}	c. 1800K	
Option	adiabatic	
Phases	gas	

Tab. 3.1(a)   Data and figures for example systems

---

1 Flash point is limit of flammability in binary fuel/oxidant
system

Fuel	Carbon Monoxide
Oxidant	Air
Inert	Water Vapour
Pressure	1 atm
Init. Temp.	100°C
Lower Flash Pt.	11.5% CO
Upper Flash Pt.	73% CO
T_L^{calc}	1374.3K
T_U^{calc}	1363.2K
T_L^{exp}	n/a[1]
T_U^{exp}	n/a
Option	adiabatic
Phases	gas

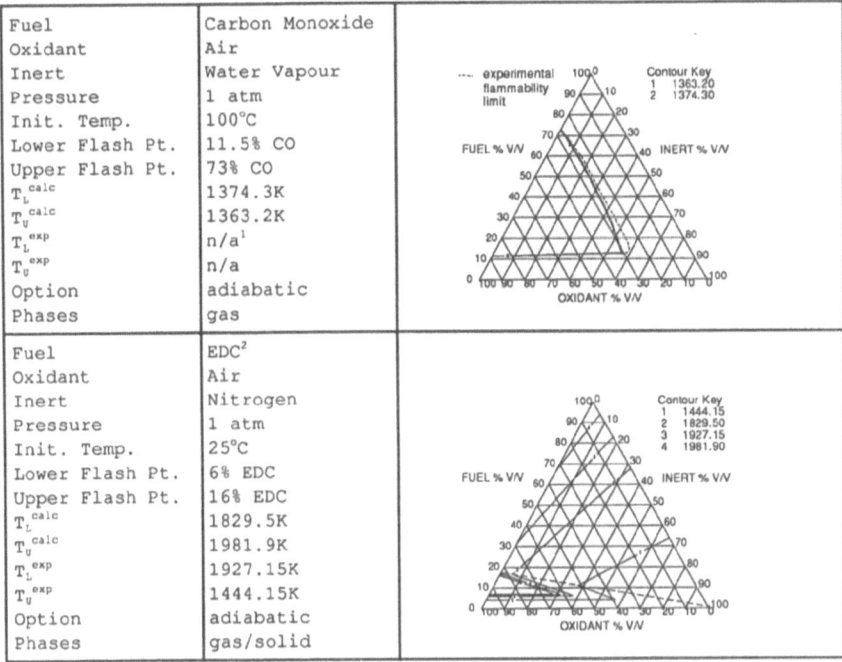

Fuel	EDC[2]
Oxidant	Air
Inert	Nitrogen
Pressure	1 atm
Init. Temp.	25°C
Lower Flash Pt.	6% EDC
Upper Flash Pt.	16% EDC
T_L^{calc}	1829.5K
T_U^{calc}	1981.9K
T_L^{exp}	1927.15K
T_U^{exp}	1444.15K
Option	adiabatic
Phases	gas/solid

Tab. 3.1(b)   Data and figures for example systems

The flame temperature contours at T_L^{calc}, T_U^{calc}, and where available T_L^{exp} and T_U^{exp} are shown for the three systems in table 3.1. When experimental flammability limits are known, these are shown with dotted lines.

In the third example, EDC/Air/Nitrogen, a solid carbon phase can be produced. Trials with the FLAME program show that the condition for this is roughly EDC:Air>0.18. This limit is shown as a dotted line in table 3.1, where the calculations have included the effect of solid carbon for this example.

4   Conclusions
This paper has described an application of mathematical modelling to a process industry problem. Using mathematical techniques, a software package has been assembled so that the

1 n/a means not available
2 EDC is ethylene dichloride

Explosion Hazards experts can estimate flammability envelopes
for fuel-oxidant-inert mixtures. The system has been delivered
on a standard microcomputer. The system can be used to reduce
pressure on experimental work because for a process which operates well away from the limit of flammability only a few experimental points would be needed to confirm the theoretical
predictions. However, for a process which is close to the limit,
detailed experimental investigation would still be necessary.

The three examples in this paper are simple cases which
illustrate the performance of the system. Reasonable agreement
between computed and experimental flammability limits is
obtained for each of the systems.

5   References

[1]   Guggenheim, E. A.: Thermodynamics, 6th ed: North-Holland
       1977.

[2]   James, G. D.; Heard, K. S.; Suttie, I. R.: Point-in-
       polygon Project Stage 1. (U.K.) Dept. of the Environment
       Research Report 2 (1975)

[3]   JANAF Thermochemical tables, 3rd ed. J Phys Chem Ref Data
       14 supplement 1 (1985)

[4]   van Zeggeren, F.; Storey, S. H.: The computation of chemi-
       cal equilibria: Cambridge 1970: pp124-127

6   Acknowledgements

The author wishes to thank the Research and Technology Manager, ICI Chemicals and Polymers Ltd for permission to publish
this paper. The motivation for this work came from Peter Doran
and Russell Greig, who also supplied the experimental data. The
advice and comments of Ian Parker are also gratefully acknowledged.

ON THE TAPER CURVES IN THE FOREST INDUSTRY

Aatos Lahtinen

Abstract. Some taper curves the forest industry uses in inventories and various processes are reviewed. A new shape preserving taper curve is presented and its properties are discussed.

1. Introduction

The forest industry must, like any other industry, take care of the availability and quality of its raw material. This means regular inventories of the forests. For instance, in Finland where the forest industry is large and surpassed only by the metal industry there exists for inventories in forests a permanent grid of sample plots covering the whole country. The measurements and observations of trees in these plots provide the material on the basis of which the state and growth of forests is estimated.

The mathematical model of a tree is a central concept in all forest mensuration. The most important primarly uses are the determination of the volume of (a certain part of) the tree and of the diameter at a given height. These make it possible to build different models to simulate certain things such as growth or harvesting strategy etc. In Finland one important task is to determine the stumpage prices and wages for harvesting work in timber lots of over 10 mill. m^3 annually.

It can be assumed that the form of the tree stem is for each species independent of the size and that the stem is a solid of revolution. Then the mathematical model of the stem can be presented by a function of one variable (relative height) which generates the stem by rotation. Such a function is called *a taper curve*. The taper curve of a regular tree is monotonically decreasing, first convex then concave, having one point of inflection. The tapering at the butt is very rapid.

2. Common taper curves

Several types of taper curves are used in the forest industry (Cf. [1]-[4], [6], [11]). The main objective in the construction is to get to certain tasks a "sufficiently accurate" taper curve by using the measured height and as few measured diameters as possible. We will mention some of the most common types here.

a. Polynomial model

Here the taper curve is a polynomial. Various degrees are used up to say 34. In a typical case the diameter p at the height x is presented by using one measured

Hj. Wacker and W. Zulehner (eds.),
Proceedings of the Fourth European Conference on Mathematics in Industry, 323–327.
© 1991 *B.G. Teubner Stuttgart and Kluwer Academic Publishers.*

diameter $d(ah)$ at a relative height $ah, 0 < a < 1$, as a parameter.

$$p(x) = d(ah) \sum_{j=j_1}^{j_n} b_j (1 - \frac{x}{h})^j, \quad 0 \le x \le h.$$

The coefficients b_j are fixed by using suitable sample tree material (Cf. [3]). These taper curves are mostly used to determine the total volume of a stand of timber.

b. Logarithmic model

Here the taper curve is a polynomial of moderate degree (e.g. 4) combined linearly with one or two powers of the logarithm. The latter terms are intended to give a natural form to the butt of the stem which tapers very quickly. Also here the diameter r at the height x is usually presented by using a measured diameter $d(ah)$ at a height ah as a parameter.

$$r(x) = d(ah)(\sum_{j=1}^{n} b_j (\frac{x}{h})^j + b_{n+1} \ln x + b_{n+2} (\ln x)^2), \quad 0 \le x \le h$$

The coefficients b_j are again fixed by using sample trees (Cf. [3]). The main uses are the determination of volume and also the lumber assortment of a stand of timber.

c. Cubic spline model

For more sophisticated use the aforementioned models are too coarse. A more accurate taper curve can be obtained by using interpolating cubic splines. For details cf. [6]. An interpolating cubic spline is uniquely determined by two initial values which can for example be two (estimated) diameters. The construction of the spline is global, i.e. a change in the diameter or initial value affects the spline at every point of definition.

The cubic spline as a rule gives a good taper curve for regular trees (Cf. [6]). Therefore a taper function construction based on cubic splines has been in standard use in Finland since 1979. However, not all trees are regular. In fact, about 25% of trees have bulges. The cubic spline has a tendency to exaggregate these bulges. In addition the cubic spline may create bulges to the taper curve also in some cases where the measured data does not indicate it. Such cases are not uncommon, about 20%. They may cause for instance over 10 % relative errors in lumber assortment and render short period growth estimates grossly inaccurate. This phenomen cannot be prevented, because it is due to the fact that the cubic spline minimises the strain energy. Thus the taper curves constructed by cubic splines are not sufficiently accurate for tasks where the shape of the stem has to be taken into account. The same is even more true for the polynomial and logarithmic model.

3. Shape preserving quadratic splines

The oscillation of the cubic spline highlights the need for a shape preserving taper curve. In order to meet this need one has at first to define the concept "shape preserving". In practice we only know the height and some diameters of the tree. Therefore we include in the shape preservation model only properties that can be estimated sufficiently accurately on the basis of this initial data.

Let L be a linear spline interpolating the data at the breakpoints. If the taper curve is monotonically increasing and decreasing on the same intervals as L and also convex and concave on the same intervals as L, then it is said that the taper curve *preserves the shape of the data* or *is shape preserving*.

For practical calculations one must suppose in addition that the taper curve is smooth, at least continuously differentiable. Thus we can state the problem as follows:

Suppose that the data $D = \{(x_i, y_i)\}_1^n$ is given. Construct a continuously differentiable function which interpolates at D and preserves the shape of D.

The solution is sought here among quadratic splines. The usual interpolating quadratic spline is not able to preserve the shape (Cf. [8]), but there are several methods which give a solution (Cf. e.g. [5], [7], [10]). The decisive step is to have a local construction where breakpoints are added between interpolating points. Perhaps the most flexible method is the one in [5] (Cf. also [10]). The essential results are:

Proposition 1. *Suppose that there are given data $D = \{(x_i, y_i)\}_1^n$, $0 < x_1 < \cdots < x_n = h$, $\{y_i\}_1^n \subset \mathbf{R}$. If $\{m_i\}_1^n \subset \mathbf{R}$ and $\{\xi_i\}_1^n$, $x_i < \xi_i < x_{i+1}$, are given, there exists on the interval $[0, h]$ a quadratic spline $s \in C^1$ so that $s(x_i) = y_i$, $Ds(x_i) = m_i$, $i = 1, \ldots, n$. All points x_i are breakpoints. On the interval $]x_i, x_{i+1}[$ there is at most one breakpoint. If such one exists, it is ξ_i.*

For the proof cf. [10]. The result means that we can locally construct an interpolating quadratic spline with additional breakpoints. Here the derivatives at interpolating points and the places of the additional breakpoints are parameters which can be freely chosen. A change in these values changes the spline only at a certain neighbourhood of the change.

It has been shown in [5] that these parameters can be chosen from certain intervals so that the resulting spline is shape preserving.

Proposition 2. *Given data D there exists intervals $A_i = [m_{Li}, m_{Ui}]$ and $B_i = [\xi_{Li}, \xi_{Ui}]$ such that the quadratic spline s, where $m_i \in A_i$ and $\xi_i \in B_i$ (whenever ξ_i is needed), preserves the shape of the data D.*

The lower and upper limits of intervals have exact expressions based on the data. The breakpoint limits depend also on the chosen derivatives.

4. Shape preserving taper curves

Shape preserving taper curves can be now formed by using the presented construction. According to Proposition 2 there is still freedom in the choice of parameters. This can be used to ensure the quality of the taper curve. The quality means here good approximations for three things, which are the *volume* of the tree or any part of it, the *lumber assortment* of the tree and the *shape* of the tree.

The most essential property is the estimation of the volume. This depends mainly on the choice of the derivatives and only to some extent on the places of the additional breakpoints. Therefore it is essential to choose the derivatives properly. One possibility is to present the derivatives as convex combinations of the slopes of the linear spline L interpolating the data. Irregularities in the data may demand on different constructions. The details are to be found in [5].

There is in [4] an algorithm for the construction of the taper curve in this way. Tests done with representative sample tree material show that the parameters of the shape preserving quadratic spline can be chosen so that the resulting taper curve gives as good volume estimates as the best earlier taper curves. The lumber assortment given by this new construction is more accurate than with the earlier taper curves.

The shape of the taper curve is to some degree fixed by the earlier demands. The graphs of the taper curves show it to be quite natural (Cf. [4]). There remains, however, some freedom in the choice of parameters which can be used for corrections if they are necessary.

It is also noted in [4] that this shape preserving construction needs for a given accuracy a smaller number of initial measurements than the standard taper curves based on cubic splines. This means in practice for instance lower costs in forest mensuration.

On the basis of test results we can say that the presented shape preserving construction of taper curves seems to be superior to the old ones for almost all uses which are connected to the shape of the tree. Better lumber assortments and more accurate growth estimates are among the direct advantages of this construction. It can also give improvements to many other models in the forest industry which need taper curves. Especially this is the case if the model uses relatively few sample trees with several measurements from each tree. We can also mention an economic model which tries to predict the optimal thinning and rotation of a forest. The present constructions are not easy to use due to the structure of some of its components (Cf. [9]). By using shape preserving taper curves one can have a less complicated optimization problem leading to a better solution in less computer time.

As a conclusion we can say it to be expected that the use of shape preserving taper curves presented will offer to the forest industry advantages which are also economically useful.

References

[1] Kilkki, P.; Varmola, M.: Taper curve models for Scots pine and their applications. Acta For. Fenn. 174 (1981) 1-60.

[2] Kozak, A.; Munro, D.D.; Smith, I.H.G.: Taper functions and their application in forest inventory. For. Chron. 45 (1969) 278-283.

[3] Laasasenaho, J.: Taper curve and volume functions for pine, spruce and birch. Commun. Inst. For. Fenn. 108 (1982) 1-74.

[4] Lahtinen, A.: On the construction of monotony preserving taper curves. Acta For. Fenn. 203 (1988) 1-34.

[5] Lahtinen, A.: Shape preserving interpolation by quadratic splines. J. Comput. Appl. Math. 29 (1990) 15-24.

[6] Lahtinen, A.; Laasasenaho, J.: On the construction of taper curves by using spline functions. Commun. Inst. For. Fenn. 95 (1979) 1-63.

[7] McAllister, D.; Roulier, J.: An algorithm for computing a shape-preserving os-culatory quadratic spline. ACM Trans. Math. Software 7 (1981) 331-347.

[8] Passow, E.: Piecewise monotone spline interpolation. J. Appr. Th. 12 (1974) 240-241.

[9] Roise J.: A nonlinear programming approach to stand optimization. Forest Sci. 32 (1986) 735-748.

[10] Schumaker, L.: On shape preserving quadratic spline interpolation. SIAM J. Numer. Anal. 20 (1983) 854-864.

[11] Sterba, H.: Stem curves - a review of the litterature. Forest Products Abstracts. Commonwealth For. Bur. 3 (1980) 69-73.

Author's adress:
University of Helsinki, Department of Mathematics
Hallituskatu 15   SF-00100   Helsinki   Finland

ANALYTICAL THREE-DIMENSIONAL STRESS DISTRIBUTION IN THE HEALTHY HIP JOINT

S. Mazzullo, T. Simonazzi, D. Tartari.

Summary: to design a total hip joint replacement, two basic ideas have been followed from the prototype of Charnley (1963) on:
1) replication of the healthy joint geometry, and
2) low friction prosthesis, on the assumption that the low friction of the healthy joint is due to the synovial liquid.
This theoretical work, on the contrary, proves that the synovial liquid contained in the intact articular capsule, is an essential "material" to achieve the proper stress distribution within the joint. The synovial liquid acts not only as a lubricant but also, and above all, it transforms the body weight compression stress on the articulation into a homogeneous radial compression over the entire acetabular surface. The future trend in total prosthesis design should, thus, be aimed at achieving this mechanical behaviour.
    The three dimensional analytical solution of the acetabular joint elastostatic problem is made possible by neglecting the geometrical details of the pelvis shape. The resulting simplified geometry is an infinite body (the pelvis bone) which contains within it a single spherical inclusion (the femur head) surrounded by a shell (cartilage and synovial liquid). The system is subjected to uniaxial compression which is uniform at infinity. The solution is obtained in terms of spherical harmonics by extension of a classical work by J.N. Goodier (1933).

1. Introduction

    Construction of a mathematical model for articulation of a healthy hip inevitably passes through simplifying hypotheses. These are, in general, dictated by the purpose for which one builds the model. In the present case the purpose is to obtain a reasonably accurate picture of the mechanical behaviour of a healthy hip so as to then imitate it with innovative prostheses both in terms of design and materials.

    The spherical geometry of the articulation is particularly simple while the shape of the pelvis, where the femur head is positioned, is quite complex (Fig.1). Many works have given priority to an accurate description of the shape of the pelvis using finite elements methods [1]. The present work, instead, describes just what takes place in the immediate vicinity of the

329

Hj. Wacker and W. Zulehner (eds.),
Proceedings of the Fourth European Conference on Mathematics in Industry, 329–335.
© 1991 B.G. Teubner Stuttgart and Kluwer Academic Publishers.

articulation, neglecting geometric details regarding the shape of
the pelvis. However by doing so it is possible to obtain a three-
dimensional analytical solution to the elasto-static problem of
the acetabular articulation.

The simplified geometry (Fig.2) of the mechanical model to
be studied here is an infinite body (pelvis bone) which contains
within it a single spherical inclusion (the femur head)surrounded
by a shell (the cartilage and synovial liquid). Such geometry is
the result of a two step pattern: 1) The first idealizes the
pelvis as a semi-infinite, three-dimensional domain containing
a hemi-sphere (the femur head). In this step the function of the
muscles and of the femur is to maintain the articular system
subjected to body weight under static equilibrium. 2) The second
is an artifice used to determine the intensity of these equilibrium
forces. In fact, an equal and opposite half-space replaces the
forces and thus an infinite, three-dimensional domain is obtained
subjected to uniaxial compression.

This work claims the fact that the articular capsule enclosing
the articulation (Fig.1b)has been given a mechanical role.
The intact capsule is a container for the synovial liquid and can
support pressure [6]. As a result, the synovial liquid acts not
only as a lubricant but also, and above all, it transforms the
body weight compression stress into homogeneous radial compression
over the entire surface of the acetabular cavity.

2) Mathematical model

The model for the articulation consists of an infinite domain
containing a shell-core spherical inclusion. The elasto-static
problem requires the solution of the equilibrium equation for a
linear elastic isotropic and heterogeneous body, undergoing
uniform uniaxial compression P.
The cartesian components of equilibrium displacement $u^{(i)}$in the
elastic solid, free from body forces, satisfy the vectorial
equation:

$$\text{grad div } \underline{u}^{(i)} + (1-2\nu_i) \text{ div grad } \underline{u}^{(i)} = \underline{0}; \quad i = 0, 1, 2,$$

where grad, div and ν are the gradient, the divergence and the Poisson ratio respectively. The meaning of the indices are (0)-core, (1)-shell, (2)-infinite domain.

Taking into account the axial symmetry of the problem, we may use zonal harmonics to obtain solutions that are symmetrical about an axis and then transform displacement $\underline{u}$ to spherical coordinates [2]. The boundary conditions, to complete the formulation of the problem are:

i) Adhesion at interfaces $r = a_i; \quad i = 0,1$

$$u_r^{(i)}(a_i) = u_r^{(i+1)}(a_i)$$

$$u_\theta^{(i)}(a_i) = u_\theta^{i+1}(a_i)$$

$$\widehat{rr}^{(i)}(a_i) = \widehat{rr}^{(i+1)}(a_i)$$

$$\widehat{r\theta}^{(i)}(a_i) = \widehat{r\theta}^{(i+1)}(a_i).$$

ii) Far field conditions: $\widehat{zz}_\infty^{(2)} = P$

iii) Regularity at the origin: $\underline{u}^{(o)}(0) = \underline{0}$.

One special aspect of this problem is the case of an infinite body containing an adhering spherical inclusion (without the shell). Such a case was analytically solved by J.N. Goodier [2]. Subsequently, V.A. Matonis et al. [3] extended this analysis and formulated the problem for the general case in which n-shells are present.

The particular case we are dealing with here, where only one shell is present, has been studied in detail [3], [5]. However, it has only been studied numerically since the infinite domain shell-core boundary value problem generates a system of 12 linear equations the analytical solution of which has, to date, been deemed unfeasible. The present contribution to this problem has been to analitically solve this system of 12 linear equations [4].

Thus, by specializing the general solution by appropriate
hypotheses, the complete analytical solution for the case in
question here has been achieved and discussed.

3) Discussion

The special hypotheses previously outlined in order to
simplify discussion of the results are that:
1) the shell consists of an ideal non viscous, compressible fluid
2) the core and the infinite body are made of the same material
3) the infinite body and the shell have the same bulk modulus.

The first hypothesis is the strongest. In symbols it reads:

$$\mu_1 \rightarrow 0; \quad \nu_1 \rightarrow 1/2$$

The formula simplification resulting from this choice is
enormous. In practice a fluid can be considered ideal as long as
it has a low viscosity and is far from the walls. Thus, in the
present case such requirements are encountered either in studying
the equilibrium system or in the presence of dynamic slow strains
for significant, non-infinitesimal shell thicknesses. Such a shell
is unable either to support shear stress or to transmit such
stresses to the core. Nor can an infinite body having such a core
support shear stresses other than zero at the infinite domain
shell interface. The core-shell structural inclusion can, therefore
only compress and expand in a isotropic manner. The calculations
indicate that the stress distribution within the infinite domain
precisely coincides with that of a byphasic composite with
spherical inclusions of only compressible liquid. In other words,
the infinite body is not affected by the fact that the inclusion
has a core. Mechanically it only "sees" the liquid. Shell and core
undergo hydrostatic compression whose intensity is 33.% of applied
stress at infinity. Stress distribution at the infinite body-
shell interface takes on the following values:
In polar position ($\theta = 0$):
radial stress: $\widehat{rr}$ $= + \dfrac{1}{3}$ P   (compression)

zenithal   stress : $\widehat{\theta\theta}$ = $-\dfrac{20}{21}$ P   (tension)

azimuthal stress : $\widehat{\psi\psi}$ = $-\dfrac{20}{21}$ P   (tension)

In an equatorial position: (θ = $\boldsymbol{\pi}$/2):

radial stress   : $\widehat{rr}$ = $+\dfrac{1}{3}$ P   (compression)

zenithal stress : $\widehat{\theta\theta}$ = $+\dfrac{40}{21}$ P   (compression)

azimuthal stress : $\widehat{\psi\psi}$ = $+\dfrac{1}{21}$ P   (compression)

The surface of acetabulum is in a homogeneous radial compression state, $\widehat{rr} = \dfrac{1}{3}$ P, whereas the tangential stresses are a function of the angular position on the interface.

If we observe the anatomy of the articulation (Fig.1b)we see that the structure of the spongeus bone in the head of femur is exactly radial. This lends support to our modelling activity which predicts a hydrostatic compression for the head of femur.

Figure 3a shows a planar map of the radial stress in the iliac zone. Radial stress are not only reduced to 0.33, of the applied compression, over the whole surface of the acetabulum, but also at the polar position the radial stress reaches a minimum of 0.20 immediately after the interface.

In then gradually rises toward the asymptotic value 1.00. Thus the presence of the liquid is such that in the most critical zone there is a significant decrease in the load applied at infinity. If we compare this map with the stress distribution in a homogeneous matrix (Fig. 3b) we may better appreciate the effect of the synovial liquid: the stress pattern is completely different and in particular in the polar position the radial stress reaches its maximum value 1.0.

4) References

[1] Bernadou M.; Christel P.; Crolet J.M.: Comportment mécanique des cupoles dans les prothéses de hanche. Interpretation des descellements. INRIA, Rapports de Recherche n.272 (1984).

[2] Goodier J.N.; Concentration of stress around spherical and

cylindrical inclusions and flaws. J. Appl. Mechs. 55 (1933) 39-44.

[3] Matonis V.A.; Small N.C.: A macroscopic analysis of composites containing layered spherical inclusions, Polym. Eng. Sci. 9, (1969) 90-104.

[4] Mazzullo S.: Stress field around a n-layred spherical inclusion. In: Mechanical behaviour of composites and laminates, W.A. Green and M. Micunovic (Eds). Elsevier Appl. Science (1987) 245-250.

[5] Ricco T.; Pavan A.; Danusso F.: Micromechanical analysis of a model for particulate composite materials with composite particles. Survey of craze initiations. Polym. Eng. Sci. 18, (1978) 774-780.

[6] Tepic S.: Maciroswki T.; Mann R.W.: Experimental temperature rise in human hip joint in vitro simulated walking. J. Orthop. Res. 3, (1985) 516-519.

S. Mazzullo, T. Simonazzi, D. Tartari
HIMONT ITALIA S.r.l.
Centro Ricerche "G.Natta"
44100 FERRARA (Italy)

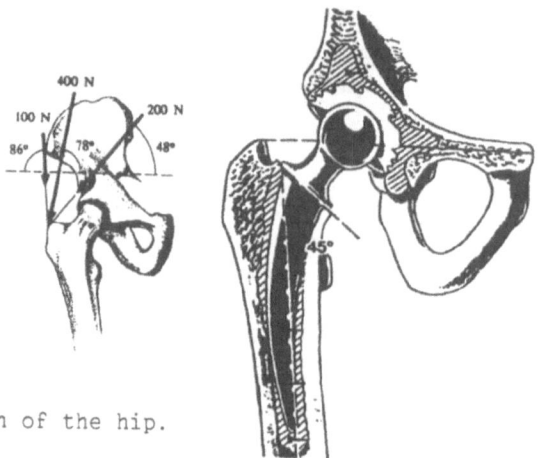

FIG. 1a- Articulation of the hip.

FIG. 1b - Hip joint showing the articular capsule

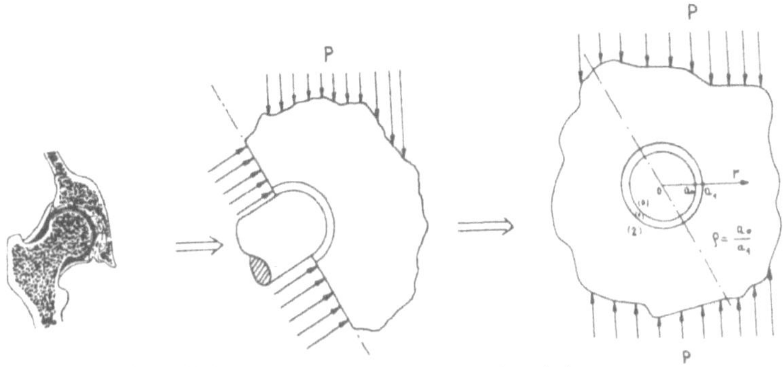

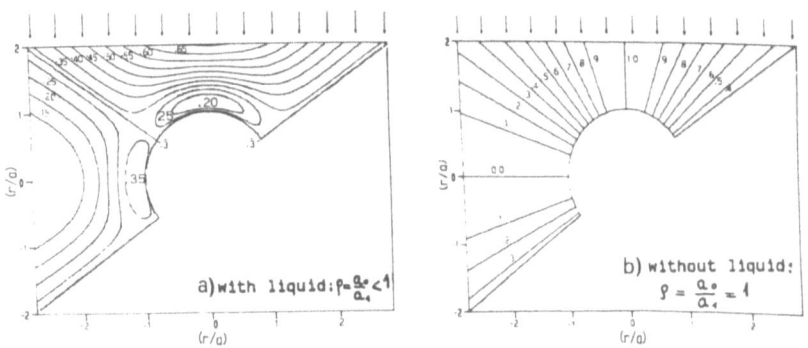

FIG. 2 - Simplified geometry of the hip-joint

FIG. 3 - Planar maps of radial stress

TORSIONAL VIBRATIONS IN A CRANK SHAFT

J. Molenaar
Institute for Mathematics Consulting
Eindhoven University of Technology
P.O. Box 513
5600 MB Eindhoven
The Netherlands

Hj. Wacker and W. Zulehner (eds.),
Proceedings of the Fourth European Conference on Mathematics in Industry, 337–345.
© 1991 *B.G. Teubner Stuttgart and Kluwer Academic Publishers.*

1. Introduction

In this paper, we analyze the torsion in the crank shaft of a gas compressor with one crank. The effect of damping will not be studied here. The geometry is as drawn in Fig. 1. Two forces are exerted on the shaft. First, it is at one end driven by a motor. We assume that this happens with uniform angular velocity ω, irrespective of the load of the crank. Second, the piston meets with the gas force, which is passed on to the shaft via the connecting rod and the crank. The geometry of the system implies, that the moments of inertia around the shaft of the moving parts may depend on the rotation angle ϕ of the shaft. The equation of motion for the torsion in this shaft has consequently periodically varying coefficients. A common approach to analyze this torsion is to average these coefficients over a period [1]. The resulting modelling equation can be dealt with in terms of eigenfrequencies, resonances and solutions, which are periodic in time. However, the solutions of the original equation of motion might be non-periodic and even show chaotic behaviour. The purpose of this study is to compare the behaviour of the reduced model with that of the original one in order to check the reliability of the reduction in case of realistic compressor data.

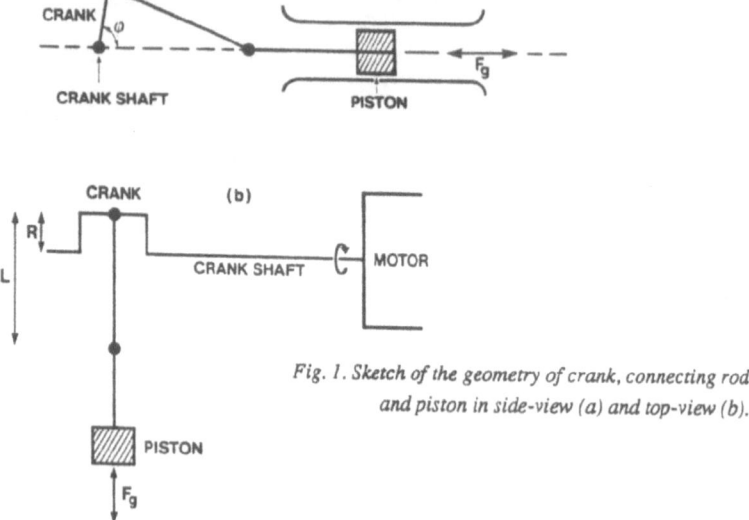

Fig. 1. Sketch of the geometry of crank, connecting rod and piston in side-view (a) and top-view (b).

2. Equation of Motion

The kinetic energy T of the system is composed of four parts:

$$T = T_{\text{crank shaft}} + T_{\text{crank}} + T_{\text{piston}} + T_{\text{connecting rod}}. \tag{2.1}$$

T can be expressed in terms of the rotational speed $\dot{\phi}$ of the shaft and the moments of inertia of the respective components:

$$T = \frac{1}{2} \left[I_{cs} + I_c + I_p(\phi) + I_{cr}(\phi) \right] \dot{\phi}^2$$

$$\equiv \frac{1}{2} I(\phi) \dot{\phi}^2. \qquad (2.2)$$

With the linear motion of the piston corresponds a moment of inertia I_p around the shaft given by

$$I_p(\phi) = m_p R^2 Z_1^2(\phi) \qquad (2.3)$$

with m_p the piston mass, R the crank radius and Z_1 a dimensionless factor. If we introduce a factor $f(\phi, \lambda)$, with $\lambda \equiv R/L$, by the definition

$$f(\phi, \lambda) = \lambda \cos \phi / \sqrt{1 - \lambda^2 \sin^2 \phi} \qquad (2.4)$$

we may write

$$Z_1(\phi) = \sin \phi (1 + f(\phi, \lambda)). \qquad (2.5a)$$

To obtain an expression for I_{cr}, we note that the motion of the connecting rod is the sum of the linear motion of its centre of mass and a rotation with respect to its centre of mass.

I_{cr} can be expressed in terms of the mass m_{cr} of the connecting rod, its moment of intertia I_{cr} with respect to its centre of mass and two dimensionless factors Z_2 and Z_3 given by

$$Z_2^2(\phi) = (1 - p)^2 \cos^2 \phi + \sin^2 \phi (1 + p \, f(\phi, \lambda))^2 \qquad (2.5b)$$

$$Z_3(\phi) = f(\phi, \lambda). \qquad (2.5c)$$

The parameter p, with $0 < p < 1$, denotes the position of the centre of mass. We have $L_1 = pL$ and $L_2 = (1 - p)L$ with L_1 and L_2 as indicated in Fig. 2.

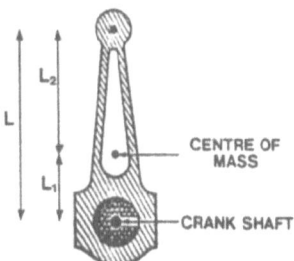

Fig. 2. The connecting rod and its centre of mass.

The expression for $I_{cr}(\phi)$ reads as

$$I_{cr}(\phi) = m_{cr} R^2 Z_2^2(\phi) + I_{cr} Z_3^2(\phi) \qquad (2.6)$$

In practice, one often has a geometry with $\lambda \ll 1$. Then, we may approximate up to first order in λ:

$$Z_1(\phi) = \sin\phi + \frac{1}{2}\,\lambda\sin(2\phi) \tag{2.7a}$$

$$Z_2^2(\phi) = 1 + p(p-2)\cos^2\phi + \lambda p\,\sin\phi\sin 2\phi \tag{2.7b}$$

$$Z_3(\phi) = \lambda\cos\phi. \tag{2.7c}$$

So, in that case we have $Z_3^2 \ll Z_2^2$.

We assume the torsion $\phi - \omega t$ to vary linearly along the crankshaft between crank and motor and take for the torsional moment at the crank position a harmonic repulsive force:

$$M_{\text{torsion}}(\phi) = -k(\phi - \omega t) \quad (k > 0). \tag{2.8}$$

With the gas force F_{gas} exerted on the piston a gas moment corresponds given by

$$M_{\text{gas}}(\phi) = F_{\text{gas}}(\phi)\,R\,Z_1(\phi). \tag{2.9}$$

Substitution of the expressions for T, M_{torsion} and M_{gas} into Lagrange's equation

$$\frac{d}{dt}\left[\frac{\partial T}{\partial\dot\phi}\right] - \frac{\partial T}{\partial\phi} = M_{\text{torsion}} + M_{\text{gas}} \tag{2.10}$$

yields the equation of motion

$$I(\phi)\,\ddot\phi + \frac{1}{2}\,I'\,\dot\phi^2 + k(\phi - \omega t) = M_{\text{gas}}(\phi) \tag{2.11}$$

with $I' \equiv dI/d\phi$. We are interested in the torsion $y(t) = \phi(t) - \omega t$ and put the equation in terms of y rather than in terms of ϕ. Furthermore, we introduce the variable τ instead of t by

$$\tau = \omega t. \tag{2.12}$$

Then, we obtain the equation

$$\omega^2 I\,\ddot y + \omega^2 I'\,\dot y + \frac{1}{2}\,\omega^2 I'\,\dot y^2 + ky = M_{\text{gas}} - \frac{1}{2}\,\omega^2 I' \tag{2.13}$$

with the notations $I' = \partial I/\partial y$ and $\dot y = dy/d\tau$. The torsion angle y is in practice very small. If we linearize this equation around $y = 0$, we arrive at the dimensionless equation

$$\ddot y + a(\tau)\,\dot y + b(\tau)\,y = F(\tau) \tag{2.14}$$

with

$$a(\tau) = I'/I$$

$$b(\tau) = (k + \frac{1}{2}\,\omega^2 I'' - M'_{\text{gas}})/\omega^2 I - \frac{1}{2}\,(I'/I)^2 \tag{2.15}$$

$$F(\tau) = (M_{\text{gas}} - \frac{1}{2}\,I'\,\omega^2)/\omega^2 I.$$

The coefficients a and b and the driving force F are periodic with, in general, period 2π. If $F_{\text{gas}} = 0$ the period is π. For convenience, we rewrite equation (2.14) into "lossless" form by means of the transformation

$$y(\tau) = x(\tau) \exp(-\frac{1}{2} \int_0^\tau a(\tau') \, d\tau'). \tag{2.16}$$

This leads to

$$\ddot{x} + d(\tau) x = G(\tau) \tag{2.17}$$

with

$$d(\tau) = b(\tau) - \frac{1}{2} \dot{a}(\tau) - \frac{1}{4} a^2(\tau)$$

$$G(\tau) = F(\tau) \exp(\frac{1}{2} \int_0^\tau a(\tau') \, d\tau'). \tag{2.18}$$

Both d and G are periodic functions. Equation (2.17) resembles a driven Mathieu equation.

3. Transition Matrix

The solution of (2.17) is given by

$$x(\tau) = \psi(\tau,0) \, x(0) + \int_0^\tau \psi(\tau,\sigma) \, G(\sigma) \, d\sigma \tag{3.1}$$

with $x = (x,\dot{x})^T$, $G = (0,G)^T$, and $\psi(\tau,\sigma)$ the transition matrix of the system. In the literature, no method is known to determine the transition matrix analytically. Instead of resorting to numerical methods, we prefer to apply a technique, which is proved to be reliable in many cases [2]. This method is based on the observation that the contributions of the Fourier components of $d(\tau)$ to the calculation of ψ strongly decrease with increasing frequency. This inspired Franks and Sandberg [3] to replace $d(\tau)$ by a more regularly shaped function with a Fourier spectrum, which closely coincides with that of $d(\tau)$ as far as the lower frequencies are concerned. We use a staircase function with N equal intervals. The height of the n-th step d_n can be calculated from the formula

$$d_n = \sum_{\substack{m \\ |m| < N/2}} \frac{(m\pi/N)}{\sin(m\pi/N)} \, e^{im\pi(2n-1)/N} \, \bar{d}_m. \tag{3.2}$$

The fourier component $\bar{d}_m$ of $d(\tau)$ is, as usually, defined by

$$\bar{d}_m = \frac{1}{2\pi} \int_0^{2\pi} d(\tau) \, e^{-im\tau} \, d\tau. \tag{3.3}$$

The transition function of the staircase function is easily obtained. At the interval $[t_{n-1}, t_n]$, with $t_n = \frac{n2\pi}{N}$, $n = 1, \ldots, N$, this matrix is given by

$$\phi_n(\tau,\sigma) = \begin{bmatrix} \cos\sqrt{d_n}\ (\tau-\sigma) & \dfrac{1}{\sqrt{d_n}}\sin\sqrt{d_n}\ (\tau-\sigma) \\[2mm] -\sqrt{d_n}\ \sin\sqrt{d_n}\ (\tau-\sigma) & \cos\sqrt{d_n}\ (\tau-\sigma) \end{bmatrix} \quad \text{if } d_n \geq 0 \tag{3.6}$$

$$= \begin{bmatrix} \cosh\sqrt{d_n}\ (\tau-\sigma) & \dfrac{1}{\sqrt{d_n}}\sinh\sqrt{d_n}\ (\tau-\sigma) \\[2mm] -\sqrt{d_n}\ \sinh\sqrt{d_n}\ (\tau-\sigma) & \cosh\sqrt{d_n}\ (\tau-\sigma) \end{bmatrix} \quad \text{if } d_n \leq 0. \tag{3.6}$$

For arbitrary arguments ψ is obtained by evaluating products of ϕ_n matrices. This simplifies the evaluation of (3.1) considerably. It remains to evaluate the integral in (3.1) numerically, because the gas force F_{gas} is not given in analytical form. A further simplification is obtained by writing it in an iterative form. If we want to evaluate the solution at grid points $\tau_i = i\,\Delta\tau$, $i = 0,1,2,...$, with $\Delta\tau$ a given increment, we may write

$$x(\tau_{i+1}) = \psi(\tau_{i+1},0)\,x(0) + \int_0^{\tau_{i+1}} \psi(\tau_{i+1},s)\,G(s)\,ds$$

$$= \psi(\tau_{i+1},0)\,[x(0) + K(0,\tau_{i+1})] \tag{3.7}$$

with K defined by

$$K(\tau,\tau') = \int_\tau^{\tau'} \psi(0,s)\,G(s)\,ds. \tag{3.8}$$

If we use that

$$K(0,\tau_{i+1}) = K(0,\tau_i) + K(\tau_i,\tau_{i+1}) \tag{3.9}$$

and

$$\psi(\tau_{i+1},0) = \psi(\tau_{i+1},\tau_i)\,\psi(\tau_i,0) , \tag{3.10}$$

we obtain the recursive formula

$$x(\tau_{i+1}) = \psi(\tau_{i+1},\tau_i)\,[x(\tau_i) + \psi(\tau_i,0)\,K(\tau_i,\tau_{i+1})]. \tag{3.11}$$

Thus, if $\psi(\tau_i,0)$ and $x(\tau_i)$ are known at a gridpoint τ_i, the calculation of $x(\tau_{i+1})$ and $\psi(\tau_{i+1},0)$ merely needs the analytical evaluation of $\psi(\tau_{i+1},\tau_i)$, the numerical evaluation of $K(\tau_i,\tau_{i+1})$, and some simple algebraic manipulations.

4. Stability Analysis

From Floquet theory we know that the stability character of the solutions of the system can be determined from the eigenvalues λ_i of the transition matrix evaluated over one period. For a second-order system the λ_i are solutions of the equation

$$\lambda^2 - S\lambda + D = 0 \tag{4.1}$$

with $S = \text{trace } \psi(2\pi,0)$ and $D = \det \psi(2\pi,0)$. For a "lossless" equation one has $D = 1$ and thus

$$\lambda_\pm = S/2 \pm \sqrt{(S/2)^2 - 1}. \tag{4.2}$$

These characteristic multipliers determine not only the stability character of solutions of the "lossless" equation but also of solutions of the original linearized equation of motion. This follows from the property

$$\int_0^{2\pi} a(\tau') \, d\tau' = 0. \tag{4.3}$$

We discern three cases

-   $|S| > 2$: λ_+ and λ_- are real and one of them has magnitude greater than one.

-   $|S| = 2$: $\lambda_+ = \lambda_-$ with unit magnitude.

-   $|S| < 2$: λ_+ and λ_- form a complex conjugate pair with unit magnitude.

The homogeneous equation has unbounded solutions if $|S| > 2$. For $|S| \leq 2$ the solutions are bounded and periodic with period $2\pi n / \omega$, with $n \in \mathbb{N}$, $n \geq 1$; it is the smallest integer for which

$$\lambda_\pm^n = 1. \tag{4.4}$$

We include in this formulation the case $n = \infty$. Some special cases are $n = 1$ if $S = 2$, $n = 2$ if $S = -2$, $n = 4$ if $S = 0$, and $n = \infty$ if $S \notin \mathbb{Q}$. The inhomogeneous or forced equation may have unbounded solutions if

a)   the unforced equation is unstable, i.e. $|S| > 2$

b)   the Fourier spectrum of the driving force contains a component with the same frequency as a periodic solution of the unforced equation.

It is not hard to deduce that the driving force $F(\tau)$, and thus $G(\tau)$, has components with frequencies which are multiples of $\omega / q\pi$ with $q = 1$ if $M_{\text{gas}} = 0$ and $q = 2$ otherwise. The unforced equation has, for $|S| \leq 2$, a periodic solution with frequency $\omega/2\pi n$. So, resonant behaviour may be expected if $q = 2$, i.e. $M_{\text{gas}} \neq 0$, and $n = 1$, i.e. $S = 2$.

Note, that these conclusions are quite different from the results of a stability analysis of a system, in which the τ-dependence of the moments of inertia, thus in the factors Z_1, Z_2 and Z_3, is averaged out. In that approach, the coefficient $d(\tau)$ is replaced by a constant $\bar{d} > 0$. The resulting system has the resonance frequencies $\pm \sqrt{\bar{d}}$.

5. Example

Here, we illustrate the conclusions in §4 for a system with parameters $I_c = 10\,kg\,m^2$, $I_{cr} = 74\,kg\,m^2$, $R = 0.15\,m$, $L = 1.0\,m$, $L_1 = 0.2\,m$, $L_2 = 0.8\,m$, $m_{cr} = 350\,kg$, $m_p = 900\,kg$, $k = 5.10^6\,N\,m$. In Fig. 3 we present calculated S-values for ω in the interval $(10,50)\,rad/s$. It appears that S as a function of ω varies in a harmonic way with nearly everywhere unit magnitude and frequency decreasing with increasing ω. We find $|S| > 2$, i.e. instability, only in some small intervals.

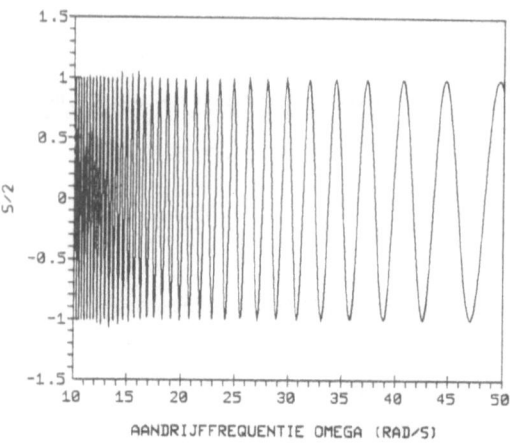

Fig. 3. Calculated S/2 values for $10 \leq \omega \leq 50\,rad/s$.'

If the gas force is applied, we may expect resonant behaviour for the ω-values with $S = 2$. To show this point, we present in Fig. 4 the solution $y(\tau)$ if $\omega = 40.65$, for which $S = 2.0$, is chosen. Fig. 4a contains the solution if no gas force is present, and Fig. 4b shows the unbounded solution if a gas force is applied. To demonstrate the sensitivity for the choice of ω, we give in Fig. 5 the solution if $\omega = 40.40$, for which $S = 0.9$, is used and with a gas force present. No resonance occurs, although the ω-value is only slightly changed in comparison with the ω-value in Fig. 4b.

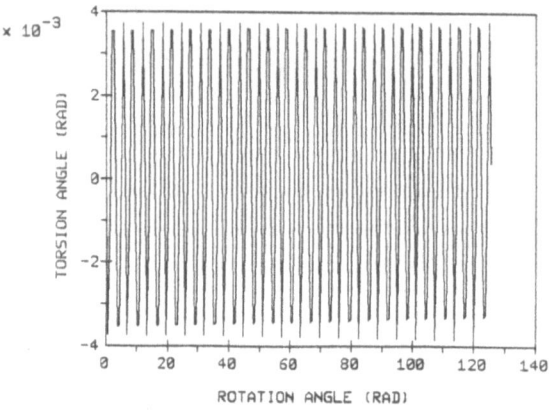

FIGURE 4.A

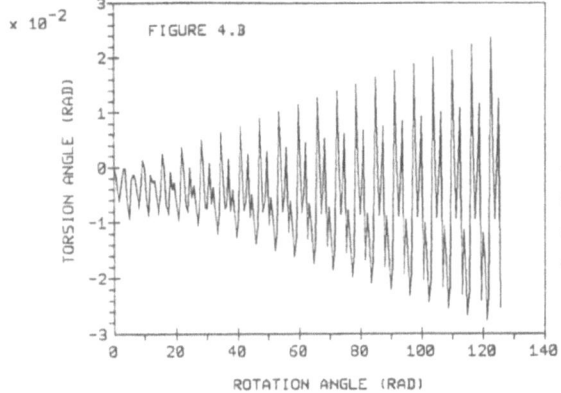

Fig. 4. Behaviour of the torsion as a function of the rotation angle τ for $\omega = 40.65$ and $M_{gas} = 0(a)$ or $M_{gas} \neq 0(b)$.

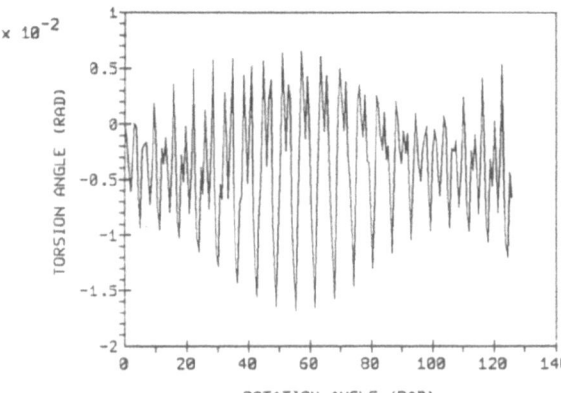

Fig. 5. Behaviour of the torsion as a function of τ for $\omega = 40.40$ and $M_{gas} \neq 0$.

1. Hafner K.E., Maas H., Torsionsschwingungen in der Verbrennungskraftmaschine, 1985, Springer Verlag Wien, ISBN 3-211-81793-X.

2. Richards J.A., Analysis of Periodically Time-Varying Systems, 1983, Springer Verlag Berlin, ISBN, 3-540-11689-3.

3. Franks L.E., Sandberg I.W., Bell Syst. Techn. J. 89, 1960, pp. 1321-1350.

4. Kibble T.W.B., Classical Mechanics, 1966, McGraw-Hill Publishing Company Limited.

Author's address: J. Molenaar, Faculty of Mathematics and Computerscience,
                  Eindhoven University of Technology, P.O. Box 513, 5600 MB Eindhoven, The
                  Netherlands

AN ADAPTIVE FINITE ELEMENT METHOD
FOR STEFAN PROBLEMS

Ricardo H. Nochetto, Maurizio Paolini, Claudio Verdi

1. Introduction. During recent years, degenerate parabolic equations have attracted the attention of both scientists and engineers mainly because of their relevance in modelling industrial processes, [3]. A common feature in dealing with such problems is the intrinsic lack of regularity of solutions across the free boundaries which, in turn, are not known in advance. For the two-phase Stefan problem, for instance, the temperature cannot be better than Lipschitz continuous and the enthalpy typically exhibits a jump discontinuity across the interface. This lack of smoothness makes piecewise linear finite elements, defined on quasi-uniform meshes, perform worse than expected according to the interpolation theory. Methods studied so far are not completely satisfactory in that they do not profit from the fact that singularities occur in a small part of the entire domain of definition, at least whenever the interface is sufficiently smooth. Consequently, a possible remedy is to be found in terms of a suitably designed adaptive algorithm. In fact, one would like to use a finer mesh near singularities in order to equidistribute the errors but still preserve the number of degrees of freedom, and thus the computational complexity. We refer to [1,6] for an account of the state-of-the-art on this topic along with numerous references.

In this light, the aim of this paper is to present a local mesh refinement method for the simplest two-phase Stefan problem in 2-D, which is based on equidistributing interpolation errors. Several numerical tests are performed on the computed solution to extract information about first and second derivatives as well as to predict discrete free boundary locations. A typical triangulation is coarse away from the discrete interface, where discretization parameters satisfy a parabolic relation, whereas is locally refined in its vicinity for the relation to become hyperbolic. The resulting scheme is stable and necessitates less degrees of freedom than previous practical methods on quasi-uniform meshes to achieve the same global asymptotic accuracy.

Hj. Wacker and W. Zulehner (eds.),
Proceedings of the Fourth European Conference on Mathematics in Industry, 347–352.
© 1991 *B.G. Teubner Stuttgart and Kluwer Academic Publishers.*

2. The adaptive algorithm. Let $\Omega \subset \mathbf{R}^2$ be a bounded and convex polygon and $T > 0$ be fixed. Consider the simplest two-phase Stefan problem which consists of finding $\{u, \theta\}$ such that

$$(2.1) \qquad\qquad u_t - \Delta\theta = 0, \quad \theta = \beta(u), \qquad \text{in } \Omega \times (0, T),$$

subject to $\theta = 0$ on $\partial\Omega \times (0, T)$ and $u(0) = u_0$ (which is assumed piecewise regular). The *strongly nonlinear* constitutive relation β is defined by $\beta(s) := (s - 1)^+ - s^-$. In the classical situation, the free boundary motion is governed by the so-called *Stefan condition* $|\nabla\theta^+(x, t) - \nabla\theta^-(x, t)| = V(x)$, where $x \in F(t) := \{x \in \Omega : \theta(x, t) = 0\}$ *(free boundary)* and $V(x)$ is the normal velocity of $F(t)$ at point x.

Denote by $\tau := T/N$ the *time step* and by S^n a partition of Ω into triangles; S^n is assumed to be *weakly acute* and *regular* uniformly in $1 \le n \le N$, [2, p. 132]. Given a triangle $S \in S^n$, h_S stands for its size and verifies $\lambda\tau \le h_S \le \Lambda\tau^{1/2}$ $(0 < \lambda, \Lambda$ fixed). Let $\mathbf{V}^n \subset H_0^1(\Omega)$ indicate the usual piecewise linear finite element space over S^n and $\Pi^n : C^0(\bar\Omega) \to \mathbf{V}^n$ the Lagrange interpolation operator. Let $U^0 \in \mathbf{V}^1$ be suitably defined in terms of u_0, [8]. The discrete scheme then reads as follows: for any $1 \le n \le N$ find $U^n, \Theta^n \in \mathbf{V}^n$ such that $\Theta^n = \Pi^n\beta(U^n)$, $\hat{U}^{n-1} := \Pi^n U^{n-1}$ and

$$(2.2) \qquad \int_\Omega \Pi^n\left(\frac{U^n - \hat{U}^{n-1}}{\tau}\chi\right) + \int_\Omega \nabla\Theta^n \cdot \nabla\chi = 0, \qquad \forall\chi \in \mathbf{V}^n.$$

Note that the integrals in (2.2) can be easily evaluated element-by-element via the vertex quadrature rule. As the resulting mass matrix is diagonal, the (strongly) nonlinear algebraic system (2.2) is efficiently solved by a nonlinear SOR method, [7,9,10].

Based on equidistributing interpolation errors, S^n is designed to be coarse away from the discrete interface $F^n := \{x \in \Omega : \Theta^n(x) = 0\}$, where the typical mesh size is $O(\tau^{1/2})$, and locally refined near F^n for triangles to reach a size $O(\tau)$. The new mesh is intended not to be altered for quite a while, say $O(N^{1/2})$ time steps, simply because remeshing is too expensive to be exercised very often.

We would like to discuss now the various tests to be performed at each time step on the computed solution Θ^n to either accept or discard the actual mesh S^n. The elements of S^n are split into three classes according to whether they belong to the interior of the refined region, to its boundary or to the rest of the domain. These regions are respectively labeled *Security Zone*, *Red Zone* and *Normal Zone*. Our algorithm checks whether the discrete interface F^n is within the Security Zone or not. In the event the Normal Zone has been reached, the computed solution is discarded and the previous one recovered; S^n must be discarded as well. To prevent the program from performing a useless time step, the Red Zone alerts that an inminent remeshing must be done.

The relation between two consecutive meshes, say S^n and S^{n+1}, is based on several tests carried out on the computed solution Θ^n to extract information about its derivatives as well as about the future location of the discrete interface F^{n+1}.

The first constraint arises from the fact that the function $\hat{U}^{n-1}$, which is used in (2.2) to determine U^n, does not coincide with U^{n-1}. This error might accumulate in time making the algorithm not convergent or, even worse, unstable. Keeping interpolation errors under control entails that the bigger second derivatives are the smaller must be the mesh-size. In addition, having control of quadrature errors leads to restrictions on triangle diameters wherever the first derivative exceeds certain tolerance. All triangles for which first or second derivatives are "bad" behaved are kept fixed. This set is denoted by S_B^n.

Of different nature is the further refinement employed near the discrete interface F^n to ensure that singularities do not escape from the *refined region* R^{n+1} during the $O(N^{1/2})$ time steps in which S^{n+1} will not be altered. The region traversed by the free boundary is predicted using a discrete version of the Stefan condition. Since the exact temperature is at most locally Lipschitz continuous, the local mesh-size in the refined region should be $O(\tau)$ to balance the interpolation error in space with the truncation error in time. Since U^n typically varies from 0 to 1 within one single element, even a slight perturbation of triangles S traversed by F^n would produce an error $\|U^n - \hat{U}^n\|_{L^\infty(S)} = O(1)$ and a subsequent lower bound $\|U^n - \hat{U}^n\|_{L^2(\Omega)} \geq C\tau^{1/2}$, which is too crude for the method to work. Consequently, we do not modify triangles crossed by F^n. This set is called S_F^n.

The mesh is then completely regenerated rather than being enriched or coarsened. Therefore, meshes S^n and S^{n+1} are *not compatible*. This reduces overhead required by enrichment-coarsening procedures as well as simplifies programming, but requires the use of a very efficient mesh generator [5, 11]. The desired local mesh-size is given by a piecewise constant function h defined on an auxiliary uniform square mesh Q of size $O(\tau)$, created and kept fixed from the beginning. For any $R \in Q$, the value $h|_R$ is the minimum of those provided by the various tests. The automatic mesh generator will then produce an adequate (weakly acute) triangulation, namely: for any $S \in S^{n+1}$, $h_S \leq \mathbf{h}(x)$ for all $x \in S$.

Our algorithm is still a *fixed domain method* because the predicted region to be invaded by the discrete interface is only used as a refinement indicator and not to solve uncoupled heat equations, as customary for front-tracking methods.

3. Analysis. Our adaptive algorithm is *stable* and *convergent*. In particular, the following error estimates have been proved, [8]:

$$(3.1) \qquad \|e_u\|_{L^\infty(0,T;H^{-1}(\Omega))} + \|e_\theta\|_{L^2(0,T;L^2(\Omega))} = O(\tau^{1/2}|\log \tau|^2).$$

The key issue here is not the rate of convergence but the fact that it holds for properly graded sets of meshes $\{S^n\}_{n=1}^N$, for which this is the first theoretical result.

4. The finite element code. Implementational details can be found in [9,11]. A schematic flow chart of the main program is as follows:

```
n = 0; do (while n < N) {
    H_DEFINE;                              (constructs the matrix h)
    if (n > 0) FIXED_ELEMENT_INSERT;       (inserts triangles of S_F^n ∪ S_B^n in S^{n+1})
    MESH;                                  (generates the triangulation S^{n+1} of Ω)
    RED_ZONE_DEFINE;                       (determines the refined region R^{n+1})
    INTERPOLATION_U;                       (calculates Û^n = Π^{n+1}U^n)
    do (while SECURITY_ZONE & n < N)       (F^n is away from the boundary of R^n)
        {n = n + 1; STEFAN}                (solves the nonlinear system (2.2))
    if (NORMAL_ZONE)                       (F^n has escaped from R^n)
        {n = n - 1; SOL_UPDATE}      }     (retrieves the previous solution)
```

5. Computational issues. The refined region R^n is typically a strip $O(\tau^{1/2})$-wide around F^n. Since the local mesh size is $O(\tau)$, the number of triangles within R^n is $O(\tau^{-3/2})$. Except possibly for a small transition region, triangles outside of R^n are $O(\tau^{1/2})$-big and so their number becomes $O(\tau^{-1})$. Since the number of time steps is $N = T/\tau$, the total number of *degrees of freedom* is $\mathrm{DOF} = O(N^{5/2})$, for a global accuracy $O(N^{-1/2})$. This quantity compares quite favorably with similar ones for practical methods involving a single quasi-uniform mesh. In fact, in that case $h_S = O(\tau)$ for all $S \in S^n$ and so $\mathrm{DOF} = O(N^3)$ for the same global accuracy [4,7,10,12].

To illustrate the superior performance of our adaptive method with respect to fixed mesh techniques, we have chosen a two-phase Stefan problem with an interface that moves up and down. This is a severe test whose exact solution is known [8,9]. Several numerical experiments were performed with both our adaptive method and the standard one with a fixed mesh [4,7,12] on a VAX/VMS 8530. Results are reported in the table below, where the following notation is employed: $DOF = N \times J :=$ number of (uniform) time steps × average number of nodes, $E_\theta^2 := \|e_\theta\|_{L^2(\Omega \times (0,T))}$, $E_\theta^\infty := \|e_\theta\|_{L^\infty(\Omega \times (0,T))}$, $CPU :=$ CPU time in seconds.

Adaptive Method				Fixed Mesh Method			
DOF	E_θ^2	E_θ^∞	CPU	DOF	E_θ^2	E_θ^∞	CPU
50×509	0.131	0.092	108	60×779	0.131	0.182	105
75×807	0.070	0.068	329	105×2318	0.071	0.114	495
100×1138	0.055	0.042	592	130×3672	0.057	0.091	951
150×1818	0.037	0.031	1694				

Our adaptive method requires less computational labor, say CPU, for a desired global accuracy. The free boundary is approximated within one single element. We thus have a practical $O(\tau)$-rate of convergence in distance for interfaces, the best one can hope for. The figures below show the locally refined initial mesh and the discrete and continuous interfaces for $N = 100$.

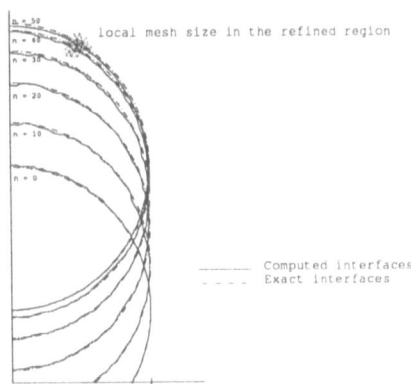

Initial mesh   Exact and discrete interfaces

References

[1] I. BABUŠKA, O.C. ZIENKIEWICZ, J. GAGO and E.R. DE A. OLIVEIRA, *Accuracy estimates and adaptive refinements in finite element computations*, John Wiley and Sons, 1986.

[2] P.G. CIARLET, *The finite element method for elliptic problems*, North Holland, Amsterdam, 1978.

[3] J. CRANK, *Free and moving boundary problems*, Clarendon Press, Oxford, 1984.

[4] C.M. ELLIOTT, *Error analysis of the enthalpy method for the Stefan problem*, IMA J. Numer. Anal., 7 (1987), 61-71.

[5] R. LÖHNER, *Some useful data structures for the generation of unstructured meshes*, Comm. Appl. Numer. Methods, 4 (1988), 123-135.

[6] R.H. NOCHETTO, *Numerical methods for free boundary problems*, in *Free Boundary Problems: Theory and Applications*, K.H. Hoffmann and J. Sprekels (eds), Pitman, 1988, to appear.

[7] R.H. NOCHETTO and C. VERDI, *Approximation of degenerate parabolic problems using numerical integration*, SIAM J. Numer. Anal., 25 (1988), 784-814.

[8] R.H. NOCHETTO, M. PAOLINI and C. VERDI, *An adaptive finite element method for two-phase Stefan problems in two space dimensions Part I. Stability and error estimates*, Math. Comp, to appear.

[9] R.H. NOCHETTO, M. PAOLINI and C. VERDI, *An adaptive finite element method for two-phase Stefan problems in two space dimensions Part II. Implementation and numerical experiments*, SIAM J. Sci. Stat. Comp., to appear.

[10] M. PAOLINI, G. SACCHI and C. VERDI, *Finite element approximations of singular parabolic problems*, Int. J. Numer. Meth. Eng., 26 (1988), 1989-2007.

[11] M. PAOLINI and C. VERDI, *An automatic mesh generator for planar domains*, Rivista di Informatica, to appear.

[12] C. VERDI, *Optimal error estimates for an approximation of degenerate parabolic problems*, Numer. Funct. Anal. Optimiz., 9 (1987), 657-670.

R. H. NOCHETTO, Department of Mathematics and Institute for Physical Science and Technology, University of Maryland, College Park, MD 20742, USA.

M. PAOLINI, Istituto di Analisi Numerica del CNR, 27100 Pavia, Italy.

C. VERDI, Dipartimento di Meccanica Strutturale, Università di Pavia, 27100 Pavia, Italy.

This work was partially supported by NSF Grant DMS-8805218 and by MPI (Fondi per la Ricerca Scientifica 40%) and CNR (IAN, Contract 880032601 and Progetto Finalizzato "Sistemi Informatici e Calcolo Parallelo", Sottoprogetto "Calcolo Scientifico per Grandi Sistemi") of Italy.

The Theoretical Ecology of Coccidia with

Reference to Poultry Production

S. Parry, M.E.J. Barratt and S. Jones
Unilever Research

S. McKee
University of Strathclyde

J.D. Murray
University of Oxford

1. Introduction

Chickens have natural parasites known generically as coccidia: these are closely adapted to their host displaying rigid host specificity and minimal pathogenicity under natural conditions. On the other hand the host has an immune system that can, given the opportunity, provide a sterilising anticoccidial immunity. To cope with this the parasite has evolved a complex, but highly well defined, life cycle.

Unfortunately, with modern husbandry practice, oocysts (the first stage in the life cycle of coccidia) can rapidly accumulate in the litter exposing a large number of young birds to infection. The delicate host-parasite relationship then breaks down under this level of exposure and drug control or vaccination become essential.

However, while potential approaches to vaccination are readily apparent, the way to achieve safe, effective and practical vaccination of large numbers of birds under commercial conditions are less well appreciated (see e.g. [1], [9]). The principal difficulty lies in devising an effective delivery system which can be easily and safely used in the real world of commercial poultry production.

In 1952 an oral vaccination was developed by Edgar [5], but this was administered through the bird's drinking water and there was real difficulty in ensuring that they received the correct dose (see e.g. [6]). Later,

353

Hj. Wacker and W. Zulehner (eds.),
Proceedings of the Fourth European Conference on Mathematics in Industry, 353–359.
© 1991 *B.G. Teubner Stuttgart and Kluwer Academic Publishers.*

following the work of Davis and Reynolds [4], the practical application of a
trickle dose, supplied orally, became achievable through incorporating
oocysts in a thick starch paste or more recently in alginate beads [1].

It was critical that such an oral delivery system should be accurate,
that is the birds must be given sufficient parasites to stimulate their
immune response, while at the same time it was crucial for the system to be
commercially viable that the birds were not given more than was minimally
necessary. These considerations coupled with the well defined self-limiting
life cycle of the parasite naturally led to the development of a mathematical
model of the ecology of coccidia. The model we shall ever so briefly
describe was the result of an interdisciplinary team consisting of three
immunologists and three applied mathematicians.

2.   The Mathematical Model

The life cycle of the coccidia is complex, and involves a number of stages.
In Fig. 1 these stages, with their multiplication rates, are set out
diagrammatically.

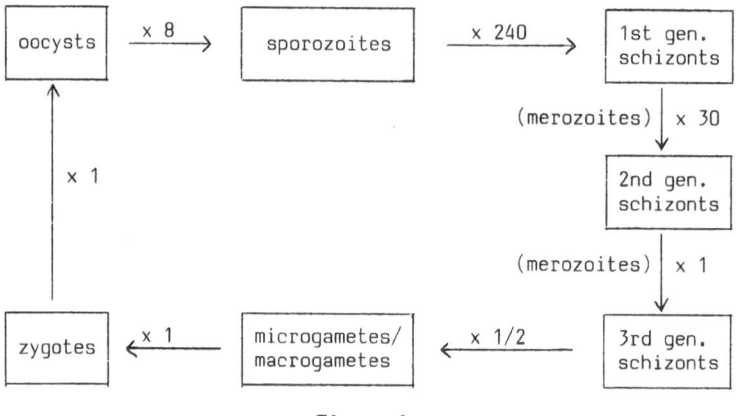

Figure 1

Firstly, for the purposes of building a model, this life-cycle can be
simplified (by lumping the three schizont stages and omitting the

micro/macrogametes stage) as these play no role in the stimulation of the

immune response. Then defining

   x_0 = the number of oocysts per bird in the litter

   x_1 = the number of oocysts inside a single bird

   x_2 = the number of sporozoites inside a single bird

   x_3 = the number of schizonts inside a single bird

   x_4 = the number of zygotes inside a single bird.

The following mathematical model of the ecology of the coccidia may be

derived:

$$\dot{x}_0 = a_4 x_4(t-t_1) - d_0 x_0 - b_0 x_0$$

$$\dot{x}_1 = F(t) + b_0 x_0 - b_1 x_1$$

$$\dot{x}_2 = a_1 x_1 - b_2 x_2$$

$$\dot{x}_3 = \frac{a_2 x_2(t-t_2)}{(1+I(t-t_2))^p} - b_3 x_3$$

$$\dot{x}_4 = a_3 x_3(t-t_3) - b_4 x_4$$

where $I(t)$ is the immunity factor, $F(t)$ denotes the bird feed intake and

$a_i (i = 1,...,4)$, $b_i (i = 0,...,4)$ and d_0 denotes rate parameters which have

to be estimated; the integer p also requires to be estimated.

   A full discussion of the derivation of these equations and parameter

estimation is not possible here; this may be found in [7]. However, it is

perhaps instructive to indicate the arguments involved in obtaining the

nonlinear equation for the schizont stage.

   Schizonts develop, as we have seen, from sporozoites, and, in turn,

develop into zygotes via the micro/macrogamete stages. So, the rate of

change in the number of schizonts per bird is equal to the sum of two terms:

   - A rate proportional to the number of sporozoites per bird present

represents the development of schizonts from sporozoites. However, this

development takes a finite time t_2 (approximately 96 hours for E. tenella) to occur. Consequently, the rate of production of the schizonts at time t

is proportional to the time delayed population $x_2(t-t_2)$.

   However, as the schizont population increases, the immune system of

the host bird is stimulated, and depresses the development of sporozoites

into schizonts. This implies that the corresponding development rate

coefficient must be some function of the schizont population. Now the effect

of the immune system depends upon both the schizont levels in the gut and

the time of exposure to the schizont population. Following Berding et al.

([1] and [2]; see also [7]), we postulate that the schizont development rate

coefficient depends upon the net schizont load in the bird over time.

Physically, we can formulate such a dependence by the following argument.

Let I(t) be an immunity factor which depresses the schizont rate of

production at time t. We postulate that the rate of production of schizonts

takes the form

$$a_2 x_2(t-t_2)/(1+I(t))^P,$$

where a_2 is the natural development rate coefficient of schizonts from

sporozoites in the absence of immunity, and p is an integer power law

coefficient. Initially, the bird's immune system is not activated against

coccidiosis, and I = 0. When the bird's immune system is exposed to

schizonts in the gut, it is stimulated to act against the infestation, and

the immunity factor I grows. Since there is a time delay in the activation

of the immune response, it seems natural to assume that the rate of change

of I is governed by the time-delay equation

$$\dot{I} = Kx_3(t-t_4)$$

where K is a rate coefficient which is a measure of the response of the

immune system, and t_4 is the immune system time delay. However, the immune

system of a bird is known to be less efficient when it is younger.

Consequently, we must expect K to be a function of time in general.

Following [1] we postulate that this time dependence can be implicitly

accounted for by expressing K as a sigmoidal function of the bird's weight

thus

$$K(t) = K_0[w(t)/(w(t) + w_c)]^2$$

where K_0 is the ideal immune response coefficient, and w_c is a critical

weight parameter.

-   A removal rate proportional to the number of schizonts per bird

present, which represents the loss of those schizonts from the schizont

population which develop into zygotes. This term takes the form $b_3 x_3$.

Adding these two contributions the schizont equation is then

$$\dot{x}_3 = a_2 x_2 (t-t_2)/(1+I(t-t_4))^P - b_3 x_3$$

where $I(t) = \int_0^t K(s) x_3(s) ds$.

3.   Numerical Method and Software

As a general rule a numerical scheme for (integro) differential equations

tends to have better stability characteristics when applied to delay

equations. With this in mind a full implicit scheme was not deemed necessary

and an order one apparently implicit scheme was employed explicitly by

solving the equations in a prescribed order (cf. Gauss Seidel for linear

equations). Experience revealed the scheme to be both robust and fast.

This code was then embedded in a large piece of software designed for

interactive use and, apart from the occyst ingestation rate and the bird

weight, which were supplied through user-created data files, the data is

entered interactively by the user via screen-driven menus. On completion of

the calculations which takes a matter of seconds the user is presented with

a choice of output plots (both linear and logarithmic) which can be produced

on the screen or sent to a hard copy plotter. Thus the resulting software

package can be used by a biologist or manager with minimal or no formal
mathematical background.

4. Results

In Fig. 2 we provide a comparison
between experimental data and the
results of the mathematical model
for oocyst output. Given the
difficulty of collecting accurate
biological information the agreement
is good.

More importantly,
experimentation with the software
package led to the realisation that
a cumulative feed programme over an
early period of the bird's life which
was related to the bird's weight
could lead to a commercially viable
product, thus saving years of
experimentation with live chicks
since each experiment takes seven
weeks - the production life of the
bird.

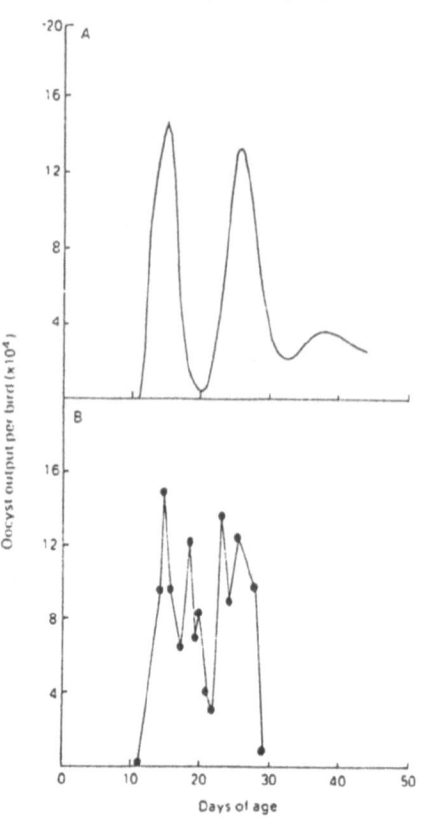

Figure 2

5. Conclusions

The widespread adaption of vaccination as a means of coccidiosis control in
the poultry industry is very appealing. It is entirely adequate, totally
natural form of control, with no problems from drug residues, toxicity (to
the bird or the consumer) or build-up of resistance. Such advantages are of

real value to the industry in these days of international "Green" politics

and their associated pressures. A mathematical model of the ecology of

coccidia has been presented. This model has provided real insight providing

a means of designing vaccination regimes in a much shorter time than would

have been taken by a typically empirical experimental approach.

References

[1] Berding, C., Keymer, A.E., Murray, J.D., Slater, A.F.G. (1986).
     The population dynamics of herd immunity to helminth infection.
     J. Theor. Biol. 122, 459-471.

[2] Berding, C., Keymer, A.E., Murray, J.D., Slater, A.F.C. (1987).
     The population dynamics of acquired immunity to helminth infection:
     experimental and natural transmission. J. Theor. Biology 126, 167-182.

[3] Davis, P.J., Barratt, M.E.J., Morgan, M. and Parry, S.H. (1985).
     Immune response of chickens to oral immunisation by 'trickle' infections
     with Eimeria.

[4] Davis, P.J. and Reynolds, F. (1982). UK patent specification
     No. 2144331A (Unilever PLC) published 6th March, 1985.

[5] Edgar, S.A. (1958). Coccidiosis of chickens and turkeys and control
     by immunisation. Proc. 11th World Poultry Congress, Mexico City,
     415-421.

[6] Long, P.L. (1984). Gordon memorial lecture. Coccidiosis control:
     Past, present and future. British Poult. Sci. 25, 3-18.

[7] Murray, J.D. (1989). Mathematical Biology. Springer-Verlag, Heidelberg.

[8] Parry, S.H., Barratt, M.E.J., Jones, S., McKee, S. and Murray, J.D.
     The development of immunity to coccidiosis in chickens (to appear).

[9] Rose, M.E. (1985). Immune responses to Eimeria infections. Georgia
     Coccidiosis Conference, Lake Lanier Islands, GA(Nov. 18-20), 449-469.

S. Parry, M.E.J. Barratt and S. Jones
Unilever Research
Colworth Laboratory
Colworth House
Sharnbrook
Bedford MK44 1LQ.

S. McKee   J.D. Murray
Department of Mathematics   Mathematical Institute
University of Strathclyde   University of Oxford
Livingstone Tower   24-29 St. Giles
26 Richmond Street   Oxford OX1 3LB.
Glasgow G1 1XH.

A dynamic two-phase flow model for a purification plant

W. PRAGER AND G. PROPST

A primary task in a class of sewage cleaning techniques is to enrich the sewage with oxygen in order to support the metabolism of microorganisms that play the key role in the biological process of purification. The purification plant that we consider, the Submers-Reaktor, is designed and built by Waagner-Biró AG in Graz. Its basic configuration consists of two basins, a pump and a siphon. Fig.1 shows a scale picture of an experimental Submers-Reaktor of about 6m height. The geometry of the siphon is idealized in the sense that the diameter of the real siphon slightly varies along its axis.

The basins and the siphon contain liquid sewage. While the pump maintains the level-difference h, the fluid flows, due to gravity, from the upper basin through the siphon into the lower one. An air induction device is placed in the upgoing part of the siphon. This device consists of three concentrically arranged annular chambers whose side screens are made of perforated caoutchuc membranes. Each of the chambers is supplied with air through pipes leading outer the siphon. The outside entrance to the pipes is controlled by three independently adjustable throttles. When a throttle is open, due to the difference of the atmospheric pressure p_0 outside and the low pressure inside the siphon, air is sucked into a chamber and penetrates the fluid through the perforations in the chamber's membranes.

As in [3] the rate of air induction q_L through one air supply pipe is modelled by

$$q_L(t) = \text{const} \sqrt{\frac{\rho_F g h_B(t) + \rho_F(\tau v_1(t))^2/2}{1 + \zeta_m + \zeta_{th}(t)}}$$

where ζ_m is the friction coefficient of the membrane and ζ_{th} is the adjustable friction coefficient of the throttle. ρ_F is the density of the fluid, g the acceleration due to gravity and v_1 the velocity of the gas-free fluid in the first segment of the siphon upstream of the membranes. τ is a factor that accounts for the velocity profile near any one of the chambers.

At the membranes the air forms bubbles that are carried along with the fluid through the siphon's bend into the lower basin. This way oxygen is submerged into the sewage.

Up to one tenth of the volume of the second segment of the siphon is occupied by bubbles of air, so a turbulent liquid/gas flow has to be modelled. We employ average velocities of the two phases in the second segment: the fluid velocity v_2 and the gas velocity v_G. In our model, they are correlated by $v_2 - v_G = v_r(q_L) > 0$, the relative velocity v_r depending solely on the rate of air induction q_L. v_r is greater than zero, because in the long downgoing part of the siphon the air bubbles, because of buoyancy, lag behind the fluid. The velocity v_1 is correlated to v_2 by a conservation of mass condition at the membranes: $\rho_F v_1(t) = \rho(t, 0) v_2(t)$.

Due to throttle adjustments and the internal dynamics of the flow, the rate of air induction varies in time, thus the density $\rho(t, x)$ of the mixture in the second segment

Supported by FWF (Austria), project S3206
A more detailed version of this paper will be submitted to Mathematical Engineering in Industry

Hj. Wacker and W. Zulehner (eds.),
Proceedings of the Fourth European Conference on Mathematics in Industry, 361–365.
© 1991 B.G. Teubner Stuttgart and Kluwer Academic Publishers.

varies in time and space. As the changes of the density of the mixture are set to be equivalent to the movements of the bubbles, the evolution of $\rho(t,x)$ is described by the continuity equation

(1) $$\rho_t(t,x) + v_G(t)\rho_x(t,x) = 0, \quad 0 \le x \le \ell_2, \ t \ge t_0,$$

with time dependent transport velocity v_G. Since per unit of time the volume q_L of air (density zero) is distributed over the cylindrical volume $v_G Q$ downstream of the membranes (Q is the cross section of the siphon), the boundary condition at $x = 0$ is

(2) $$\rho(t,0) = \rho_F \left(1 - \frac{p_0}{p_2(t)} \frac{q_L(t)}{Q v_G(t)}\right), \quad t \ge t_0,$$

where p_2 is the average pressure in the second segment of the siphon. We do not model the expansion and contraction of the air bubbles travelling along the siphon but in (2) we apply Boyle's law in a global sense multiplying the induced volume of air by the quotient p_0/p_2 of the pressures inside and outside the siphon.

The change of the level-difference h reflects the balance of the volume of fluid that enters or leaves each of the two basins with cross sections F_1 and F_2 respectively:

(3) $$\frac{d}{dt}h(t) = \frac{F_1 + F_2}{F_1 F_2} q_P(h(t)) - \frac{Q}{\rho_F}\left(\frac{\rho(t,0)}{F_1} + \frac{\rho(t,\ell_2)}{F_2}\right)v_2(t), \quad t \ge t_0.$$

The delivery of the pump q_P depends nonlinearly on h. By equating the time derivative of the overall momentum of the flow with the sum the forces involved we obtain

(4) $$\frac{d}{dt}v_2(t) = \frac{p_{dr}(t) - p_{dyn}(t) - p_{tfr}(t) - p_{pfr}(t) - p_{md}(t)}{\ell_1 \rho(0,t) + \ell_2 \rho_2(t)}, \quad t \ge t_0,$$

ρ_2 being the average density of the mixture in the second segment. The numerator of the right hand side of (4) represents the balance of driving and inhibiting pressures. The driving pressure $p_{dr} = \rho_F gh$ in the case of air induction has to be modified according to the density distribution and the geometry of the siphon. The pressure drop p_{dyn} takes into account approximations for the acceleration of fluid at the entrance of the siphon and at the membranes. For the friction term p_{tfr} of the two-phase flow through the tube we use the Lockhart-Martinelli correction factor ([1] p.749ff) that in terms of the model is $(\rho_F/\rho_2)^{7/4}$. The friction between the two phases in the up– and downgoing parts of the siphon is simply set equal to the buoyancy of the gas relative to the fluid yielding a pressure drop p_{pfr} that depends on $\rho(t,x)$ and the geometry of the siphon. Finally p_{md} summarizes the terms that arise from the derivative of the time dependent mass and density. Equation (4) is an insteady Bernoulli equation including terms that model an averaged interaction of the two phases.

Given initial conditions at time t_0, the model equations are solved numerically. Fixing a grid of points of distance Δx apart in the second segment of the siphon, we compute the solution of (3) and (4) by a Runge–Kutta procedure with variable time step $(\Delta t)_k = \Delta x / v_G(t_k)$, $k = 0, 1, \ldots$. Making the time stepping depend on the variable v_G in this way, the solution of the transport equation (1) is simulated by shifting the density by one grid point every Runge–Kutta step. In order to compute equilibrium states of the model corresponding to a fixed throttle coefficient $\overline{\zeta_{th}}$ we transform the model equations using $h \equiv \dot{v}_2 \equiv 0$ and $\rho(t, x) \equiv \rho_2$ into a fixed point problem for ρ_2 that can be solved iteratively. The resulting density $\rho_2(\overline{\zeta_{th}})$ yields the equilibrium values of all the other state variables.

We investigated three series of experiments with an experimental machine (Fig.1), each one subdivided by abrupt adjustments of just one of the three throttles. After each throttle adjustment the system passes through a transient and converges to another equilibrium state. We use the average values of the noisy numerical data during the equilibrium states, in order to adapt the model to the experimental plant.

For each chamber, τ can be reconstructed from pressure measurements in the air induction pipes while the throttles are closed. The overall friction coefficient of the tube was found by use of q_P- and h-measurements during gas-free equilibriums. Then we looked for the functional form of v_r and ζ_m. It turns out that, at least for higher fraction of gas, $v_r(q_L)$ is well approximated by $v_r(q_L) = \alpha q_L$ with some characteristic $\alpha > 0$ for each one of the three chambers. Furthermore, the formula $\zeta_m = (\varepsilon_0 + \varepsilon_1 q_L)/q_L^2$ that was found during independent experiments with perforated caoutchuc membranes corresponds very well to the estimated values of ζ_m that yield correct equilibrium states of the Submers-Reaktor model. We finally applied the Levenberg-Marquardt procedure ([2] p.82ff, we used the IMSL routine ZXSSQ) minimizing the deviation of the model states h and q_L from measured equilibriums, with respect to the three scalars $\alpha, \varepsilon_0, \varepsilon_1$.

The parameters of the model thus being determined for each chamber, we use functions $q_P(h)$ fitted to the q_P-measurements, for the numerical simulation of the experiments. In Fig.2 simulated trajectories are plotted onto a copy of the curves that were continuously recorded during the experiments with the outermost chamber of membranes. The upper curve shows the level difference, the lower one reflects the rate of air induction. Shifting the zero level of both curves upward by the same amount, the simulated trajectories are plotted above the graphic experimental record. The elapsed time in minutes is given at the horizontal axis (from right to left). Starting from gas free equilibrium the throttle is opened and closed twice ($t = 1.2 - 2.7, t = 4.25 - 7.3$); then the throttle is opened in a series of five steps ($t = 9.75, 15., 18., 20., 22.$). The model–input $\zeta_{th}(t)$ used for the simulations is a step function, whose values are determined from measurements at pressure gauges in the air supply pipe inside and outside the throttle. The data for this experiment cover 23 minutes. The simulations carried out on a conventional PC take less time. The quality of the model-trajectories during transients can be judged from the graphics, numerical values for the equilibrium states are given in Tab.1. In particular for higher rates of air induction ($> 60 m^3/h$) the correspondence of the simulations and the experimental data is satisfactory, while for lower fraction of gas the model's friction terms are too large, yielding level-differences which are too large. This also holds for the experiments with the middle and with the innermost chamber of membranes.

[1] H. Brauer, "Grundlagen der Einphasen- und Mehrphasenströmungen", Sauerländer, Aarau, 1971.

[2] R. Fletcher, "Practical Methods of Optimization", J. Wiley, Chichester, 1980.

[3] S. Thaller, *Submersreaktor IV*, Institut für Mathematik der Universität Graz, 1988.

Address: Institut für Mathematik, Universität Graz, Elisabethstraße 16, 8010 Graz, Austria

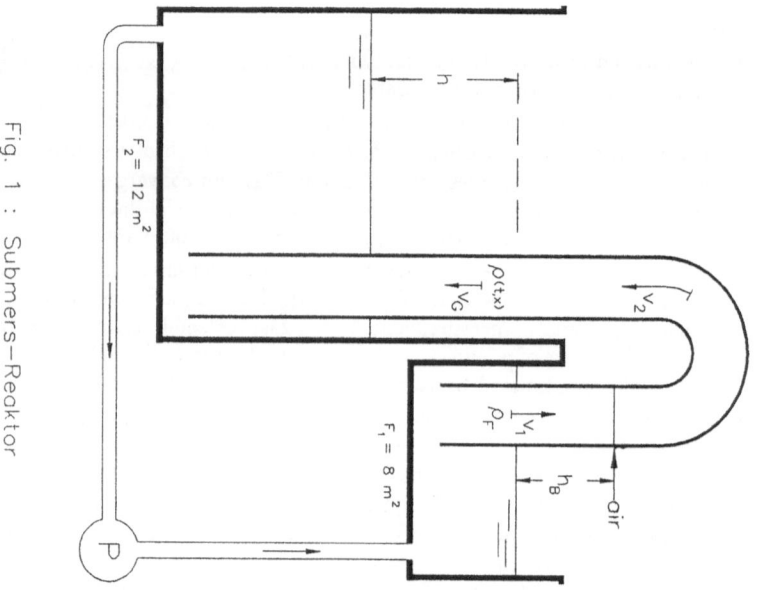

Fig. 1 : Submers-Reaktor

	1000/zth	h m	v2 m/s	vG m/s	rho2 kg/m3	qL m3/h	qP m3/h	zm
measured		.605				.00	1294.	
computed	.0000	.598	1.89	.00	998.	.00	1275.	★★★★
measured		.771				28.45	1272.	
computed	.1985	.834	1.92	1.82	975.	27.80	1268.	1743.8
measured		.874				37.70	1267.	
computed	.3787	.917	1.93	1.79	966.	37.16	1266.	1021.7
measured		1.013				56.70	1263.	
computed	.9747	1.085	1.96	1.75	950.	55.44	1261.	499.3
measured		1.193				69.50	1256.	
computed	1.7015	1.212	1.98	1.72	938.	68.82	1258.	343.3
measured		1.402				87.10	1250.	
computed	3.2959	1.387	2.01	1.68	921.	86,62	1253.	232.7
measured		1.583				105.50	1247.	
computed	6.4032	1.577	2.04	1.64	903.	105.25	1247.	169.1
measured		1.706				115.10	1246.	
computed	9.6288	1.692	2.06	1.62	892.	116.16	1244.	144.3

Tab. 1 : Equilibrium States

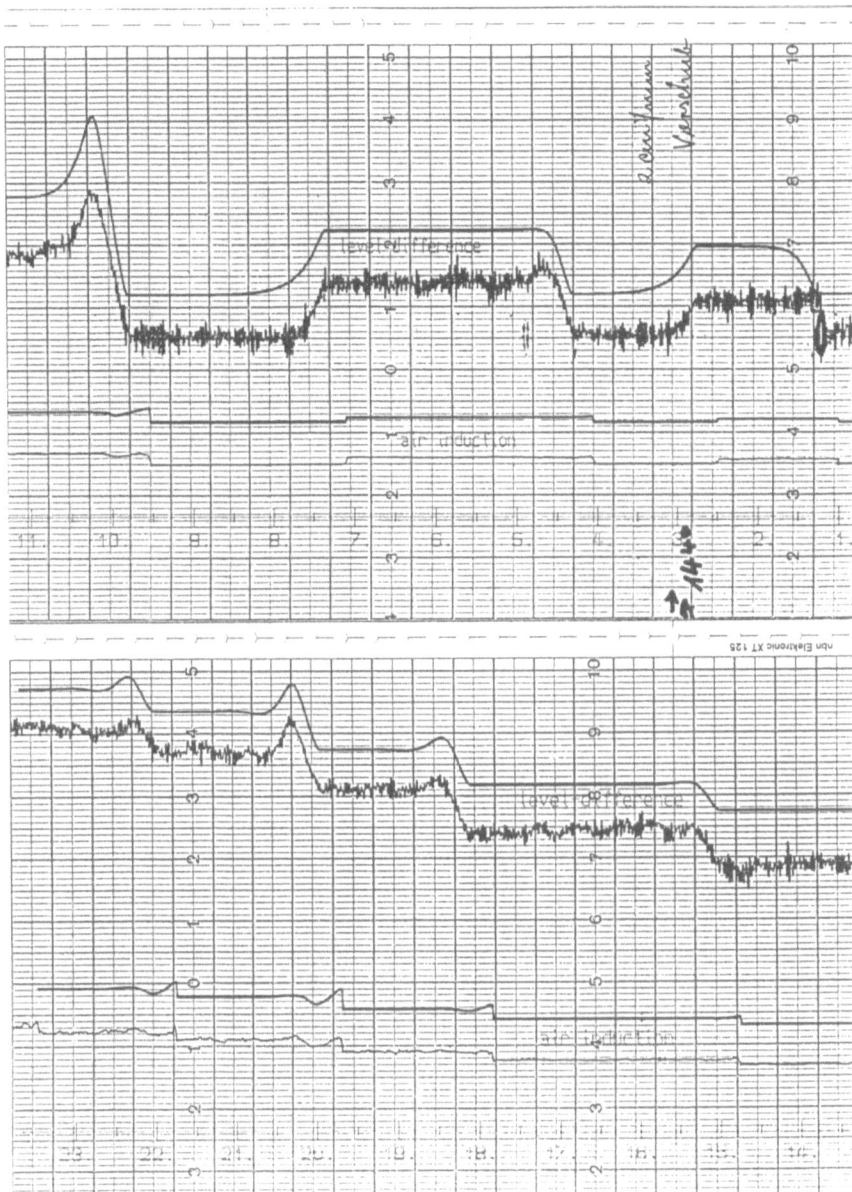

Fig. 2 : Experimental and Simulated Trajectories

EXCHANGE OF PARAMETRIC CURVE AND SURFACE GEOMETRY
BETWEEN COMPUTER AIDED DESIGN SYSTEMS

M J Pratt, R J Goult & M A Lachance, Cranfield, England.

1   Introduction

The transfer of geometric data between dissimilar computer aided design (CAD) systems is a problem of great concern to manufacturing industry.   It arises when a single organisation uses different systems at different stages of the design/ manufacturing cycle, and also in contractor/subcontractor situations where the two organisations do not use the same system. In either case it is clearly preferable to transfer the data on magnetic tape and have it read automatically than to generate drawings using the sending system and to re-enter them manually into the receiving system.   However, different CAD systems use different internal formats for data storage, and the idea of a 'neutral format' has arisen to facilitate the interchange.   In a draughting context, for example, such a medium provides a neutral, system-independent, means for representing lines, circular arcs and various types of annotation commonly used on drawings. Native information from the sending system is written in this form by a translator known as a preprocessor.   A postprocessor is then required to translate the neutral file data into the native internal format of the receiving system.   Thus just two translators are required for every CAD system, one for input and one for output. This method of data transfer is already in widespread use, the most commonly used format being IGES [10].   However, this will be superseded in due course by a new international standard known as STEP (STandard for the Exchange of Product data), the first draft proposal for which has recently been published [11].

2   Curves and Surfaces in CAD

While neutral file transfer is proving successful in many cases involving simple part geometry, difficulties arise in the transmission of the parametric curve and surface geometry widely used, for example, in the aerospace and automotive industries. The types of mathematical representations used in existing

Hj. Wacker and W. Zulehner (eds.),
Proceedings of the Fourth European Conference on Mathematics in Industry, 367–378.
© *1991 B.G. Teubner Stuttgart and Kluwer Academic Publishers.*

systems are basically of the following types:

	Curves	Surfaces

Polynomial: $\sum_{i=0}^{n} \underline{a}_i f_i(t)$ $\qquad$ $\sum_{i=0}^{n} \sum_{j=0}^{n} \underline{a}_{ij} f_i(u) f_j(v)$

Rational: $\sum_{i=0}^{n} \underline{a}_i w_i f_i(t) / \sum_{i=0}^{n} w_i f_i(t)$ $\quad$ $\sum_{i=0}^{n} \sum_{j=0}^{n} \underline{a}_{ij} w_{ij} f_i(u) f_j(v) / \sum_{i=0}^{n} \sum_{j=0}^{n} w_{ij} f_i(u) f_j(v)$

Procedural: Either of the above   Specified boundary data with associated 'blending' procedure.

Here the $\underline{a}_i$ and $\underline{a}_{ij}$ are vector-valued coefficients and the $f_i(*)$ polynomial basis functions of degree n. Curves are parametrised in terms of t, surfaces in terms of u, v; the range of these parameters is often [0,1]. The w_i and w_{ij} are scalar weights which may be used to vary the shape or the parameter distribution of rational curves and surfaces. More generally, the degree of surface representations may differ in the u and v directions.

It should be noted that the polynomial form is a subset of the rational form (obtained when the denominator of the latter sums to unity). Procedural surface definitions are sometimes expressible exactly in one of the other two forms, but this is usually not so. In the polynomial and rational cases different choices of basis functions give rise to the various popular formulations (Bezier,B-spline,Ferguson/Hermite) described in Faux & Pratt [5] or Pratt [17].

3   Curve and Surface Data Transfer

To provide examples for further discussion some commercial CAD systems are listed below, together with the curve and surface representations they use:

CATIA (Dassault Systemes, marketed by IBM)
                            Polynomial, maximum degree 15
GMS (Graftek)   Polynomial, maximum degree 3
EMS (Intergraph)   Rational, maximum degree 15

PDGS (Ford Motor Co., marketed by Prime)
Cubic polynomial curves,
procedural surfaces

Exact surface data transfer is possible in principle between the following pairs of systems:

CATIA → EMS

GMS → CATIA

GMS → EMS.

The first is possible since a degree 15 polynomial is a special case of a degree 15 rational function, the second and third since cubic polynomials are a subset of polynomials of degree ≤ 15. In all other cases exact transfer is impossible; PDGS uses a highly non-standard procedural surface representation termed a 'chord height blend' with no explicit polynomial or rational formulation [13], and in all other possible combinations of the remaining three systems the receiver can only handle a subset of the geometry transmitted by the sender. In such situations it is necessary to approximate the transmitted surfaces in terms of entities available within the receiving system.

It should be remarked in passing that whether the data transfer is approximate or whether it is in principle exact significant computation may be involved. For example, it may be required to pass information from a Bezier-based system into a B-spline-based system via a neutral format which represents polynomials in terms of the power or monomial basis. This requires two changes of basis, and inevitably computational errors will arise. It is important that these errors have a minimal effect on the geometry as reconstructed in the receiving system. This topic will not be further dwelt upon here, but recent work by Farouki & Rajan [4] and Lachance [12] have shown that there is an advantage in terms of numerical stability in using the Bezier basis wherever possible, and especially in the specification of neutral formats for transfer of this type of information.

4   Approximation of Parametric Curves and Surfaces

Clearly, when any approximation of engineering data is
performed the user must be able to specify a tolerance on the
accuracy of the results.   In practice this means that a high-
degree curve segment or surface patch will have to be split
into several segments or patches on transfer into a system
limited to a lower degree.   Subdivision achieves the added
flexibility needed in the receiving system for accurate representation of the transmitted data.   With this strategy in mind,
two methods for generating parametric approximations will now
be described.

4.1   Degree Reduction for Pure Polynomial Systems

Approximation algorithms for parametric curves and surfaces can be based on classical methods, though certain modifications are necessary. A given scalar-valued polynomial $p_n(x)$
of degree n has a best uniform approximation $p_{n-1}(x)$ of degree
(n - 1) on the interval [-1,1] which may be found by Chebyshev
economisation [3].   The approximation is derived by subtracting
from p (x) an appropriate multiple of the Chebyshev polynomial
of degree n so that the terms of degree n cancel.   This minimises the maximum value of the quantity

$$e(x) = |p_n(x) - p_{n-1}(x)|$$

on the interval [-1,1], and distributes the error over the
interval in such a way that it oscillates between maximum and
minimum values having the same magnitude, the number of oscill-
ations depending on the degree.

Since a parametric polynomial curve has two or three
components which are themselves scalar polynomial functions
of the parameter, the method of Chebyshev economisation may
be applied directly to each component to achieve a degree reduction.   Since parametric curves and surfaces often use the
parameter range [0,1] it is more convenient to work in terms
of the shifted Chebyshev polynomials defined on this interval
by a simple linear transformation.   The major problem stems
from the fact that the Chebyshev polynomials are non-zero at

the ends of the interval concerned. In geometric terms this
implies that the end-points of the original and the economised
curves do not coincide. Then if it is desired to approximate
a piecewise polynomial curve the economised segments resulting
from this procedure will in general not even be connected to
each other. In practice the original curve may well have continuity up to first or second derivatives at its junctions,
and a satisfactory approximation algorithm should be able to
reproduce these conditions if desired as well as mere posit-
ional continuity.

Lachance [14] has proposed a method for achieving the
aims described above, based on the use of constrained Chebyshev
polynomials. These share the property known as equioscillation
with the standard and shifted Chebyshev polynomials, but are
constrained so that their values and those of their first m
derivatives are zero at the end points of the parameter range
of interest, usually [0,1]. A different set of constrained
polynomials results for each choice of m, though for practical
purposes it is not usually necessary to go beyond m = 3. For
m > 0 the coefficients of the polynomials were computed numerically using a modified Remes algorithm. Use of these polynomials gives a realistic upper bound on the error of the approximation for each curve segment; if the error estimate is greater
than some predefined tolerance then the segment being approximated may be subdivided at its parametric mid-point and the
procedure applied separately to the two resulting smaller segments. This subdivision may be repeated recursively until
the desired tolerance is achieved over the entire length of
the segment. The method has the advantage that it approximates
not only the geometry of the curve but also its parametrisation,
which may be important for some applications.

Lachance's paper also describes the extension of his
method to the economisation of polynomial surface patches,
which may be of different degrees in their two parameters.
In this case the degree is reduced by one in each parametric
direction.

Multiple applications of the Chebyshev economisation

method are needed to achieve degree reductions of more than
one.   In this case it can no longer be guaranteed that the
resulting approximation is best possible in the uniform sense.
Nevertheless the procedure has been found to work well in practice;   the use of subdivision as described above permits the
reliable approximation of the original geometry to any desired
lower degree and to within any specified tolerance.

4.2   A More General Approximation Algorithm

The Chebyshev-based method described above is restricted
to degree reduction of polynomial curves and surfaces. However,
a polynomial approximation of some specified degree is frequently required to an entity originally represented in non-
polynomial terms, usually rational or procedural.   For this
purpose Goult [6] has devised a parallel approach based on
least-squares rather than uniform approximation.   What is now
minimised is the integral of the square of the distance between
corresponding points on the original and the approximating
curve or surface.   Classically, this is achieved most efficiently in terms of Legendre polynomials [3].   In a CAD context,
for reasons mentioned in the last section, it proves necessary
to define and work in terms of shifted, constrained Legendre
polynomials.   These were generated for degree up to 20 and
for $m = 0,1,2,3$ using Gram-Schmidt orthogonalisation.   Goult's
method requires the evaluation of integrals, for which Gaussian
quadrature may be used. Then the curve or surface being approximated need only be sampled at a set of discrete points.   Provided this is possible the actual nature of the curve or surface
being approximated is immaterial.   Whereas Lachance's algorithm
reduces the degree by one on each application, a single application of Goult's method will generate a polynomial approximation of any specified degree.   As with the former method,
the parametrisation as well as the geometry of the original
curve or surface is approximated.   It is worth reiterating
that the operation of Goult's method is independent of the
nature of the entity being approximated. This being so, extended applications of this technique include
(1)   approximation of procedurally defined elements such as
          parallel offset curves and surfaces, and

(2)   degree elevation, i.e. the approximation of piecewise
        curves and surfaces composed of low-degree elements in
        terms of a smaller number of higher-degree elements.

5   Other Curve and Surface Approximation Methods

        Related techniques have been described by other research-
ers. Watkins & Worsey [18] achieve degree reduction (for Bezier
curves only) using unconstrained Chebyshev polynomials.   This
has the effect of moving the end points as pointed out above;
they overcome this by simply moving the end points back to
their original positions after the degree reduction.   An error
analysis indicates that the results are acceptable for single
segments, but it is not possible to obtain continuity of first
or higher derivatives when piecewise curves are dealt with
in this manner.   Dannenberg & Nowacki [2]   describe a more
general method for curve and surface approximation based on
a subdivision strategy and a knot placement algorithm due to
Holzle [7].   Hoschek [8] and Hoschek & Wissel [9] describe
a related technique for lowering or raising the degree of spline
curves whilst retaining derivative continuity, using modification of the curve parametrisation to achieve geometrically
optimal results.   It has already been mentioned, however, that
significant changes in the parametrisation may not be acceptable for certain applications.   Patrikalakis [16], Bardis &
Patrikalakis [1] describe techniques for the approximate conversion of curves and surfaces from rational to non-rational
B-spline form, based on iterative adjustment of knot vectors
and control points.   Lyche & Mørken [15] use an approach based
on an initial dense sampling of the entity to be approximated.
In the curve case, this gives a piecewise linear approximation
which is regarded as a B-spline curve of degree 1.   Then the
degree of the B-spline curve is progressively raised and data
points discarded in such a way that the curve always remains
within some specified tolerance of the data.   The process can
be stopped at whatever degree is required for the final approximation.   It is believed, however, that the techniques of
Lachance and Goult are optimal in being   computationally
efficient, general in application and based on classical principles giving 'best' or 'near-best' results.

6   Examples

Two examples are presented. Figure 1 shows a patch taken from an automobile design generated on the system STRIM (developed and marketed by CISIGRAPH). This is a polynomial-based system with maximum degree 20. The chosen patch is of degree 13 x 14, and Lachance's constrained Chebyshev method has been used to approximate it to within a tolerance of 0.01 mm by six bicubic patches joining with first derivative continuity. In the Figure, isoparametric lines are plotted at intervals of 0.1 in both parameters. Note that subdivision has only been necessary in one parametric direction.

Figure 2   illustrates the use of Goult's constrained Legendre   method. A curve is defined procedurally as a normal offset from another explicitly specified curve. The offset distance varies linearly with parameter value along the original curve. Note that whereas the latter is a single segment, the approximate offset curve is composed of seven segments having second derivative continuity at their junctions.

7   Acknowledgements

The work reported on here was partially funded by the European Commission under ESPRIT Project 322 (CAD Interfaces). One of the authors (M.A. Lachance) developed the Chebyshev-based algorithm whilst receiving support from the UK Science and Engineering Research Council as a Visiting Fellow at Cranfield Institute of Technology. Both sources of funding are gratefully acknowledged. The authors would also like to thank the BMW car company, Munich, FRG, for supplying realistic test data, and both BMW and Cisigraph, Vitrolles, France for testing the algorithms developed.

8   References

[1]   Bardis, L.; Patrikalakis, N.M.: Approximate Conversion of Rational B-spline Patches. Computer Aided Geometric Design 6 (1989) 189-204.

[2]   Dannenberg, L.; Nowacki, H.: Approximate Conversion of Surface Representations with Polynomial Bases. Com-

puter Aided Geometric Design 2 (1985) 123-131.

[3]   Davis, P.J.: Interpolation and Approximation. Dover 1963.

[4]   Farouki, R.T.; Rajan, V.T.: On the Numerical Condition
        of Polynomials in Bernstein Form. Computer Aided Geomet-
        ric Design 4 (1987) 191-216.

[5]   Faux, I.D.; Pratt, M.J.:   Computational Geometry for
        Design and Manufacture. Ellis Horwood 1979.

[6]   Goult, R.J.: Parametric Curve and Surface Approximation.
        In D.C. Handscomb (ed.): The Mathematics of Surfaces
        III. Oxford University Press (1990).

[7]   Holzle, G.E.:   Knot Placement for Piecewise Polynomial
        Approximation of Curves.   Computer Aided Design 15 (1983)
        295-296.

[8]   Hoschek, J.:   Approximate Conversion of Spline Curves.
        Computer Aided Geometric Design 4 (1987) 59-66.

[9]   Hoschek, J.; Wissel, N.:   Optimal Approximate Conversion
        of Spline Curves and Spline Approximation of Offset Curves.
        Computer Aided Design 20 (1988) 475-483.

[10]   IGES: Initial Graphics Exchange Specification, Version
        4.0.   National Institute of Standards and Technology
        (NIST), Gaithersburg, Maryland, 1988.

[11]   ISO: First Draft Proposal for the STEP Version 1.0 CAD/
        CAM Product Data Exchange Standard, NIST.   Gaithersburg,
        Maryland, 1989.

[12]   Lachance, M.A.   Some Experiments on Transferring Poly-
        nomial Data.   In I. Bey & J. Leuridan (eds.), ESPRIT
        Project 322:   CAD*I (CAD Interfaces) Status Report 3.
        Report No. KfK-PFT-132, Kernforschungszentrum Karlsruhe,
        F.R.G.,   March 1987.

[13]   Lachance, M.A.:   Data Proliferation in the Exchange of
        Surfaces.   In R.R. Martin (ed.) The Mathematics of Surf-
        aces II. Oxford University Press, 1987.

[14]   Lachance, M.A.:   Chebyshev Economisation for Parametric
        Surfaces. Computer Aided Geometric Design 5(1988) 195-208.

[15]   Lyche, T.; Mørken, K.:   Knot Removal for Parametric B-
        spline Curves and Surfaces.   Computer Aided Geometric
        Design 4 (1987) 217-230.

[16]   Patrikalakis, N.M.:   Approximate Conversion of Rational

Splines. Computer Aided Geometric Design <u>6</u> (1989) 155-165.

[17] Pratt, M.J.:   Parametric Curves and Surfaces as used
in Computer Aided Design.   In J.A. Gregory (ed.)   The
Mathematics of Surfaces. Oxford University Press, 1986.

[18] Watkins, M.A.; Worsey, A.J.:   Degree Reduction of Bezier
Curves. Computer Aided Design <u>20</u> (1988) 398-405.

Department of Applied Computing & Mathematics
Cranfield Institute of Technology
Cranfield, Bedford MK43 0AL, England.

STRIM + User Defined Tolerance + Target Degree
Surface Data 3X3

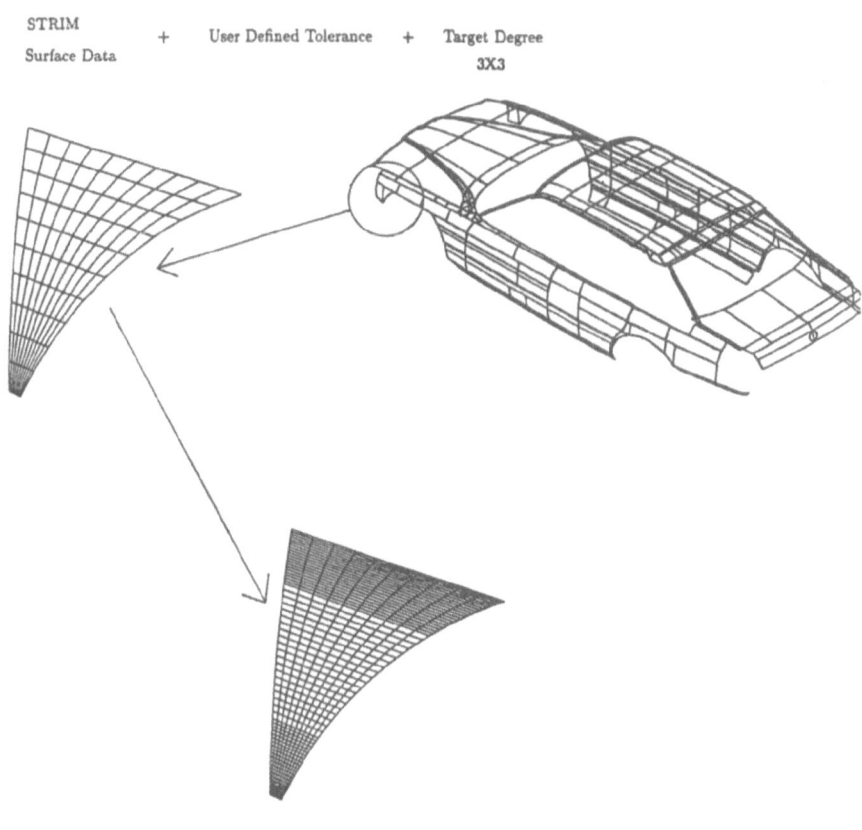

Figure 1 - Constrained Chebyshev Approximation of a 13 x 14 Patch
as an Assembly of Tangent-continuous Bicubic Patches.

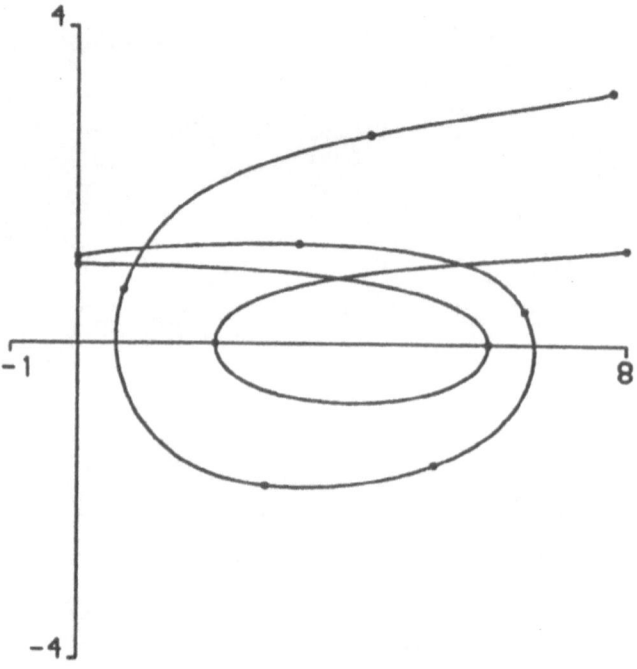

Figure 2 - Constrained Legendre Approximation to a Curvature-
continuous Tapered Offset Curve.

Fairing of ship lines with mechanical splines

Klaus-Dieter Reinsch

Introduction

The mathematical model for ship design and its local characterisation is presented together with two special cases:

- the spline through frictionslessly rotating slides,

- the spline loaded with point forces only.

The shape of a ship can, for example, be graphically presented by a plan of smooth lines consisting of plane intersections through the ship surface in three directions - broadwise and lengthwise intersections perpendicular to the watersurface and intersections parallel to the watersurface (so called waterlines).

These lines should be equivalent to mechanical elastic splines — thin rods with circular round cross section — which pass through given frictionslessly rotating slides and which are loaded with given forces.

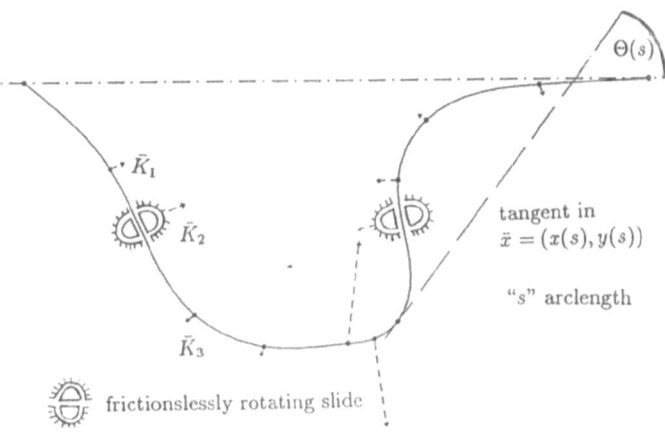

Figure 1

Hj. Wacker and W. Zulehner (eds.),
Proceedings of the Fourth European Conference on Mathematics in Industry, 379–383.
© 1991 B.G. Teubner Stuttgart and Kluwer Academic Publishers.

1   Mathematical Model and Local Characterisation of the Elastic Line

Mathematically, such a line can be characterised by a curve $\bar{x}(s)$ which minimises the following functional, (Theorem of Minimum Potential Energy [4])

$$
\begin{aligned}
I[\bar{x}] &= \text{strain energy} - \text{outer work} \\
&= \frac{1}{2}\int_0^L \|\ddot{\bar{x}}(s)\|_2^2 \, ds - \sum_{i=0}^n \bar{K}_i^T(\bar{x}(s_i) - \bar{x}_i),
\end{aligned}
$$

(1.1)

where $\bar{x}(s)$ is piecewise twice continuously differentiable, $\bar{K}_i$ are the given point forces or the a priori unknown forces in the slides.
Let

$$
(F_x^{(i)}, F_y^{(i)})^T := -\sum_{j=0}^i \bar{K}_j, \quad j = 0, \ldots, n-1,
$$

then from the calculus of variation we obtain the following results:
• the curvature $\kappa(s)$ of $\bar{x}(s)$ is continuous,
• the Euler-Lagrange-differential equation is given by

(1.2) $\dot{\kappa}(s) = F_x^{(i)}\,\dot{y}(s) - F_y^{(i)}\,\dot{x}(s) \quad s \in]s_i, s_{i+1}[.$

Depending on the specific problem special boundary conditions are obtained which define a unique solution of (1.2). Integration of (1.2) $\times \kappa(s)$ yields

(1.3) $\frac{1}{2}\kappa^2(s) = -F_x^{(i)}\,\dot{x}(s) - F_y^{(i)}\,\dot{y}(s) + c_i, \quad s \in [s_i, s_{i+1}].$

Equation (1.3) has the analytic solution

(1.4) $\kappa(s) = 2\sqrt{f_i}\,k_i\,cn(\sqrt{f_i}(s - \sigma_i), k_i),$

$f_i = \|F_i\|_2, \quad k_i^2 = \dfrac{c_i + f_i}{2\,f_i}, \quad \sigma_i \text{ constant of integration,}$

where $cn(z, k)$ is the Jacobian elliptic function *cosine amplitudinis* with argument z and modulus k. In [3] it was shown that $\bar{x}(s)$ is uniquely determined by $(\Theta, \kappa, E, x, y)|_{s=s_i}$, $\Theta(s)$ is a continuous function and denotes the angle between x-axis and the tangent at $\bar{x}(s)$. $E(s)$ is defined by $E(s) = \int_{s_0}^s \kappa^2(\bar{s})\,d\bar{s}$. Furthermore, it was shown that these quantities satisfy a system of nonlinear equations. In order to solve this system numerical values of the elliptic integrals of the first and second kind have to be calculated. Here

(1.5) $el2(x, k_c, a, b) = \displaystyle\int_0^x \frac{a + b\xi^2}{(1+\xi^2)\sqrt{(1+\xi^2)(1+k_c^2\,\xi^2)}}\,d\xi$

$k_c^2 = 1 - k^2$

denotes the elliptic integral of the second kind and $el1(x, k_c) = el2(x, k_c, 1, 1)$ the elliptic integral of the first kind. These integrals can be calculated very quickly via Landen transformation (see [1]).

2   Spline through rotating slides (open nonlinear spline)

Let $Q_i = (x_i, y_i)$, $i = 0, \ldots, n$, be $n+1$ given points. There are rotating slides without friction, i.e. $(s_{i+1} - s_i)$, $i = 0, \ldots, n - 1$ and, in addition, the quantities $\bar{K}_i$, $i = 0, \ldots, n$ are unknown a priori. In $(Q_i)_i$ only normal forces are acting.

From the calculus of variation we obtain the "natural" boundary conditions $\kappa_0 = \kappa_n = 0$, and all constants of integration c_i are vanishing, i.e. $k = k_c = \frac{1}{\sqrt{2}}$ in all intervalls $[s_i, s_{i+1}]$.

Fig. 2 shows the big difference between the open nonlinear spline and the linearised form, the so called cubic spline. Cubic splines yield good initial data for solving the nonlinear equation system mentioned in chapter 1.

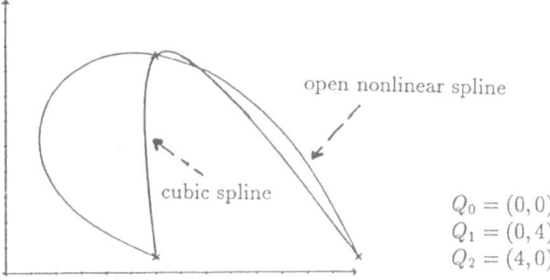

open nonlinear spline

cubic spline

$Q_0 = (0,0)$
$Q_1 = (0,4)$
$Q_2 = (4,0)$

Figure 2

If there are acute angles in the polygon $(Q_i)_i$, then it can be difficult to find good initial data. In order to overcome this a new set of points $(Q_i^{(0)})_i$ is calculated, where the angles are obtuse and a homotopy path is constructed (see [3]) to the given set of points (see Fig. 3).

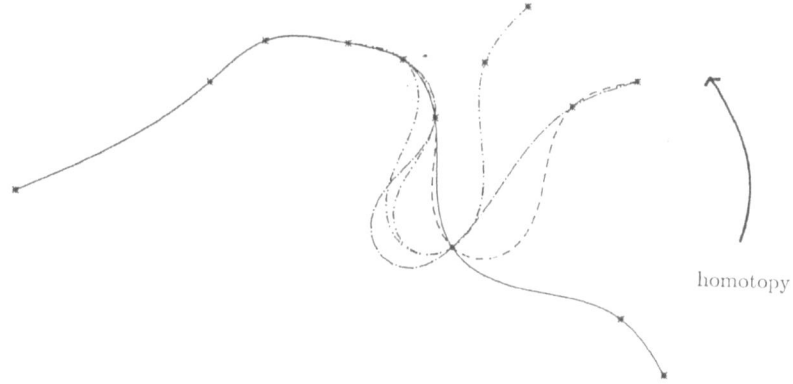

homotopy

Figure 3

3   Spline loaded with point forces only

At given points $s_i, i = 0, \ldots, n$, 0 or n are eventually omitted depending on the problem, point forces $\bar{K}_i$ are prescribed together with certain boundary conditions "bc", for example x_0, y_0, y_n fixed and x_n free (see Fig. 4).

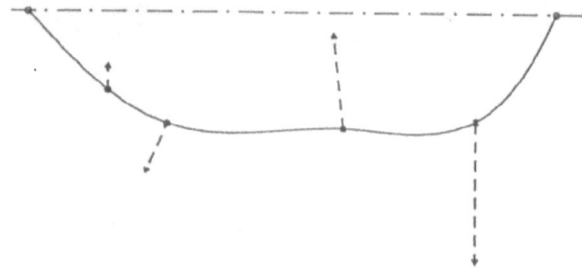

Figure 4

$(\Theta_i, \kappa_i, E_i, x_i, y_i)$, $i = 0, \ldots, n$ can be computed from the nonlinear system (1.2). As described in chapter 2 we start the algorithm by solving a reduced problem $\bar{K}_i^{(0)} = (0, \nu\, K_y^{(i)})$, $0 < \nu \leq 1$ and possibly altered boundary conditions "$bc^{(0)}$". From the linear beam equation new initial data are obtained which are used as starting values for the construction of a homotopy path $(\bar{K}_i^{(0)}, bc^{(0)}) \Longrightarrow \ldots \Longrightarrow (\bar{K}_i^{(l)}, bc^{(l)}) = (\bar{K}_i, bc)$ to find the true solution (see [3]).

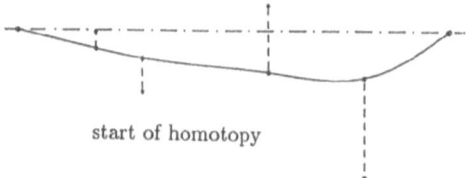

start of homotopy

Figure 5a

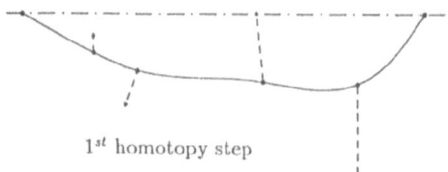

1st homotopy step

Figure 5b

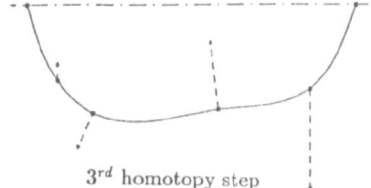

3rd homotopy step

Figure 5c

4 References

[1] Bulirsch, R.
     Numerical Calculation of Elliptic Integrals and Elliptic Functions
     Num. Math. **7** (1965)

[2] Reinsch, K.-D.
     Nonlinear Spline Functions
     Diplomarbeit, Technische Universität München (1977)

[3] Reinsch, K.-D.
     Numerische Berechnung von Biegelinien in der Ebene
     Doktorarbeit, Technische Universität München (1981)

[4] Sokolnikoff, I. S.
     Mathematical Theory of Elasticity
     McGraw-Hill Book Comp. Inc. (1956)

Klaus-Dieter Reinsch
Mathematisches Institut
Technische Universität München
Postfach 20 24 20
D-8000 München 2

Numerical simulation of the stretching behavior of polymeric liquids by

nonlinear Volterra integrodifferential equations with boundary conditions

W. Schmidt, TU München

1 A problem from polymer rheology

In 1978 Lodge, McLeod, and Nohel [3] described the nonlinear viscoelastic stretching of a filament of a molten polymer by the following equation:

$$(1.1) \quad \mu y(t) + \int_0^t a(t-s) \left\{ \frac{y^3(t)}{y^2(s)} - y(s) \right\} ds = f(t) y^2(t) \ , \ 0 \leqq t < \infty \ .$$

Equation (1.1) models the length $y(t)$ of a polymer melt which is totally relaxed at $t = 0$. Then it is stretched by a time dependent force $f(t) \geqq 0$. The parameter μ is a nonnegative material constant denoting a Newtonian contribution to the viscosity, and the memory kernel a has often the form of an exponential sum:

$$(1.2) \quad a(t) = \sum_{k=1}^r a_k \exp\left(\frac{-t}{\tau_k}\right) \quad ; \quad a_k > 0 \ , \quad \tau_k > 0 \ .$$

$f(t)$

Because of the term $y^3(t)$ under the integral, equation (1.1) has not the standard form of a Volterra integral equation. By some simple calculations and with the new variable $y_3(t) := y(t)$ (1.1) can be written as a system of three Volterra integral equations:

$$y_1(t) = \int_0^t a(t-s) \, y_3^{-2}(s) \, ds$$

$$(1.3) \quad y_2(t) = \int_0^t a(t-s) \, y_3(s) \, ds$$

$$y_3(t) = y_3(0) + \frac{1}{\mu} \int_0^t \left\{ f(s) y_3^2(s) - y_1(s) y_3^3(s) + y_2(s) \right\} ds \ .$$

Hj. Wacker and W. Zulehner (eds.),
Proceedings of the Fourth European Conference on Mathematics in Industry, 385–389.
© 1991 B.G. Teubner Stuttgart and Kluwer Academic Publishers.

2   Formulation of a boundary value problem

In the form (1.3) the problem is a initial value problem and can be solved numerically by standard algorithms, like collocation, Runge-Kutta, or extrapolation methods. In practical cases sometimes a parameter appears which can be fixed only by boundary conditions. Therefore it seems to be useful to introduce the following boundary value problem instead of an initial value problem. We consider the Volterra integral equation of the second kind:

$$(2.1)\ y(t) = y_0 + \int_0^t k(t,s,y(s))ds \quad , \quad 0 \le t \le T$$

with the nonlinear boundary condition

$$(2.2)\ r(y(0),y(T)) = 0 \ .$$

Here $k : \{(t,s) \in \mathbb{R}^2 : 0 \le s \le t \le T\} \times \mathbb{R}^n \to \mathbb{R}^n$ and $r : \mathbb{R}^n \times \mathbb{R}^n \to \mathbb{R}^n$ are continuous functions. The unknown n-dimensional vector y_0 is fixed by the boundary condition (2.2).

3   Numerical solution

Volterra integral equations and ordinary differential equations are very closely related. Thus the well known multiple shooting techniques (1) can be applied to integral equations. This leads to an algorithm with the following steps:

1)   The integration interval is divided into $m-1$ subintervals:

$$0 = x_1 < x_2 < x_3 < \ldots < x_{m-1} < x_m = T \ .$$

2)   At each node x_j $(1 \le j \le m)$ an initial guess $s_j \in \mathbb{R}^n$ is prescribed.

3)   In each subintervall $x_j \le t \le x_{j+1}$, $1 \le j \le m-1$ a solution Y_j of the integral equation at the point x_{j+1} must be computed. This means in detail: In every interval $x_j \le t \le x_{j+1}$ an initial value problem of Volterra integral equations with the initial value s_j must be solved.

Due to the integral term the solution $Y_j(x_{j+1})$ does not only depend on the initial value s_j but, however, on all former values $s_1, \ldots, s_{j-1}$. Thus, we write more precisely $Y_j(x_{j+1}; s_1, s_2, \ldots, s_j)$ instead of $Y_j(x_{j+1})$.

4)   Since the solution of (2.1) should be continuous the jump discontinuities must vanish. Together with the boundary condition (2.2) this leads to the following nonlinear equation for the vectors $s_1, s_2, \ldots, s_m$:

$$(3.1) \quad F(s_1, \ldots, s_m) := \begin{pmatrix} Y_1(x_2; s_1) & - s_2 \\ Y_2(x_3; s_1, s_2) & - s_3 \\ \cdot \\ \cdot \\ \cdot \\ Y_{m-1}(x, s_1, \ldots, s_{m-1}) & - s_m \\ r(s_1, s_m) \end{pmatrix} = 0$$

Remarks to step 3) and 4) :

Equation (3.1) can be solved with a damped Newton method. This requires the computation of the Jacobi-matrix, namely a matrix of the following form:

$$(3.2) \quad DF(s) := \begin{pmatrix} G_{1,1} & -I & 0 & \cdots & & 0 \\ G_{2,1} & G_{2,2} & -I & & & \cdot \\ \cdot & & & \ddots & & \cdot \\ \cdot & & & & & 0 \\ \cdot & & & & & \\ G_{m-1,1} & G_{m-1,2} & & & G_{m-1,m-1} & -I \\ A & 0 & \cdots & & 0 & B \end{pmatrix}$$

$s := (s_1, \ldots, s_m)^T$ is a vector and the $n \times n$-matrices are defined by:

$(3.3) \quad G_{i,j} := \dfrac{\partial F_i(s_1, \ldots, s_m)}{\partial s_j}$, where F_i is the i-th component of F and

$(3.4) \quad A := \dfrac{\partial r(s_1, s_m)}{\partial s_1}$, $B := \dfrac{\partial r(s_1, s_m)}{\partial s_m}$

In the differential case the Jacobi-matrix is a block-diagonal matrix with $-I$ in its upper diagonal. Normally this one is computed by numerical differentation. This means that $n(m-1)$ ordinary differential equations must be solved per Newton-iteration step. In the integral case the amount increases rapidly: Each computation of the Jacobi-matrix requires about $\frac{n}{2}(m-1)(m-2)$ solutions of Volterra integral equations. Futhermore these equations must be solved very accurate to avoid cancellation. Therefore another way must be taken: The initial value problems of step 3) can be solved by an extrapolation method, based on the mid-point-rule. This algorithm was introduced by Hock (2) in 1979. With several modifications it can be transformed to a so called differentiated midpoint-rule. This means that the components of the Jacobi-matrix are simultaneously computed with the trajectories of step 3). Thus the matrices $G_{i,j}$ can be calculated without increasing costs.

4   Numerical results

Finally the developed algorithm is tested at some problems from polymer rheology. Concrete data for equation (1.1) were given by Meißner (5) and Wagner (6). They investigated a low-density-polyethelene melt, called ' melt I '.

In figure 1 an elongation-relaxation experiment with melt I is simulated. In a time period t_s this rubberlike liquid is stretched to 4-times its initial length by the stretching force $F(t) = F_c t$. Then the force is cut off and the fila-ment can relax: if the material is stretched very rapidly only small relaxation can be observed; slow elongation causes large recovery. The unknown parameter F_c is calculated by coupling the 'trivial' equation $\dot{F}_c = 0$ to the integral equation system (1.3) and introducing a new boundary condition:

$$(4.1) \quad y_3(t_s) = 4 \cdot y_3(0).$$

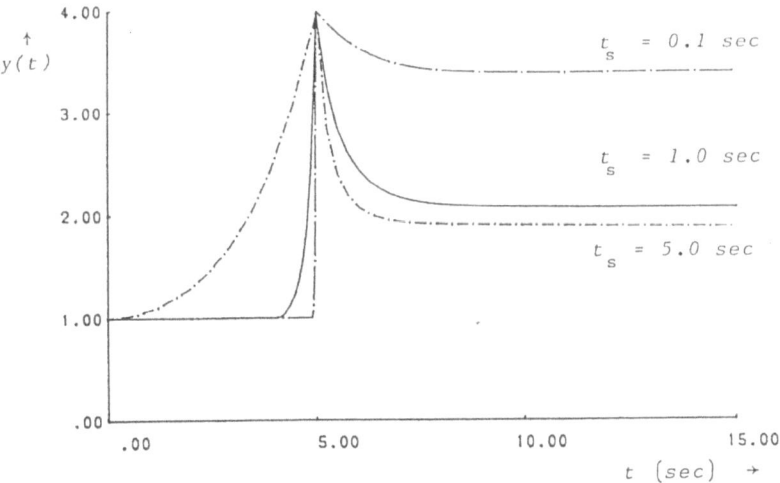

Fig. 1:   Elongation-relaxation behaviour of melt I

5   References

[1]   Bulirsch, R.: Die Mehrzielmethode zur numerischen Lösung von nichtlinearen Randwertproblemen und Aufgaben der optimalen Steuerung. Report der Carl-Cranz-Gesellschaft: Heidelberg 1971

[2]   Hock, W.: Asymptotic Expansions for Multistep Methods Applied to Nonlinear Volterra Integral Equations. Num. Math. 33 (1979) 77 - 100

[3]   Lodge, A. S.; McLeod, J. B.; Nohel, J. A.: A nonlinear singularly pertubed Volterra integrodifferential equation occurring in polymer rheology. Proc. Roy. Soc. Edinburgh 80A (1978) 99 - 137

[4]   Markowich, P.; Renardy, M.: A Nonlinear Volterra Integrodifferential Equation Describing the stretching of Polymeric Liquids. SIAM. J. Math. Anal. 14 (1983)

[5]   Meißner, J.:Dehnungsverhalten von Polyäthylen-Schmelzen. Rheol. Acta 10 (1971) 230 - 242

[6]   Wagner, M. H.: Analysis of stress-growth data for simple extension of a low-density branched polyethylene melt. Rheol. Acta 15 (1976), 133 - 135

Dr. W. Schmidt

Mathematisches Institut

Technische Universität München

D-8000 München 2

Postfach 202420

3-D-FITTING OF BLADES AND VANES

P. Schmidtke, Hamburg, and D. Windelberg, Hannover

Summary: For the repair of jet-engine blades and vanes it is
often necessary to apply material by welding and to dress the
build-up areas back to the original parts contour.
  Both operations are normally performed by manual-welding and
-benching. Automation of these operations is a challenge for the
coming years, and for this a geometrical description of the
existing parts-surface is desirable.
  The repair-specification gives only some dimensional
requirements for final inspection but no information about the
3-D-surface.
  In general, such kind of surfaces are not Bezier-surfaces,
because a vane or a blade has no "natural" coordinate-system;
especially after some years of use. On the other hand, it is
necessary to describe the different 3-D-surfaces.
  In this lecture a possible description of an idealized
(special) vane is given. It will be shown, that there exists for
each of the (repairable) vanes a translation and a torsion for a
`good fitting´ into the idealized vane. Criterias for the
quality of fitting are introduced. An automatic method of
fitting for this kind of problems will be discussed.

1 Problem

  We start with a 2-D-fitting problem: Let us introduce two
ordered sets of points NOM ("nominal") and MEA ("measured")
describing a closed area A_{NOM} resp. A_{MEA}. Then we want to find a
motion of MEA, that A_{MEA} "fits best" to A_{NOM} - or to find a
"degree of disparity", which describes the disparity by a number
between 0 and infinity.

  In 4 v.Hemdt and Pfeifer give an algorithm to calculate the
components of such a motion. They assert, that their algorithm
would find a "best fitting" (fig. 1): At first they calculate
the (geometric) center c_{NOM} of A_{NOM} resp. c_{MEA} of A_{MEA}. Then a
translation moves c_{MEA} into c_{NOM}. In a second step they calculate the main axis of inertia and rotate A_{MEA} with center c_{MEA}
so that the pair of axis of A_{MEA} correspond with those of A_{NOM}.

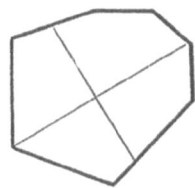 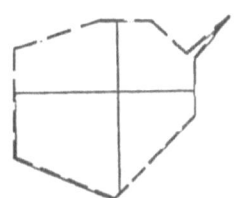 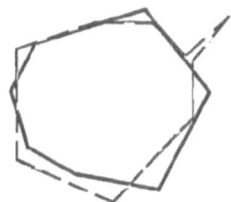

Fig.1: A_{NOM} A_{MEA} Best fit of MEA in NOM ?

Hj. Wacker and W. Zulehner (eds.),
Proceedings of the Fourth European Conference on Mathematics in Industry, 391–396.
© 1991 B.G. Teubner Stuttgart and Kluwer Academic Publishers.

We can ask for a second motion for better fitting. But we do´nt have a definition of "disparity" for describing the quality of fitting, and there are many possibilities for this definition. Following are some examples:

1. For each point of NOM let d_i be the distance to the nearest point of MEA. Then the disparity is the mean of these distances.

2. For each point of NOM let d_i (1 <= i <= NOM) be the distance to the nearest point of MEA. Then the disparity is the maximum of these distances d_i.

3. In fig. 1 we have connected the points of NOM resp. MEA by straight lines. Therefore we can define the disparity as the area between A_{NOM} and A_{MEA}.

4. If the points of NOM resp. MEA are connected by straight lines as in fig. 1 we can define "inside" and "outside" of A_{NOM} for getting defined areas. Then the disparity may be the area between A_{NOM} and A_{MEA}.

5. If the points of NOM resp. MEA are connected by curves, we also can define "inside" and "outside" of A_{NOM} for getting defined areas. Then the disparity may be the area between A_{NOM} and A_{MEA}. - But there is no reason why we should take an interpolation or approximation by curves, because we only have sets of points.

6. If a^+ is the outside resp. a^- the inside difference area, disparity is equal to the sum $(a^+)^2 + (a^-)^2$.

7. If we can describe an inside and an outside "tolerance surface" related to A_{NOM} with distances d^+ resp. d^-, disparity is equal to the sum $(d^+)^2 + (d^-)^2$. An algorithm for 3-D-splines (as used in the car-industry 5) finding such tolerance surfaces is given by 3 .

Other authors (see 1 and 2) developped an algorithm to calculate the translation from MEA to NOM by determining the center of the smallest circle of all points of NOM resp. MEA. If A_{NOM} has a well-defined coordinate-system which can be related to a coordinate-system of A_{MEA} it is shown that for some 2-D-fitting-problems this algorithm works quite well. Here we are looking for a 3-D-algorithm, which also works if there are no comparable coordinate-systems.

2 Disparity

In a 3-D-case we also start with two sets NOM and MEA of
points. We want to have a mathematical definition of
"the disparity of MEA in NOM is equal to a real number x".

At first we have to find a system for putting these points in
order. Therefore we examine the origin of the point-datas (see
also 7): A coordinate measuring machine touches the blades or
vanes in many contour-points by steps in a given width (perhaps
1 mm) and in a given plane (f.i. z = const.). So we get an
ordered set of points for each z. The measurements in different
planes are also assortable, because the planes can be ordered.

To describe the covering contour for these sets of points, we
need to connect them with lines. In similarity to the 2-D-case
we have to decide between a linearisation and a smoothened line
(or surface). The only informations available are the exact measured points - we don't know, whether the surface is convex or
concav in certain areas. That is why we chose a linearisation,
programmed by 6 , using quadrangles (triangles are special
quadrangles) for the description.

In fig. 2 the construction of quadrangles by a special type
of connections between points of different planes is shown. It
is necessary to define "inside" and "outside" by positive or
negative direction of the normal vector from the greatest triangle which is contained in the quadrangle.

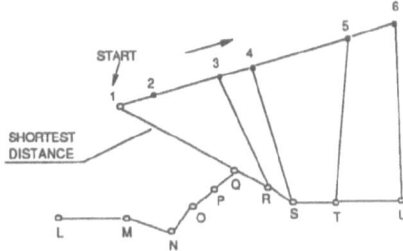

Fig.2:   Construction of quadrangles between given polygons

Now we have to define a "distance" between A_{NOM} and A_{MEA}. We
do this in two steps:

1. For each quadrangle QN of the surface A_{NOM} we have three
points a, b and c which represent the quadrangle. In this step,

let us look for one quadrangle QM of A_{MEA}: For point a of QN
and for QM let d_a be the (weighted) distance between point a
and the plane generated by the points of QM. The distances d_b
and d_c are defined the same way. The longest (weighted) distance
in the set d_a, d_b, d_c is defined as d_{QM}.

2. If A_{MEA} contains more than one quadrangle, we have to find a
distance $d_{QN > MEA}$ between the quadrangle QN and the complete
surface A_{MEA}: In the set of all d_{QM} with QM A_{MEA} we elect the
smallest (weighted) distance. On the other hand, the distance
$d_{QN > MEA}$ defines for each quadrangle QN a vector $v_{QN > MEA}$.

Now we are able to define the "disparity" of MEA in NOM: Let
$$D^+ := d_{QN > MEA} ; d_{QN > MEA} > 0 \text{ and QN } A_{NOM}$$
and $d^+_{NOM > MEA}$ the mean of all elements of D^+, weighted by
the corresponding areas of the quadrangles QN. In the same way
we define $d^-_{NOM > MEA}$ and get the disparity
$$DIS_{NOM > MEA} := (d^+_{NOM > MEA})^2 + (d^-_{NOM > MEA})^2.$$
Different definitions have to be used for special applications.

3 Automatic fitting

Now we will present an algorithm to find a translation and a
torsion for a good fitting in relation to the defined disparity.

The definition of disparity gives for each qadrangle QM A_{MEA}
the vector $v_{QM > NOM}$. The mean of these vectors, weighted by
the areas and the distances of QM, leads by dynamical optimation to a translation T.

We have to find the torsion for the translated surface A^T_{MEA}.
Again we are looking for the set of $v_{QM > NOM}$; QM A^T_{MEA} .
The mean of these vectors (weighted by the areas of QM), and the
angles between $v_{QM > NOM}$ and the perpendicular to the corresponding plane of A_{NOM}, leads by dynamical optimation to a
torsion.

The result of this fitting-process for a HPC-Blade is shown
in fig.3 and fig.4: If A_{NOM} is the top of an idealized (special)
blade and A_{MEA} the real (repairable) blade, then the fitting-
process shows the "difference" between both.

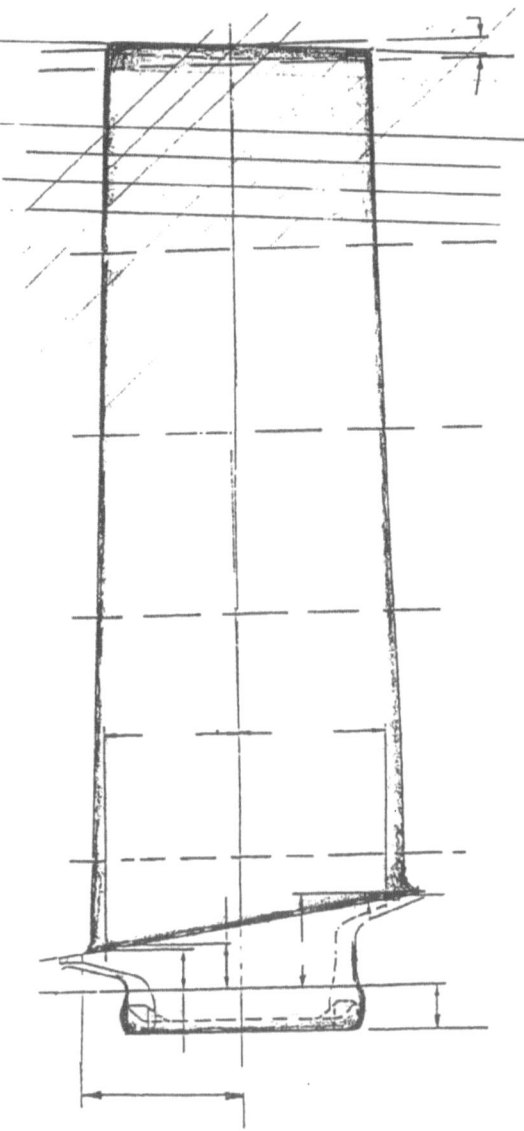

Fig.3:   Measured points of a HPC-blade

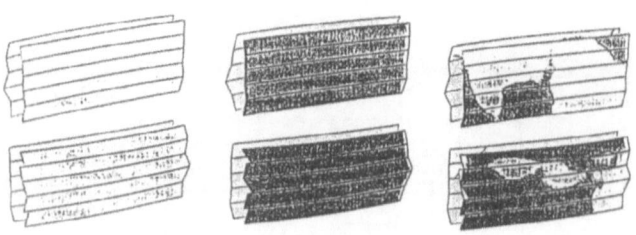

Fig.4: A_{NOM} A_{MEA} Best fit of MEA in NOM ?

4 References

1 Biehlmayer, W.; Forst,W.: Approximation by Circles.
   Erscheint demnächst in ZOR.

2 Biehlmayer, W.; Forst, W.; Weckenmann, A.: Numerische Verfahren zur Prüfung von Werkstücken unter Berücksichtigung
   der Maximum-Material-Bedingung im ebenen Fall. Technisches
   Messen tm 56 (1989) 23 - 31.

3 Hoschek, J.; Schneider, F.-J.: Spline Approximation of
   Offset Curves and Offset Surfaces. Erscheint demnächst.

4 Pfeifer, T.; v.Hemdt, A.: Lageprüfung an Freiformkurven und
   -flächen in der Koordinatenmeßtechnik. Lehrstuhl für
   Fertigungsmeßtechnik und Qualitätssicherung am WZL der RWTH
   Aachen. September 1988.

5 VDA-Flächenschnittstelle (VDAFS). Verband der Automobil-
   industrie e.V., Westendstraße 61, D-6000 Frankfurt.

6 Viergutz, H.; Windelberg, D.: Einpass. Geometrisches
   Verfahren zur Bestimmung einer optimalen Lage einer
   Ist-Kontur zu einer vorgegebenen Soll-Kontur. Erscheint
   demnächst.

7 Weckenmann, A.; Gawande, B.: Prüfen von Werkstücken mit
   gekrümmten Flächen auf Koordinatenmeßgeräten. Technisches
   Messen tm 54 (1987) 277 - 284.

Authors:
    Peter Schmidtke, Section Manager Workplanning Engine Blades
    and Vane Repair, Lufhansa German Airlines, Department WT 71,
    P.O.Box 300, D-2000 Hamburg 63 - Airport.
    Dr. Dirk Windelberg, Institut für Mathematik der Universität,
    Welfengarten 1, D-3000 Hannover 1.

Development of a method for electrical load forecasting and its implementation using Expert System Techniques

Ann Scully

1. Introduction: ESB system

ESB operates an isolated network with a peak load of 2350 MW and transmission voltages of 400, 220 and 110 kV. There is a wide variety of generation types, including coal, oil and peat-firing, conventional steam generation, dual-firing (oil/gas), combined and open cycle turbines, hydro and pumped storage. There is a total of 62 generators and 99 transmission stations on the network, connected by 5,737 km of lines and cables.

Increasing fuel prices and system complexity have increased the demand for software to help optimise dispatch of generation. Economic dispatch software optimises from minute to minute, unit commitment optimises over a week.

2. Requirement for a load forecasting program

It is intended to commission a unit commitment software package in June 89. It is estimated that the potential benefit of unit commitment to ESB is of the order of 0.5% of fuel costs i.e. one million pounds/year. Unit commitment requires an estimate of the system load for one week. In order to achieve the full benefit of this software, it is necessary to develop a load forecasting method with a high degree of accuracy. There is substantial recent interest in such developments c.f. (1), (2), (3), (4), (5).

3. Methods used at present

At present, in projecting the system load for plant scheduling the dispatching engineers search for a comparable day e.g. previous day, same day last week or last year. They then make intuitive adjustments for differences in weather conditions and then for any special events which may occur

397

Hj. Wacker and W. Zulehner (eds.),
Proceedings of the Fourth European Conference on Mathematics in Industry, 397–402.
© 1991 B.G. Teubner Stuttgart and Kluwer Academic Publishers.

An experienced operator will achieve an accuracy of 40-50 MW in estimating the peak demands, with a lead time of 3-4 hours.

A programmed approach is therefore required to produce longer term forecasts (one week), to allow a range of forecasts to be easily produced and to increase consistency in forecasts. It is hoped that the accuracy of the forecast can match operator accuracy over the short term i.e. 2-3 days.

4. Data required to forecast load

ESB has archived system load data for the past several years. Weather data, temperature, light, wind speed and direction and humidity are currently measured in Dublin, and have been for the past three years. A significant industrial load, an electric arc furnace, has been recorded separately for the past 18 months. Weather data at other sites around the country are available from the Meteorological Office.

5. Factors which affect electricity demand

Electricity demand, although highly variable, is largely predictable with a small degree of randomness. The variations in the demand for power can be attributed to several factors including, periodic changes, weather influences, special events and other random factors.

Periodic changes include hour of day, day of week, season of year and recurring events e.g. Christmas, New Year, holidays. Weather effects also influence electricity demand. By observation, the most important factors are, in decreasing order of importance, ambient temperature, light intensity, wind velocity and humidity/ precipitation. Wind direction does not appear to directly influence load. The initial analysis of the data also revealed a time lag effect due to temperature. Special events include major sporting events, elections or any social or political events which may electricity demand.

The distribution of load across the country is very uneven, with the Dublin area accounting for 60-70% of the total system demand. Since changes in the weather will affect one part of the country before the others, it may be necessary to include parameters of weather measured around the country. Only two areas will be included initially, Dublin and the rest of the country. This will be modified later if necessary.

6. Load model

A linear relationship between total load and the above factors is assumed. Using historical records of actual load and weather conditions, the parameters for models of load behaviour can be calculated. These models can then be used to forecast future loads.

The total load of the system at time t can be expressed as follows:

$$Y(t) = N(t)+W(t)+S(t)+R(t) \quad (1)$$

where

$Y(t)$ = total load at time t

$N(t)$ = normal load at time t

$W(t)$ = weather-sensitive load component at time t

$S(t)$ = load increment, due to special events, at time t

$R(t)$ = random load component, at time t

At a particular hour k this is:

$$Y(k) = N(k)+W(k)+S(k)+R(k) \quad (2)$$

The first step to define the model is collection of historical data of actual system loads and weather conditions. Special loads then need to be removed from the load curve, as they are not part of normal load behaviour. This includes significant industrial loads which are not weather dependent e.g. the arc furnace, events such as elections etc.

Equation (2) then reduces to:

$$Y(k) = N(k)+W(k)+R(k) \quad (3)$$

To identify the parameters of the weather sensitive model, periods where the normal load is approximately constant will be selected. This is reasonably valid for times such as mid-morning or mid-afternoon on a week day, over a period of several weeks. Applying multiple linear regression analysis to the measurement sample for this hour will give estimates, based on the least squares criterion, of the normal load and also of the coefficients of the weather model parameters.

At this stage, it will be necessary to re-evaluate the choice of weather factors to see if some are unnecessary and/or if other factors should influence the model. Having chosen the necessary and sufficient parameters to represent the effect of the weather, this procedure will be used to identify the parameters. Initially parameters will be identified on a 2 monthly basis i.e. 6 sets of parameters/year.

For each period we can then derive an equation for the normal load:

$$N(k) = Y(k)-W(k)-R(k) \quad (4)$$

$Y(k)$ is available for each hour and $W(k)$ can be estimated from the weather model. We will assume that random effects, $R(k)$, can be modelled as zero mean uncorrelated errors with constant variances. We can then calculate the normal load for each hour of each day in the year, including weekdays, weekends and special days, by averaging $N(k)$ over the appropriate sample for the period.

We will then combine the results of the normal load model with estimates of weather sensitive loads produced from the weather model using weather forecats for the desired period.

7. Expert system model

An alternative to the model described above is to use an expert system to search a load and weather database and select days which have similar weather to the forecast, at the same time of day and during the same time period. "Similar" here would imply defining a mathematical criterion to rate a comparison of forecasts. The system would then select a pre-determined number of days where the forecasts were within this criterion. The weather dependent load for each selection can be computed as the actual load minus the normal load. The average of these weather dependent loads can be used as the forecasted weather dependent load for the time in question.

This method closely follows the method dispatching engineers use already i.e. search for a similar day. The expert system interface would allow the engineer to interrogate the system and display the days selected, the criterion of selection, normal load at this hour etc. It would also allow the engineer to specify days, if known in advance e.g. an election, and to input an estimate of the change in load. Rules to cover extreme weather conditions e.g. high summer temperature or snow could also be added. Disadvantages include the computational overhead of the database search and the need to keep large amounts of data on-line.

8. Time scale

The project commenced March 1989 in the National Institute of Higher Education, Limerick. It is intended to have weather dependent and normal load models developed and tested by April 1990 as a stand alone program. In the following year, these models will be incorporated into an expert system.

9. Future development

Automatic scheduling of the program using a link to the Meteorological Office data to produce a revised forecast at a specified interval will be investigated later.

References

(1) "Modern Power Systems Control and Operation" A.S. Debs. Kluwer Academic Publishers 1988

(2) "ALFA - Automated Load Forecasting Assistant" K. Jabbour et al. IEEE Trans. on Power Systems Vol. 3 No. 3 1988

(3) "Adaptive Forecasting of hourly loads based on Load Measurements and Weather Information"

D.P. Lijesen & I. Rosing.

IEEE Trans. PAS-90 1971

(4) "Applied Regression Analysis" 2ed

N. Draper & H. Smith.

John Wiley & Sons 1981

(5) "Forecasting of Hourly Load by

Pattern Recognition in a Small

Area Power System"

A. Dehdashti-Shahrokh.

PhD Thesis. University of

Missouri, Columbia 1982

Ann Scully, Operations Dept., E.S.B.,

27 Lr. Fitzwilliam St., Dublin 2.

Fibre Drying

Louise Terrill

Abstract

A mathematical model for the folding of fibres on a drum drier is formulated, simplified and solved.

§1 Introduction

A problem which arises in the industrial process of fibre drying is to model the folding of a broad band of fibres which is fed continuously from a faster to a more slowly rotating cylindrical drum. The drums are perforated and the fibres are held onto them by air suction through these perforations. Fixed baffles inside each drum ensure that the fibres feed from one to the other.

Experimental observations indicate that the folding occurs almost entirely in the gap between the two drums and so it was decided to model this aspect of the process.

§2 A model for the transfer of fibres between the drums

The inertia of the fibres is small and so can be neglected, giving a quasi-steady problem. Experimental observations lead us to assume that the fibres always remain in contact with the first drum until after the end of the baffle on this drum. Thus, the suction pressure, which acts normally to the fibres, is always of one sign. For simplicity, this pressure was assumed to be constant.

As indicated in Figure 1, ψ is the angle that the fibre makes with the horizontal and s is arc length along the fibre made dimensionless with $L = [\sigma/p]^{1/3}$, where σ is the flexural stiffness of the fibres and p is the suction pressure. β is the angle at which the end of the fibre would lie if the length of fibre between $s = 0$ and $s = s_2$ were laid on the second drum. The overfeed, due to the differential speeds of the drums, implies that

$$\beta = \beta_0 + \frac{\Delta U}{R} t \tag{2.1}$$

where β_0 is a constant, ΔU is the difference between the speeds of the two drums and R is the drum radius. This relation is valid until a fold is formed on the second drum. For $s \leq s_1$, $s \geq s_2$, the fibre is in contact with a drum. Hence,

$$\psi = -\alpha + \frac{s}{R^*} \quad \text{for} \quad s \leq s_1, \qquad \psi = \beta - \frac{s}{R^*} \quad \text{for} \quad s \geq s_2,$$

where $R^* = R/L$

For $s_1 < s < s_2$, the fibre is assumed to satisfy the nonlinear beam equation,

$$\psi''' + \lambda^2 \cos \psi \psi' + 1 = 0 \tag{2.2}$$

403

Hj. Wacker and W. Zulehner (eds.),
Proceedings of the Fourth European Conference on Mathematics in Industry, 403–407.
© 1991 B.G. Teubner Stuttgart and Kluwer Academic Publishers.

where $' = d/ds$ and $F = (\sigma^2 p)^{1/3}\lambda$ is the unknown compressive force acting on the fibre. The points s_1, s_2 are unknown. The appropriate boundary conditions there are that the angle and the curvature of the fibre are continuous. Thus,

$$\left.\begin{array}{ll} \psi(s_1) = -\alpha + \dfrac{s_1}{R^*}, & \psi'(s_1) = \dfrac{1}{R^*} \\[2mm] \psi(s_2) = \beta - \dfrac{s_2}{R^*}, & \psi'(s_2) = -\dfrac{1}{R^*} \end{array}\right\} \tag{2.3}$$

Two further constraints are given by the requirement that the fibre lies on the second drum at $s = s_2$. Hence,

$$\left.\begin{array}{l} \displaystyle\int_{s_1}^{s_2} \cos\psi\, ds = -R^*[\sin\psi(s_1) + \sin\psi(s_2)] \\[3mm] \displaystyle\int_{s_1}^{s_2} \sin\psi\, ds = -h^* - R^*[2 - \cos\psi(s_1) - \cos\psi(s_2)] \end{array}\right\} \tag{2.4}$$

where $h = h^* R$ is the minimum distance between the two drums.

Equations (2.3) and (2.4) constitute a set of six conditions for the third order differential equation (2.2). In principle, it is thus possible to determine the six unknowns of the problem; the three constants of integration, s_1, s_2 and λ. These will depend parametrically on t, because equation (2.1) holds.

§3 Simplified Model

A simple experiment was performed in which paper was dispensed from a roll at a faster rate than the roll was translated in the opposite direction at a short distance above the ground. This procedure, which is shown in Figure 2, produced a folding of the paper similar to that observed with the fibres on the drums.

Neglecting the curvature of the drums and taking the point of departure of the fibres from the first drum to be pinned, this problem is equivalent to the fibre-drum problem except that whilst gravity acts vertically on the paper, suction was assumed to act normally to the fibres. If $|\psi| \ll 1$, however, equation (2.2) can be linearized and the problems are then exactly equivalent. For simplicity, it was decided to study this linear differential equation first, although the problem is a free boundary problem and, therefore, inevitably nonlinear.

By suitably rescaling time, the dispenser can be assumed to move with unit speed in the positive x direction. The linear beam equation is

$$\frac{d^4 y}{dx^4} + \lambda^2 \frac{d^2 y}{dx^2} + 1 = 0. \tag{3.1}$$

where y is scaled with h, x is scaled with $(h\sigma/p)^{1/4}$ and $\lambda^2 = F(h/\sigma p)^{1/2}$.

The boundary conditions are

$$y = \frac{dy}{dx} = \frac{d^2 y}{dx^2} = 0 \qquad \text{at} \quad x = d \tag{3.2}$$

$$y = 1, \quad \frac{d^2 y}{dx^2} = 0 \qquad \text{at} \quad x = t \tag{3.3}$$

The overfeed condition implies that the length of fibre between $x = d$ and $x = t$ is equal to t. Hence,

$$\int_d^t \left[1 + \left(\frac{dy}{dx} \right)^2 \right]^{1/2} dx = t$$

Linearizing this gives

$$\int_d^t \left(\frac{dy}{dx} \right)^2 dx = 2t \tag{3.4}.$$

The solution of (3.1)-(3.4) is

$$y = \frac{1}{\lambda^4} \left[- \cos \mu \xi - \tan \frac{\mu}{2} \sin \mu \xi + \mu \tan \frac{\mu}{2} \xi + 1 - \frac{\mu^2 \xi^2}{2} \right]. \tag{3.5}$$

where,

$$\xi = \frac{x - d}{t - d}, \qquad \mu = \lambda(t - d),$$

and λ, μ satisfy the transcendental equations

$$\lambda^4 = \mu \left(\tan \frac{\mu}{2} - \frac{\mu}{2} \right), \tag{3.6}$$

$$2t\lambda^7 = \frac{\mu}{2} - \frac{1}{4} \sin 2\mu + \frac{\mu^3}{3} - 2 \sin \mu + 2\mu \cos \mu$$

$$- \tan \frac{\mu}{2} (\sin \mu - \mu)^2 + \tan^2 \frac{\mu}{2} \left(\frac{3\mu}{2} + \frac{1}{4} \sin 2\mu - 2 \sin \mu \right), \tag{3.7}$$

Equations (3.6) and (3.7) were solved numerically for a range of t and the solutions for y plotted. In each interval $0 \leq \mu < \pi$, $(2n - 1)\pi \leq \mu < (2n + 1)\pi$, $(n \geq 1)$, there is a critical time, t_n, before which no non-trivial solutions of (3.6) and (3.7) exist. In each interval for μ, the function (3.6) at $t = t_n$ was chosen as the initial condition for the overfeed. After this time, one solution exists for $0 \leq \mu < \pi$, and two solutions exist for $(2n - 1)\pi \leq \mu < (2n + 1)\pi$ where $n \geq 1$. Equations (3.6) and (3.7) are sketched for $t > t_0, t_1, t_2$ in Figure 3. The α_i's are the roots of $\tan x = x$. The arrows indicate the direction in which the points of intersection of the curves 1 and 2 move as time increases.

As $t \to \infty$, two possibilities can occur;

a) $\lambda \to \infty$, $\mu \to (2n - 1)\pi$, $t - d \to 0$, (as in the case of A,B,D),

b)$\lambda \to 0$, $\mu \to 2\alpha_n$, $t - d \to \infty$. (as in the case of C,E).

It can be shown that the solution with $0 < \mu < \pi$ has least energy and is, therefore, the stable mode. It evolves schematically as shown in Figure 4.

§4 Conclusions

The initial stages in our simplified model are predicted. From Figure 4, it is clear, however, that steepening of the profile causes the nonlinear terms neglected in (3.1) to become important, and hence control the amplitude and wavelength of the folds which are ultimately formed.

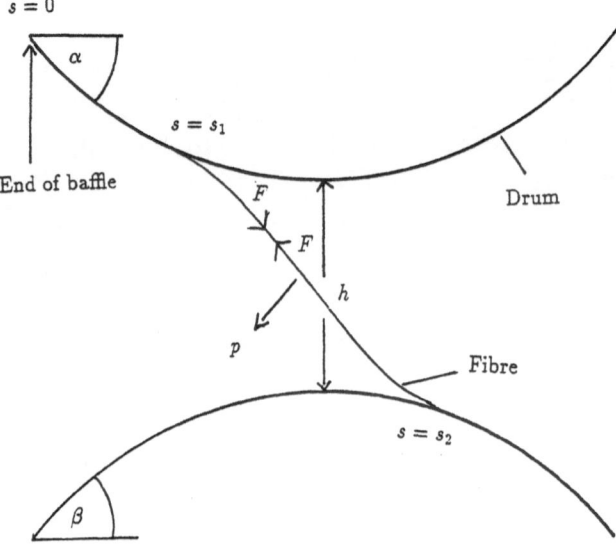

Figure 1 The Drum Model

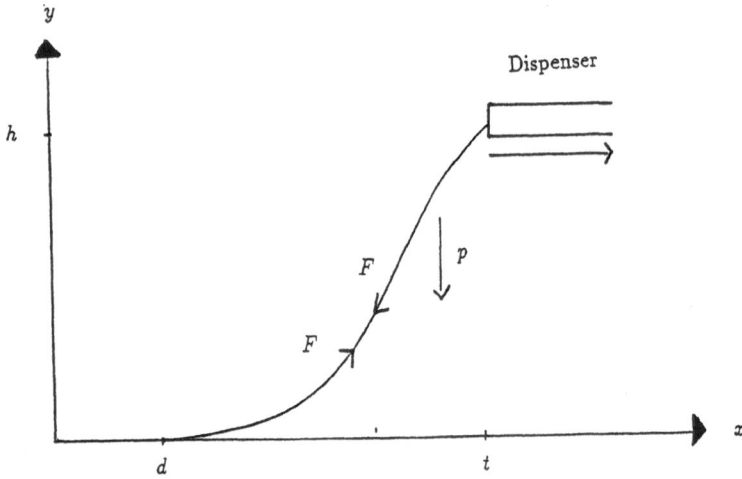

Figure 2 The Simplified Model

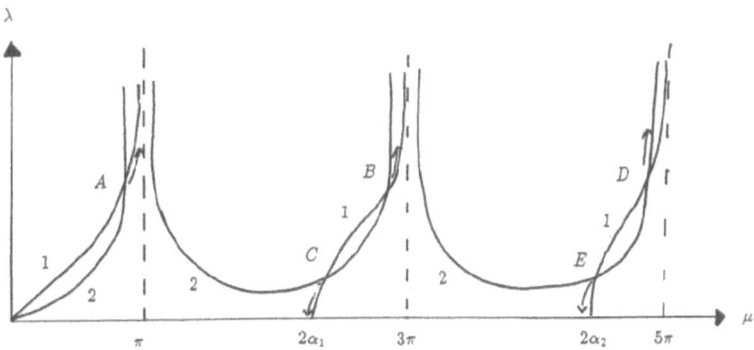

Figure 3   Curve 1: $\lambda = \mu^{1/4}\left(\tan\dfrac{\mu}{2} - \dfrac{\mu}{2}\right)^{1/4}$

Curve 2: $\lambda = \left(\dfrac{1}{2t}\right)^{1/7}\left[\dfrac{\mu}{2} - \dfrac{1}{4}\sin 2\mu + \dfrac{\mu^3}{3} - 2\sin\mu + 2\mu\cos\mu\right.$

$\left. - \tan\dfrac{\mu}{2}(\sin\mu - \mu)^2 + \tan^2\dfrac{\mu}{2}\left(\dfrac{3\mu}{2} + \dfrac{1}{4}\sin 2\mu - 2\sin\mu\right)\right]$

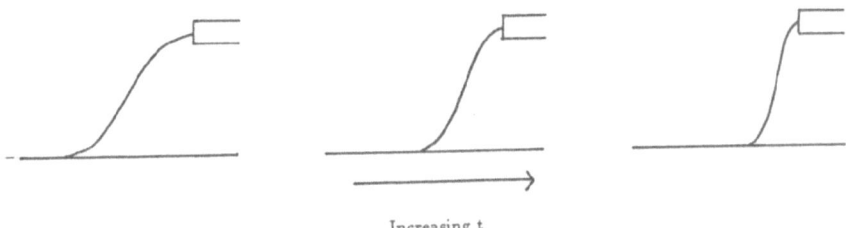

Increasing t

Figure 4 Evolution of the Stable Mode

Louise Terrill,
Mathematical Institute,
24-29 St. Giles',
Oxford,
U.K.

OPTIMAL EXTENSION OF A SYSTEM

OF HYDRO ENERGY STORAGE PLANTS

Hj.Wacker*, E.Czapka*, A.Gangl*, J.Gutenberger*

1. Introduction

In 1988 a local Austrian Electricity Board asked for advice concerning the extension of a system of two medium sized storage plants. (see fig. 1) As both the influx and the demand are random variables stochastic optimization would be a possibility (see e.g. [2]). W.r.t. the high nonlinearity of our model, however, we prefer a deterministic approach.

fig. 1: Scheme of the system (present state)

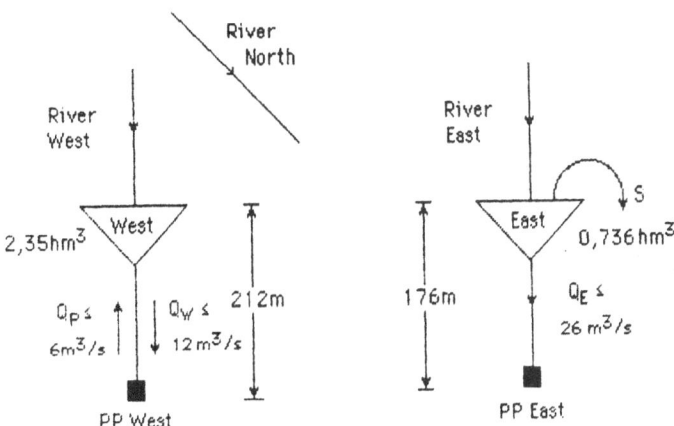

At the present time there are the following complaints concerning the system:

I) At East: After heavy rain the influx may increase up to $100m^3/s$ (average: $8.73m^3/s$) Then the capacity of the turbines resp. the capacity of the reservoir is too small to avoid spillage water.

II) At West: Austria is offered to trade peak power against base load at a favourable rate of up to 1 : 4 or even 1 : 5. Therefore one would like to produce more peak power but the capacities of the turbines both for producing and pumping and the influx to the reservoirs are too small.

---

* University of Linz, Mathematical Department, 4040 LINZ, AUSTRIA

Hj. Wacker and W. Zulehner (eds.),
Proceedings of the Fourth European Conference on Mathematics in Industry, 409–414.
© 1991 B.G. Teubner Stuttgart and Kluwer Academic Publishers.

To increase the capacity of the system (i.e. more peak power during day, more supply of energy in high winter and early spring) the Electricity Board proposed that water should be transferred from East to West. To cope with this larger influx it would be necessary to increase the capacity of the system West:

* For the transferred water the head increases from 176 [m] to 212 [m]
* Less spillage water at East
* More peak power at West because the capacity of both the turbines and the storage increases: $\overline{Q} : 69 m^3/s$, $\overline{Q}_p : 51 m^3/s$; $V_{max} : 61 hm^3$

In additon we have a larger degree of freedom for optimization.

2. Two Possibilities for Extension

Extension I: Construction of a Storage Plant North

fig. 2:

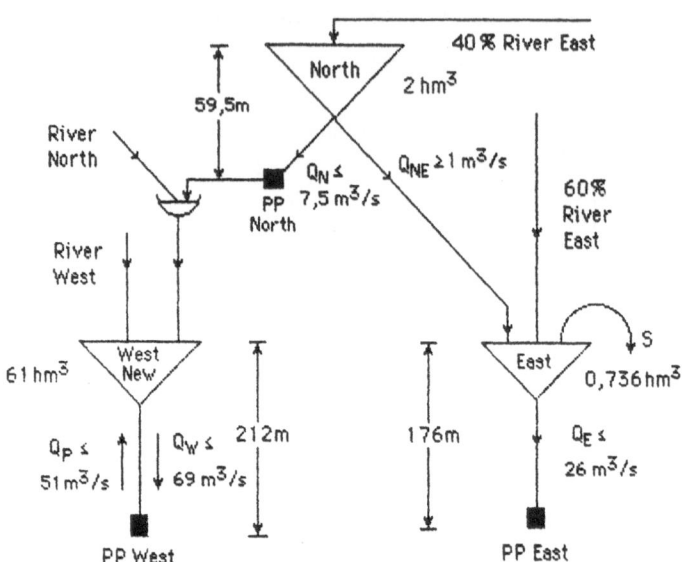

Up to 40% of the former influx to reservoir East is transferred to a newly to be built reservoir North ($2hm^3$). Water from reservoir North can be released either to the reservoir East and/or via a pipe system with a capacity of $7.5m^3/s$ to a small Plant North: PN with a head of 59.5 m. This water and the water of River North is then collected in a small pool ($0.1hm^3$) and afterwards led to reservoir West New.

Extension II: Construction of a Reservoir North East

fig. 3:

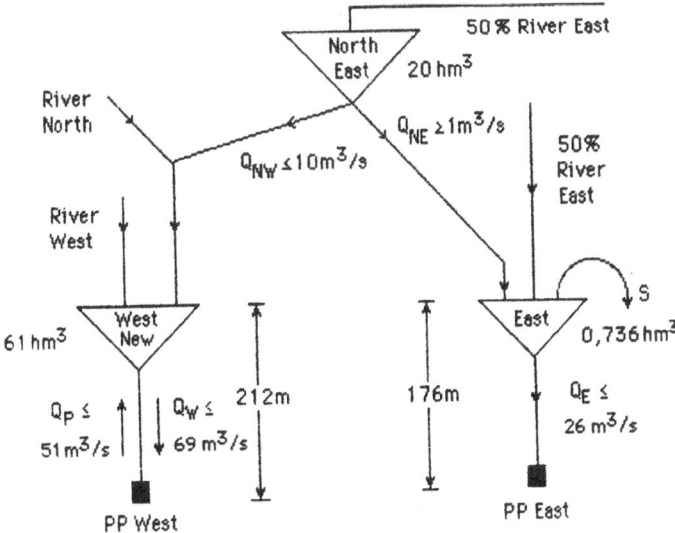

A new reservoir of $20hm^3$ is built to catch 50% of the influx of River East. Water from this reservoir can be released then either to West - by a tunnel with a capacity of $10m^3/s$ - or back to East. River North is again collected to be led to West.

The management wanted us to analyse both extensions to have a solid base for comparison - both alternatives resulted from a careful planning done by the engineers.

It was proposed to optimize the value of the production of both extensions within a given period of time. As a long term optimization could not easily cope with spillage water situations both a yearly and a weekly optimization was performed. (Compare [1]) To have a base for comparison we first optimized the old system.

3. The Mathematical Models

We confine ourselves to give a model (M) for Extension I only (yearly optimization):

$(M1)$ We want to maximize the value of the produced energy in $[o, T]$

$$E(T) = E_N(T) + E_E(T) + E_W(T) - E_P(T)$$

$$E = \frac{1}{3.6.10^6} \; g \cdot \int_v^T a(t) \underbrace{f(V(t))}_{head} \; \underbrace{Q(t)}_{discharges} \cdot \underbrace{\eta(V,Q)}_{efficiency} \; dt \qquad [ATS]$$

E_P: costs for pumping

The tariff $a(t)$ variies both with season and day/night.

(M2)   Constraints

2.1. for each reservoir there holds the continuity equation, e.g. West:

$$\dot{V}_W(t) = Z_W(t) + Z_N(t) + Q_N(t) + Q_P(t) - Q_W(t) \wedge V_W(0) = V_W(T) = V_{W,max}$$

2.2. Box constraints both for the state variable V (volume) and the control variable Q (discharge), Q_P (pumped water) and S (spillage water).

2.3. Peak Power Demands at West: $P_W(t) \geq \overline{P}_W(t)$

4. Numerical Solution-Results

To solve (M) numerically we use a Decomposition/Convexification technique proposed by Gfrerer [3],[4]. This method is a dual method, its efficiency is based on an artificial splitting of the problem into a separated one. It is ideally suited for parallel computation. The dual functional for N time steps reduces to the computation of N subproblems of dimension 9 only (splitting w.r.t. time) or dimension 4 (splitting w.r.to both time and plants).

Peak power production (expressed by average discharges at high- tariff periods):

Sept.	Oct.	Nov.	Dec.	Jan.	Febr.	Mar.	Apr.	May	June	
12	9	11	18	19	18	13	9.0	7.	6	$[m^3/s]$

(The time varying height of Reservoir West is already respected here.)

Results

EAST	WEST	NORTH	PUMPING	TOTAL	

1. The old system without peak power demands

| 45,64 | 24,21 | - | - | 69,85 Mio ATS | |

2.1. Extension I with peak power production

32,55	51,24	4,64	- 18,20	76,23 Mio ATS	= 100%
without peak power production:				84,83 Mio ATS	+11,3%
Reduction of capacity of Reservoir North from 2 to 1 $hm^2(*)$					− 0,04%

2.2. Extension II with peak power production

| 28,12 | 51,85 | – | – | 79,97 Mio ATS | = 100% |
| Without peak power production | | | | 82,96 Mio ATS | = 102,1% |

Sensibility Analysis for Extension II:

I) No peak power demands	+2.1 %
II) No water sent East ($Q_{NE} = 0$)	+4.5%
III) Reduction of tunnel capacity $6m^3/s$ (instead of $10m^3/s$)	− 0.1 %

This gives a slight advantage in favor of Extension II which in addition is both more stable w.r.t. peak power demands.

As for the long term analaysis we have worked only with averaged influces in addition a one weeks optimization was performed. We include running times of the water (e.g. from North to East: 8 hs etc.), spillage water and realistic influces. ($\Delta t = 2h$)

Example 1:
Comparison: Extension I (without peak power demands) - present state: $+ 18.2\%$
(water transfer East $\rightarrow$ West: 5.8%, plant North: 6.6%, River North: 4%, higher efficiency at West: 1.8%)

Example 2:
Nonlinear dependence of production on the influx (no spillage water) - Extension I:
  i) 50% increase of influx: 68.4% increase of production
  ii) 30% decrease of influx: 51.4% decrease of production

Example 3:
The capacity of the pool can be neglected: an increase of the 0.1 hm^3 capacity of the pool to 0.25 hm^3 gives only $+0.07$ %.

Example 4: Spillage water analysis (Extension I)
The same weekly influx from River East: 33.5 m^3/s is averaged (I), slightly peaked (II) and heavily peaked (III) at the middle of the week. For the spillage water we get: (in percentage of the influx)

	I	II	III
Without water transfer from East to West	22.4 %	26.9 %	37.6%
With transfer, no capacity at Reservoir North	0.0%	14.2%	26.3 %
Capacity North: 1 hm^3	0.0%	9.5 %	21%
Capacity North: 2 hm^3	0.0%	5.1 %	16.6 %
Capacity North: 3 hm^3	0.0%	5 %	12.1 %

These results refine decisively the results of the yearly optimization (compare 2.1. $(*)$).
In addition a thorough analysis of the pumping strategies was performed.(in [5])

References:

[1]   W.Bauer, H.Gfrerer, E.Lindner, A.Schwarz, Hj.Wacker: Optimization of
      the Hydro Energy Power Plant System Gosau-Gosauschmied-Steeg,
      Math.Eng.Ind., Vol. 1, No. 3, pp 169 - 190 (1987)

[2]   J.Dodu, M.Goursat, A.Hertz, J.P.Quadrat, M.Viot: Methodes de gradient
      stochastique pour l'optimisation des investissements dans un reseau electrique,
      EDF, Bulletin de la Direction des Etudes et Recherches, Serie C - Math,
      Inf. no.2, 1981, pp. 133 - 164

[3]   H.Gfrerer: Globally Convergent Decomposition Method for Nonconvex
      Optimization Problems, Computing 32, pp 199 - 227 (1984)

[4]   H.Gfrerer: Optimization of Storage Plant Systems by Decomposition,
      in "Appl. Opt. Techn. in Energy Probl." (ed. Hj. Wacker), Teubner,
      pp 215 - 226 (1985)

[5]   Hj.Wacker, E.Czapka, A.Gangl, J.Gutenberger: Analysis of the System
        "Oberes Mühlviertel" (Research Report for the OKA Power Supply Company),
        pp 1 - 30 (1989) (confidential/in German)

Modelling Separated Flows

P. Wilmott

Summary

There exist two very simple classical models for separated flow behind a bluff body - Helmholtz/Kirchhoff and Prandtl/Batchelor. These both have a pleasant simplicity but agreement with experiment is only qualitative. A simple combination of these two is presented for flow down a step which gives better quantitative results. The techniques of thin airfoil theory are applied when the aspect ratio of the separated region is large.

1) Introduction

There exist two simple models for separated regions in inviscid flows (i) the Helmholtz/Kirchhoff model where the region is treated as a stagnant zone with constant pressure and (ii) the Prandtl/Batchelor model where the separated region consists of a recirculating zone of constant vorticity. Unfortunately, these models do not agree well with experiment. Here we present a model for flow down a step which combines the basic features of both classical models based upon the characteristics of the flow seen in experiment.

From flow visualizations [1] the typical streamline pattern is as shown in fig. 1.

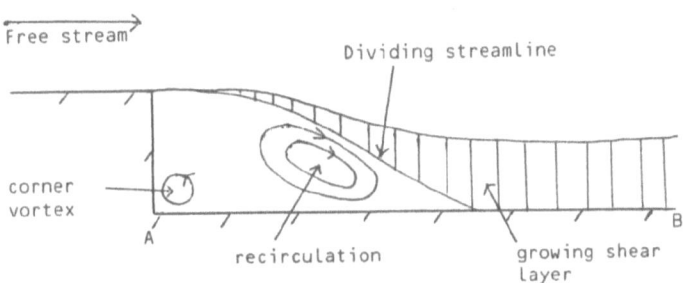

Figure (1): General description of mean flow field.

415

Hj. Wacker and W. Zulehner (eds.),
Proceedings of the Fourth European Conference on Mathematics in Industry, 415–421.
© 1991 B.G. Teubner Stuttgart and Kluwer Academic Publishers.

With pressure distribution on the wall AB as shown in fig.2.. These observations were taken from experimental work [1] and are in accord with other authors [2,3,4,5,6].

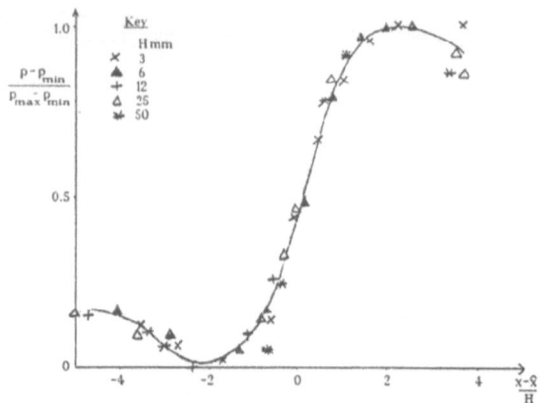

Figure (2): The results of Naranyanan *et al* [4].

The key points to note are

(i) the region of almost constant pressure immediately behind the step and

(ii) the region of recirculation further downstream.

Here we shall consider the case where the aspect ratio of the separated zone is large so that we may apply the techniques of thin airfoil theory. This corresponds to making assumptions about the magnitudes of the pressure behind the step and the vorticity as we shall see shortly.

2) Combination model

The experimental observations lead us to propose the following model, see fig.3.

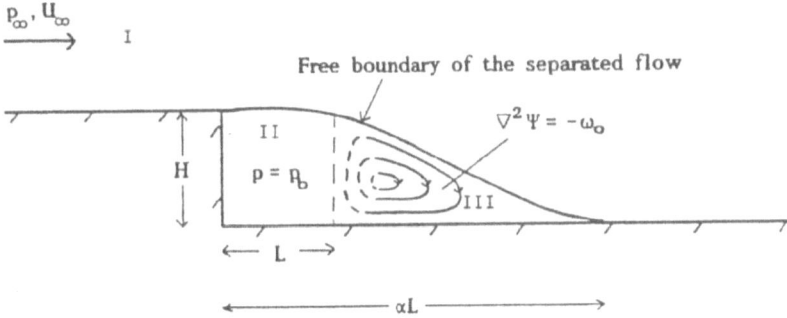

Figure (3): Diagram of modelled flow field

Define $\varepsilon = (p_\infty - p_0)/(\frac{1}{2}\rho U_\infty^2)$. We find that when $\varepsilon \ll 1$ and $L\omega_0/U_\infty =$ $O(\varepsilon^{-1/2})$ then $H/L = O(\varepsilon)$, i.e. the aspect ratio is large. Our methodology will be to derive expressions for the pressure on the streamline $y = S(x)$ when approached from regions I, II and III. Equating these pressures will then yield an equation for the streamline shape. Note that in the absence of the constant pressure region the slender Prandtl/Batchelor model was analysed by Childress [7].

First, in region I the flow is inviscid, incompressible and irrotational so that $\phi = U_\infty x + \bar{\phi}$ where $\nabla^2\bar{\phi} = 0$ with, to leading order, $\bar{\phi}_y = U_\infty S'(x)$ on $y = 0$ (here the boundary condition has been transferred to the horizontal axis). Since the x and y first derivatives of a harmonic function are related by the Hilbert transform we have

$$\bar{\phi}_x = -\frac{U_\infty}{\pi}\int_0^{\alpha L} \frac{S'(t)}{t - x}\, dt$$

and since, again to leading order, $p = p_\infty - \rho U_\infty \bar{\phi}_x$ we have

(2.1) $$p = p_\infty + \frac{\rho U_\infty^2}{\pi}\int_0^{\alpha L} \frac{S'(t)}{t - x}\, dt.$$

Trivially, in region II, we have

(2.2) $p = p_0.$

Finally, in region III, we have a streamfunction ψ such that

$$\nabla^2 \psi = -\omega_0 \text{ with } \psi = 0 \text{ on } y = 0 \text{ and } S(x) \text{ so that}$$

$$\psi = -\frac{\omega_0}{2} y \ (y - S(x)) \qquad \text{(where we have used the fact that}$$

y derivatives dominate in the Laplacian). Therefore, to leading order,

(2.3) $p = p_\infty - \rho \ (h + \frac{\omega_0^2}{8}S(x)^2)$.

We have included h, a constant, since a jump in the Bernoulli constant is permitted.

3) The integral equation

Equating pressures between regions I, II and III we find that

(3.1) $\displaystyle \frac{1}{\pi}\int_0^{\alpha L} \frac{S'(t)}{t-x}\,dt = \begin{cases} -\dfrac{\varepsilon}{2} & 0<x<L \\[2mm] -\dfrac{h}{U_\infty^2} - \dfrac{\omega_0^2}{8\,U_\infty^2}S(x)^2 & L<x<\alpha L \end{cases}$

with p continuous at $x = L$ so that

(3.2) $\displaystyle \frac{\varepsilon}{2} = \frac{h}{U_\infty^2} + \frac{\omega_0^2}{8\,U_\infty^2}S(L)^2$

and continuity and smoothness conditions

(3.3) $S(0) = H, \ S'(0) = 0 = S(\alpha L) = S'(\alpha L)$.

N.B. By multiplying (3.1) by $S'(x)$ and integrating from zero to αL we can derive the following relationship between the parameters

(3.4) $\displaystyle \frac{\omega_0^2\,S(L)^3}{3\,U_\infty^2 H \varepsilon} = 1$.

A similar relationship was derived (using a force balance argument) by Childress for his model.

4) Results

The numerical solution of (3.1) will not be discussed here in any depth. In brief, we invert the finite Hilbert transform (3.1) and integrate once. This new form of the integral equation is then iterated numerically. For

$O(1)$ values of α the convergence was found to be insensitive to the initial guess for $S(x)$.

In order to predict the pressure on the wall in dimensional form we need to prescribe three of ε, L, H, α (all of which can be measured experimentally).

Case (i): prescribed ε, L, H.

With ε, L and H given we can predict a value for α. From the experimental work [1] we find that our predicted value for α is larger than expected. To try to account for this we considered

case (ii): prescribed ε, L, α.

We can now predict an apparent value for H. This is found to be between one third and one half of the true height. This corresponds physically to the step height perceived by the flow, since far downstream the flow sees a shear layer with width of the same order as the step.

In fig. 4 are shown the comparisons between the experimental and theoretical results for $\bar{C}_p$ against x/L where

$$\bar{C}_p = \frac{C_p - C_{pmin}}{1 - C_{pmin}} \qquad \text{and} \qquad C_p = \frac{p(x) - p_\infty}{1/2 \, \rho \, U_\infty^2} .$$

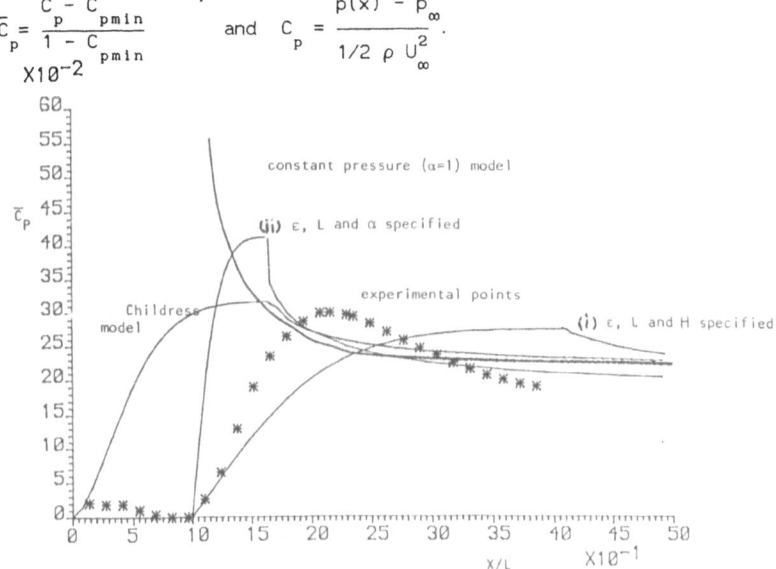

Figure (4): Comparison between theory and experiment.

5) Conclusion

From fig.4 we can see that both the cases (i) and (ii) of our proposed model give better agreement with experiment than either of the two classical models. Indeed, the true $\bar{C}_p$ vs. x/L curve lies midway between our two cases. Since the downstream shear layer appears to play such a large role in modifying the perceived step height this suggest that an improvement to our model would be to incorporate such a region.

Acknowlegements

This work was carried out jointly between the author A.D.Fitt, T.V.Jones, J.R.Ockendon and K.O'Malley.

References

[1] O'Malley,K., Fitt,A.D., Jones,T.V., Ockendon,J.R. and Wilmott,P. Models for high Reynolds number flow down a step. Submitted to J.Fluid Mech.

[2] Moore,T.W.F. (1960) Some experiments on the reattachment of a laminar boundary layer separating from a rearward facing step on a flat plate aerofoil. J.Royal Aeron. Soc. **64** 668-672

[3] Moss,W.D. and Baker,S. (1980) Recirculating flows associated with two-dimensional steps. Aero.Q. **31** 151-172

[4] Naranyanan,M.A.B., Khadgi,Y.N. and Visvanath,P.R. (1974) Similarities in pressure distribution in separated flow behind backward facing steps. Aero.Q. **25** 305-312

[5] Roshko,A. and Lau,J.C. (1965) Some observations on transition and reattachment of a free shear layer in incompressible flow. Proc.1965 Heat Transfer and Fluid Mechanics Institute, Stanford University Press.

[6] Tani,I., Iuchi,M. and Komoda,H. (1961) Experimental investigation of
      flow separation associated with a step or a groove. Aero. Res.
      Inst., Univ. of Tokyo, rept. 364

[7] Childress,S. (1966) Solutions of Euler's equations containing finite
      eddies. Phys. Fluids 9 860-872

Paul Wilmott

Mathematical Institute,

University of Oxford,

Oxford, U.K.

On the Impact of Rigid Bodies with Small Deadrise Angle onto Inviscid, Incompressible Fluid

S. K. Wilson

Abstract

In this paper a uniformly valid leading order solution for the impact of a two-dimensional rigid body with small deadrise angle onto an inviscid, incompressible fluid is obtained.

1 Introduction

The problem of the impact of a rigid body onto an inviscid, incompressible half-space of quiescent fluid is a formidable free boundary problem with many industrial applications, and has been a subject of mathematical interest since the pioneering paper by Wagner [4]. Neglecting the effects of gravity and surface tension, the simplest geometry is that of the entry of a semi-infinite wedge. Despite progress by a number of authors, including Mackie [3], no exact solutions are known even for this problem, and no existence/uniqueness theory has been developed. However, many numerical investigations, such as those by Greenhow [1], have been performed, and an excellent review of recent developments has been provided by Korobkin & Pukhnachov [2].

2 Bodies with Small Deadrise Angle

We restrict ourselves to two-dimensions, and consider only the vertical impact of symmetric, rigid bodies at constant velocity. Working in dimensionless variables, we take cartesian axes (x, y) with the y-axis vertically upwards along the axis of symmetry and the x-axis in the undisturbed fluid surface, and $t = 0$ denotes the instant of impact. In order to make progress, we consider only bodies with *small deadrise angle*, in the form $y = f(\epsilon x)$ where $\epsilon \ll 1$.

The governing equations are Euler's equations, and since the fluid is irrotational, the flow is described by a velocity potential $\phi(x, y, t)$ which satisfies Laplace's equation, $\nabla^2 \phi = 0$. We impose a normal velocity matching condition on the body, $y = f(\epsilon x) - t$, and the usual kinematic and pressure matching conditions on the unknown free surface, $y = h(x, t)$.

3 Outer Problem

The natural way to approach the problem is as a perturbation problem in the small parameter ϵ. The profile of the body changes by an $O(1)$ quantity over a $O(1/\epsilon)$ length scale, and so to investigate the problem in the *outer region* we introduce scaled outer variables X and Y defined by $X = \epsilon x$ and $Y = \epsilon y$, an outer velocity potential $\Phi(X, Y, t) = \epsilon \phi(x, y, t)$ and an outer free surface elevation $H(X, t) = h(x, t)$. To perform a systematic expansion of the boundary conditions they must be expressed in terms of quantities evaluated on their basic positions, corresponding to setting $\epsilon = 0$. However, experimental and numerical studies indicate the formation of a long, thin, fast-moving jet close to the body, and so we expect the free surface to become double valued.

423

Hj. Wacker and W. Zulehner (eds.),
Proceedings of the Fourth European Conference on Mathematics in Industry, 423–427.
© *1991 B.G. Teubner Stuttgart and Kluwer Academic Publishers.*

Hence, when linearizing onto $Y = 0$ we neglect the jet and apply the body boundary condition on $Y = 0$, $|X| < d(t)$ and the free surface condition on $Y = 0$, $|X| > d(t)$, where $d(t)$ is an unknown function. Expanding the dependent variables as asymptotic series in powers of ϵ, the leading order outer problem is

$$\nabla^2 \Phi_0 = 0 \quad \text{in} \quad Y < 0, \tag{1}$$

$$\frac{\partial \Phi_0}{\partial Y} = -1 \quad \text{on} \quad Y = 0, \quad |X| < d(t), \tag{2}$$

$$\frac{\partial \Phi_0}{\partial Y} = \frac{\partial H_0}{\partial t} \quad \text{on} \quad Y = 0, \quad |X| > d(t), \tag{3}$$

$$\Phi_0 = 0 \quad \text{on} \quad Y = 0, \quad |X| > d(t), \tag{4}$$

together with the initial conditions $H_0(X, 0) = 0$, $\Phi_0(X, Y, 0) = 0$ and the far-field condition $|\nabla \Phi_0| \to 0$ as $(X^2 + Y^2)^{\frac{1}{2}} \to 0$. This problem, which is shown in Figure 1, is mathematically equivalent to the normal impact with unit speed of a flat plate of width $2d(t)$ onto the half-space $Y < 0$, and has solution

$$\Phi_0(X, Y, t) = -\left[Y + \Re(d(t)^2 - Z^2)^{\frac{1}{2}} \right], \tag{5}$$

where $Z = X + iY$ and $\Re(\cdot)$ denotes the real part of a complex quantity. From equation (3) the leading order free surface elevation for $|X| > d(t)$ is given by

$$H_0(X, t) = -t + \int_0^t \frac{X}{(X^2 - d(\tau)^2)^{\frac{1}{2}}} \, d\tau, \tag{6}$$

and from Bernoulli's equation the leading order pressure on the body $Y = 0$, $|X| < d(t)$ is

$$\frac{1}{\epsilon} \frac{d(t) d'(t)}{(d(t)^2 - X^2)^{\frac{1}{2}}}. \tag{7}$$

The function $d(t)$ is determined by the matching condition that the leading order outer free surface, given by (6), meets the body at $X = d(t)$. (This condition is equivalent to the statement that the volume of fluid in the jet is small compared with the total volume displaced, which can be verified once the solution has been determined.) Hence, $d(t)$ satisfies the Abel integral equation

$$d(t) \int_0^t \frac{d\tau}{(d(t)^2 - d(\tau)^2)^{\frac{1}{2}}} = f(d(t)), \tag{8}$$

which has the solution

$$t(d) = \frac{2}{\pi} \int_0^d \frac{f(\xi)}{(d^2 - \xi^2)^{\frac{1}{2}}} \, d\xi. \tag{9}$$

4   Inner Solution

The leading order outer solution is singular at $|x| = d(t)/\epsilon$ and hence fails to represent the flow accurately near these points. In order to obtain the correct *inner problem* we introduce scaled inner variables $\hat{x}$ and $\hat{y}$, defined by

$$x = \frac{d(t)}{\epsilon} + \epsilon^n \hat{x}, \quad y = f(d(t)) - t + \epsilon^n \hat{y},$$

where the exponent $n > -1$ is unknown. Since the fluid velocity must be $O(1/\epsilon)$, we define a scaled inner velocity potential $\hat{\phi}(\hat{x}, \hat{y}, t)$ by $\phi(x, y, t) = \epsilon^{n-1}[d'(t)\hat{x} + \hat{\phi}(\hat{x}, \hat{y}, t)]$, and an inner free surface elevation $\hat{h}(\hat{x}, t)$ by $h(x, t) = f(d(t)) - t + \epsilon^n \hat{h}(\hat{x}, t)$. Expanding the dependent variables as asymptotic series in powers of ϵ, we obtain the leading order problem

$$\nabla^2 \hat{\phi}_0 = 0 \quad \text{in the fluid,} \tag{10}$$

$$\frac{\partial \hat{\phi}_0}{\partial \hat{y}} = 0 \quad \text{on } \hat{y} = 0, \tag{11}$$

$$\frac{\partial \hat{\phi}_0}{\partial \hat{x}} \frac{\partial \hat{h}_0}{\partial \hat{x}} - \frac{\partial \hat{\phi}_0}{\partial \hat{y}} = 0 \quad \text{on } \hat{y} = \hat{h}_0(\hat{x}, t), \tag{12}$$

$$\left(\frac{\partial \hat{\phi}_0}{\partial \hat{x}}\right)^2 + \left(\frac{\partial \hat{\phi}_0}{\partial \hat{y}}\right)^2 = d'(t)^2 \quad \text{on } \hat{y} = \hat{h}_0(\hat{x}, t). \tag{13}$$

This is a Helmholtz cavity flow with a jet whose asymptotic thickness, $h(t)$, is unknown, and is summarised in Figure 2. The problem is steady and can be solved by the standard conformal mapping techniques. The scaled complex potential $\hat{w}_0(\hat{z}) = (\hat{\phi}_0(\hat{z}) + i \hat{\psi}_0(\hat{z}))/d'(t)$, where $\hat{z} = \hat{x} + i\hat{y}$, satisfies the non-linear differential equation

$$\hat{w} = k - \frac{2h}{\pi} \left[\frac{2\hat{w}'}{(\hat{w}'+1)^2} + \log \left| \frac{\hat{w}'-1}{\hat{w}'+1} \right| \right], \tag{14}$$

where k is an unknown function of time. Without solving equation (14), we can determine the shape of the free streamline in terms of the intermediate variable $-1 < \xi < 1$ to be

$$\hat{x} = \frac{h}{\pi} \left[\log \left| \frac{1+\xi}{1-\xi} \right| - \frac{2\xi}{1+\xi} \right], \quad \hat{y} = \frac{4h}{\pi} \left[1 - \left(\frac{1-\xi}{1+\xi} \right)^{\frac{1}{2}} \right]. \tag{15}$$

The leading order pressure on the body, $\hat{y} = 0$, is expressed as a function of ξ for $\xi < -1$ and $\xi > 1$ by

$$\frac{d'(t)^2}{2\epsilon^2} \left[1 - (\xi \pm (\xi^2 - 1)^{\frac{1}{2}})^2 \right], \tag{16}$$

where we take the positive root for $\xi < -1$ and the negative one for $\xi > 1$. $\hat{x}$ is given in terms of ξ by

$$\hat{x} = \frac{h}{\pi} \left[\log \left| \frac{\xi+1}{\xi-1} \right| - 4 \left(\frac{\xi-1}{\xi+1} \right)^{\frac{1}{2}} + \frac{6+4\xi}{1+\xi} \right], \tag{17}$$

and $\xi = \pm\infty$ corresponds to the local stagnation point. Matching the far behaviour of the inner solution with the outer solution determines the unknown exponent to be $n = 1$, and the thickness of the jet, $h(t)$, to be

$$h(t) = \frac{\pi}{8} \frac{d(t)}{d'(t)^2}. \tag{18}$$

5 Jet Problem

The third flow region is the thin, fast-moving jet, as least partially attached to the body, and emanating from the inner region. To lowest order the jet is described by zero gravity shallow water theory, and the governing equations for the tangential velocity $u_0(X, t)/\epsilon$ and jet thickness $\epsilon h_0(X, t)$ in $|X| > d(t)$ are therefore

$$\frac{\partial u_0}{\partial t} + u_0 \frac{\partial u_0}{\partial X} = 0, \quad \frac{\partial h_0}{\partial t} + \frac{\partial}{\partial X}(u_0 h_0) = 0, \tag{19}$$

together with the boundary conditions $u_0(d(t), t) = 2d'(t)$ and $h_0(d(t), t) = h(t)$. The leading order pressure on the body is $\epsilon \kappa(X) h_0(X, t) u_0(X, t)^2$, where $\kappa(X)$ is the curvature of the body. The fact that the volume of the jet is only $O(1)$ confirms the mass conservation argument leading to the matching condition (8).

6   Uniformly Valid Composite Solution

The uniformly valid composite solution can now be constructed from the leading order solutions in their respective regions, and the leading order term of the total force on the body can easily be shown to be $\pi d(t)d'(t)/\epsilon^2$.

7   Examples

The simplest impacting body is a *wedge*, $f(x) = |x|$. Solving (9) gives $d(t) = \pi t/2$ and so the leading order outer free surface elevation is

$$H_0(X,t) = -t + \frac{2X}{\pi} \sin^{-1}\left(\frac{\pi t}{2X}\right),\tag{20}$$

and the solution in the jet region $\pi t/2 < X < \pi t$ is

$$u_0(X,t) = \pi, \quad h_0(X,t) = \frac{1}{4}\left[t - \frac{X}{\pi}\right].\tag{21}$$

When the impacting body is a *parabola*, $f(x) = x^2$, solving (9) gives $d(t) = (2t)^{\frac{1}{2}}$ and so the leading order outer free surface elevation is

$$H_0(X,t) = -t - X(X^2 - 2t)^{\frac{1}{2}} + X^2,\tag{22}$$

and the solution in the jet region $(2t)^{\frac{1}{2}} < X < \infty$ is

$$u_0(X,t) = \frac{X}{t}, \quad h_0(X,t) = 2\pi\frac{t^4}{X^5}.\tag{23}$$

The leading order force on a wedge is $\pi^3 t/4\epsilon^2$, and π/ϵ^2 on a parabola, and typical pressure histories are shown in Figures 3 and 4 respectively.

8   Conclusions

The present work subsumes and extends the intuitive ideas of Wagner [4]. It can easily be modified to include the effects of gravity, variable impact velocity and surface tension, and extended to fluid-fluid impacts. The same approach can be applied to three-dimensional bodies, although less analytical progress is possible. Further details are given by Wilson [5].

9   References

[1] Greenhow, M. *Appl. Ocean Res.* 1987 **9** (4) : 214-223
[2] Korobkin, A.A., Pukhnachov, V.V. *Ann. Rev. Fluid Mech.* 1988 **20** : 159-185
[3] Mackie, A.G. *Quart. Journ. Mech. and Applied Math.* 1969 **22** (1) : 1-17
[4] Wagner, H. *Z.A.M.M.* 1932 **12** (4) : 193-215 (Also *N.A.C.A. Translation 1366*)
[5] Wilson, S.K. *Oxford University D.Phil. Thesis* 1989

Stephen K. Wilson,
Mathematical Institute,
24-29 St.Giles',
Oxford.
GREAT BRITAIN.
OX1 3LB.

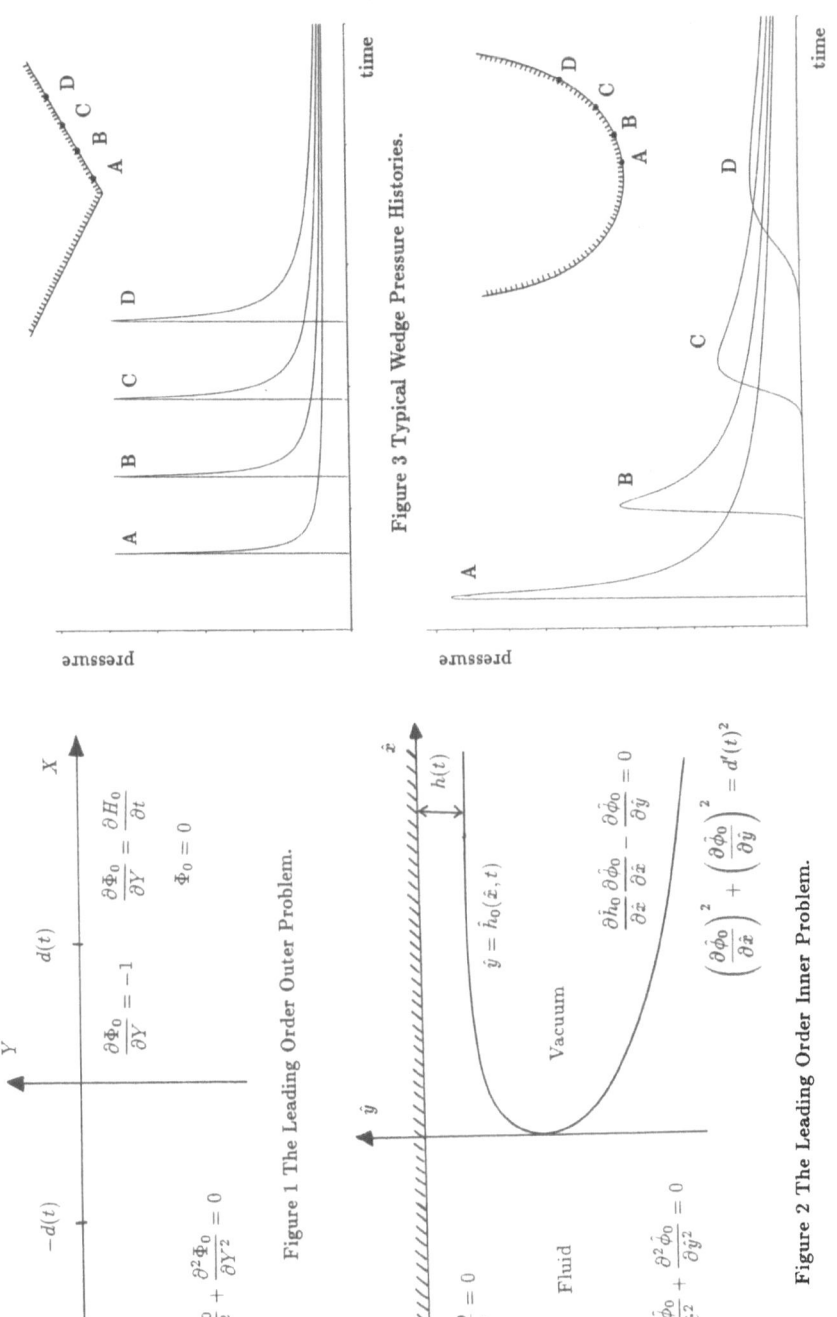

Figure 1 The Leading Order Outer Problem.

$$\frac{\partial^2 \Phi_0}{\partial X^2} + \frac{\partial^2 \Phi_0}{\partial Y^2} = 0$$

$$\frac{\partial \Phi_0}{\partial Y} = -1 \qquad \frac{\partial \Phi_0}{\partial Y} = \frac{\partial H_0}{\partial t}$$

$$\Phi_0 = 0$$

Figure 2 The Leading Order Inner Problem.

$$\frac{\partial \hat{\phi}_0}{\partial \hat{y}} = 0$$

$$\frac{\partial^2 \hat{\phi}_0}{\partial \hat{x}^2} + \frac{\partial^2 \hat{\phi}_0}{\partial \hat{y}^2} = 0$$

$$\hat{y} = \hat{h}_0(\hat{x}, t)$$

$$\frac{\partial \hat{h}_0}{\partial \hat{x}} \frac{\partial \hat{\phi}_0}{\partial \hat{x}} - \frac{\partial \hat{\phi}_0}{\partial \hat{y}} = 0$$

$$\left(\frac{\partial \hat{\phi}_0}{\partial \hat{x}}\right)^2 + \left(\frac{\partial \hat{\phi}_0}{\partial \hat{y}}\right)^2 = d'(t)^2$$

Figure 3 Typical Wedge Pressure Histories.

Figure 4 Typical Parabolic Pressure Histories.

B-SPLINE LIBRARY FOR MATHEMATICAL MODELLING AND SOLVING OF SCIENTIFIC AND INDUSTRIAL PROBLEMS.

By Anne Marie Ytrehus, Senter for Industriforskning (SI), Oslo.

SPLINE-MATHEMATICS AT SI:

Center for Industrial Research (SI) in Oslo is a non-profit research institute, doing projects paid by the industry. The staff numbers 340, and covers
the fields of Robotics and Automation, Information Technology, Industrial
Chemistry and Materials Research.

Within Information Technology, SI concentrates on: CAD/CAM and Geometric
Modelling, Knowledge Based Systems, and Distributed and Integrated
Information Systems. The Geometric Modelling department works in close
contact with the Spline-group of Prof. Tom Lyche at the Institute for
Computer Science at the University of Oslo.

We try to be a link between the mathematics and the industry, providing
the University of Oslo with interesting industrial problems, and the
industry with advanced mathematical solutions. The philosophy of our work
is to utilize advanced mathematics in practical applications. The activities
in the geometric modelling department is mainly concentrated on mathematics
research in the spline-area, and development of software for industrial
applications based on these mathematical results. We are constantly looking
into new applications.

We define Geometric Modelling as mathematical design of three-dimensional
objects, based on free form curves and surfaces. When the objects are
created, mass and area properties may be calculated, and simulation
processes may be performed. In addition, the geometry defined may be
transformed to other CAD, CAM or FEM-systems, for further calculation and

429

Hj. Wacker and W. Zulehner (eds.),
Proceedings of the Fourth European Conference on Mathematics in Industry, 429–433.
© 1991 *B.G. Teubner Stuttgart and Kluwer Academic Publishers.*

use.

THE APS-SS B-SPLINE LIBRARY:

APS-Sculptured Surfaces is a Fortran subroutine library based on B-spline
mathematics. The library is used for modelling sculptured surfaces and
complicated mathematical functions. A broad range of applications exist.
Some of these are mentioned later. An extended version, written in C is now
finished.

The work on a B-spline based sculptured surface modeller started in the
Inter-Nordic cooperation project GPM (Geometric Product Models, 1978-1981),
continued in the German-Norwegian CAD/CAM project APS (Advanced Production
Systems, 1981-1987), and will go on in the ESPRIT project IMPPACT (1989-
1991).

The B-spline theory describes mathematical methods which guarantee that all
solutions are found in geometry calculations and geometry operations, and
enables the user to influence the accuracy of calculations. The B-spline
format is efficient in calculations, gives stable algorithms and can be
stored compactly. The subroutine package contains functions for creating
and editing various types of curves and surfaces, operate on them and do
calculations such as mass and area properties.

THE MATHEMATICS OF B-SPLINES:

A curve can be defined as a piecewise polynomial of order k, on a nondecreasing knot-set. The order of the curve or surface is independent of
the number of datapoints used for the definition. This implies that e.g. a
curve interpolating 100 data-points, can be expressed by cubic splines.

The basis functions of this polynomial description have some nice properties: They are always non-negative, and they are positive only over a small area. Thus local changes doesn't influence the entire shape.

The use of multiple or single knot-values in the knot-vector gives the user the possibility of controlling the continuity of an object. This is important when surfaces or curves are combined.

The object (curve or surface) is always lying inside a polygon through the controlpoints (polynom coefficients). This property can be used for intersection-tests and collision-control (NC), and during calculations.

The recurrence relation formula for B-splines and the Oslo-algorithm (developed by Cohen, Lyche and Riesenfeld in 1979) ensure accurate, stable and effictive calculations.

APPLICATIONS OF THE APS-SS SUBROUTINE LIBRARY:

The development of the system is based on product studies and continous feedback from the industrial use of applications built on the B-spline based modeller. The system is portable, and can easily be integrated into CAD/CAM-systems or interfaced to other related systems. Today it is integrated into the following systems: SESAM, Det Norske Veritas (Norway), Technovision, Norsk Data (Norway & West-Germany), GeoRec, SysScan (Norway & West-Germany), Proren, Isykon GmbH (West-Germany) and Moldflow (West-Germany & Australia).

The APS-SS library is being used and tested in universitites and research institutes as well. Some of them are: SI, Sintef and CMI in Norway, Cranfield Institute of Technology in England, the Royal Institute of Technology in Sweden, the Technical University of Berlin and the University of Karlsruhe in West-Germany.

The main applications of the APS-SS B-spline library have been integration
into commersial CAD-systems and interfaces between different CAD/CAM-systems.

The interfaces between CAD/CAM-systems are developed as modules converting
the mathematical description from standard formates for exchange of geometric
data such as IGES and VDAFS into the B-spline description and vice versa.
This is an area of growing interest. The need for integration of several
different systems to describe the process all the way from design to
machining, puts requirement on the possibilities of exchanging geometric data.

Our participation in industrial projects has focused on a variety of indust-
rial problems that are solved by the use of the APS-SS subroutine library.
Some of them are:
Design and engineering: Water turbines, aircrafts, cars, ship propellers.
Simulation: Spray-painting robot for the car industry, robot movements,
plastic molding.
Quality control: Systems for scanning and optical meassurements, regeneration of curves, regeneration of surfaces.
Terrain modelling: Our terrain modelling system supplies tools needed for
construction and planning of built-up areas, analysis of snow-avalanche and
rock-fall hazard, subsea mapping, etc.

Data-reduction: This is a function used in applications like curve and
surface regeneration and terrain modelling. The importance of data-reduction
or data-compression is growing, because of the need for transforming,
storing and doing calculations on huge amounts of data.

NC-preparation: Calculation of machining paths, collision control, rough
machining.

FURTHER DEVELOPMENT:

The further development will be directed towards extensions of the functionality and the applications of the mathematical package. In addition, we test the feasibility of the new spline theory that can be the basis for the next generation of spline based systems for complex geometry handling. Key-words here are Non-Uniform Rational B-Splines (NURBS) and multivariate splines (Box and Simplex splines).

In the area of extended functionality we are working on regeneration of curves and surfaces, shape control and design and calculation of volumes.